Autonomous Flying Robots

Kenzo Nonami • Farid Kendoul • Satoshi Suzuki
Wei Wang • Daisuke Nakazawa

Autonomous Flying Robots

Unmanned Aerial Vehicles and Micro Aerial Vehicles

Springer

Kenzo Nonami
Vice President, Professor, Ph.D.
Faculty of Engineering
Chiba University
1-33 Yayoi-cho, Inage-ku
Chiba 263-8522, Japan
nonami@faculty.chiba-u.jp

Farid Kendoul
Research Scientist, Ph.D.
CSIRO Queensland Centre
for Advanced Technologies
Autonomous Systems Laboratory
1 Technology Court
Pullenvale, QLD 4069, Australia
Farid.Kendoul@csiro.au

Satoshi Suzuki
Assistant Professor, Ph.D.
International Young Researchers
Empowerment Center
Shinshu University
3-15-1 Tokida, Ueda
Nagano 386-8567, Japan
s-s-2208@shinshu-u.ac.jp

Wei Wang
Professor, Ph.D.
College of Information
and Control Engineering
Nanjing University of Information Science
& Technology
219 Ning Liu Road, Nanjing
Jiangsu 210044, P.R. China
wwcb@nuist.edu.cn

Daisuke Nakazawa
Engineer, Ph.D.
Advanced Technology R&D Center
Mitsubishi Electric Corporation
8-1-1 Tsukaguchi-honmachi, Amagasaki
Hyogo 661-8661, Japan
Nakazawa.Daisuke@df.MitsubishiElectric.co.jp

ISBN 978-4-431-53855-4 ISBN 978-4-431-53856-1 (eBook)
DOI 10.1007/978-4-431-53856-1
Springer Tokyo Dordrecht Heidelberg London New York

Library of Congress Control Number: 2010931387

Front cover: The 6-rotor MAV which Chiba university MAV group developed is shown here and its size is one meter diameter, 1 kg for weight, 1.5 kg for payload and 20 minutes for flying time. In order to achieve a fully autonomous flight control, the original autopilot unit has been implemented on this MAV and the model based controller has been also installed. This MAV will be used for industrial applications.

Printed on acid-free paper

Springer is part of Springer Science+Business Media (www.springer.com)

Preface

The advance in robotics has boosted the application of autonomous vehicles to perform tedious and risky tasks or to be cost-effective substitutes for their human counterparts. Based on their working environment, a rough classification of the autonomous vehicles would include unmanned aerial vehicles (UAVs), unmanned ground vehicles (UGVs), autonomous underwater vehicles (AUVs), and autonomous surface vehicles (ASVs). UAVs, UGVs, AUVs, and ASVs are called UVs (unmanned vehicles) nowadays. In recent decades, the development of unmanned autonomous vehicles have been of great interest, and different kinds of autonomous vehicles have been studied and developed all over the world. In particular, UAVs have many applications in emergency situations; humans often cannot come close to a dangerous natural disaster such as an earthquake, a flood, an active volcano, or a nuclear disaster. Since the development of the first UAVs, research efforts have been focused on military applications. Recently, however, demand has arisen for UAVs such as aero-robots and flying robots that can be used in emergency situations and in industrial applications. Among the wide variety of UAVs that have been developed, small-scale HUAVs (helicopter-based UAVs) have the ability to take off and land vertically as well as the ability to cruise in flight, but their most important capability is hovering. Hovering at a point enables us to make more effective observations of a target. Furthermore, small-scale HUAVs offer the advantages of low cost and easy operation.

The Chiba University UAV group started research of autonomous control in 1998, advanced joint research with Hirobo, Ltd. in 2001, and created a fully autonomous control helicopter for a small-scale helicopter for hobbyists. There is a power-line monitoring application of UAV called SKY SURVEYOR. Once it catches power line, regardless of the vibration of the helicopter, with various onboard cameras with a gross load of 48 kg for a cruising time of 1 hour, catching of the power line can be continued. In addition, it has a payload of about 20 kg. Although several small UAVs are helicopters — Sky Focus-SF40 (18 kg), SST-eagle2-EX (7 kg), Shuttle-SCEADU-Evolution (5 kg), and an electric motor-based Lepton (2 kg) for hobbyists, with gross loads of 2–18 kg — fully autonomous control of these vehicles is already possible. Cruising time, depending on the helicopter's class, is about 10–20 min, with payloads of about 800 g – 7 kg. These devices are what automated the commercial radio-controlled helicopters for hobbyists, because they can be flown freely by autonomous flight by one person, are cheap and simple systems, and can apply chemical sprays, as in orchards, fields, and small-scale gardens. In the future they can also be used for aerial photography, various kinds of surveillance, and rescues in disasters.

GH Craft and Chiba University are conducting further research and development of autonomous control of a four-rotor tilt-wing aircraft. This QTW (quad tilt wing)-UAV is about 30 kg in gross load; take-off and landing are done in helicopter mode; and high-speed flight at cruising speed is carried out in airplane mode. Bell Helicopter in the United States completed development of the QTR (quad tilt rotor)-UAV, and its first flight was carried out in January 2006; however, the QTW-UAV had not existed anywhere in the world until now, although the design and test flight had been attempted. The QTW-UAV now is already flying under fully autonomous conditions. Moreover, Seiko Epson and Chiba University tackled autonomous control of a micro flying robot, the smallest in the world at 12.3 g, with the micro air vehicle (MAV) advantage of the lightest weight, and have succeeded with perfect autonomous control inside a room through image-processing from a camera. The XRB by Hirobo, Ltd., about 170 g larger than this micro flying robot, has also successfully demonstrated autonomous control at Chiba University. Flying freely with autonomous control inside a room has now been made possible.

We have also been aggressively developing our own advanced flight control algorithm by means of a quad-rotor MAV provided by a German company (Ascending Technologies GmbH) as a helicopter for hobbyists. We have chosen this platform because it offers good performance in terms of weight and payload. The original X-3D-BL kit consists of a solid airframe, brushless motors and associated motor drivers, an X-base which is an electronic card that decodes the receiver outputs and sends commands to motors, and an X-3D board that incorporates three gyroscopes for stabilization. The total weight of the original platform is about 400 g including batteries, and it has a payload of about 200 g. The flight time is up to 20 min without a payload and about 10 min with a payload of 200 g. The X-3D-BL helicopter can fly at a high speed approaching 8 m/s. These good characteristics are due to its powerful brushless motors that can rotate at very high speed. Furthermore, the propellers are directly mounted on the motors without using mechanical gears, thereby reducing vibration and noise. Also, our original 6-rotor MAVs for industrial applications such as chemical spraying have been developed, and their fully autonomous flight has already been successful.

For industrial applications, a power-line monitoring helicopter called SKY SURVEYOR has been developed. A rough division of the system configuration of SKY SURVEYOR consists of a ground station and an autonomous UAV. Various apparatuses carry out an autonomous control system of a sensor and an inclusion computer, and power-line monitoring devices are carried in the body of the vehicle. The sensors for autonomous control are a GPS receiver, an attitude sensor, and a compass, which comprise the autonomous control system of the model base. The flight of the compound inertial navigation of GPS/INS or a 3D stereo-vision base is also possible if needed. The program flight is carried out with the ground station or the embedded computer system by an orbital plan for operation surveillance, if needed. For attitude control, an operator performs only position control of the helicopter with autonomous control, and so-called operator-assisted flight can also be performed. In addition, although a power-line surveillance image is recorded by the video camera of the UAV loading in automatic capture mode and is simultaneously transmitted to the ground station, an operator can also perform posture control of the power-line monitoring camera and zooming at any time.

We have been studying UAVs and MAVs and carrying out research more than 10 years, since 1998, and we have created many technologies by way of experimental work and theoretical work on fully autonomous flight control systems. Dr. Farid Kendoul worked 2 years in my laboratory as a post-doctoral research fellow of the Japan Society for the Promotion of Science (JSPS post-doctoral fellow), from October 2007 to October 2009. He contributed greatly to the progress in MAV research. These factors are the reason, the motivation, and the background for the publication of this book. Also, seven of my graduate students completed Ph.D. degrees in the UAV and MAV field during the past 10 years. They are Dr. Jinok Shin, Dr. Daigo Fujiwara, Dr. Kensaku Hazawa, Dr. Zhenyo Yu, Dr. Satoshi Suzuki, Dr. Wei Wang, and Dr. Dasuke Nakazawa. The last three individuals — Dr. Suzuki, Dr. Wang, and Dr. Nakazawa — along with Dr. Kendoul are the authors of this book.

The book is suitable for graduate students whose research interests are in the area of UAVs and MAVs, and for scientists and engineers. The main objective of this book is to present and describe systematically, step by step, the current research and development in, small or miniature unmanned aerial vehicles and micro aerial vehicles, mainly rotary wing vehicles, discussing integrated prototypes developed within robotics and the systems control research laboratory (Nonami Laboratory) at Chiba University. In particular, this book may provide a comprehensive overview for beginning readers in the field. All chapters include demonstration videos, which help the readers to understand the content of a chapter and to visualize performance via video. The book is divided into three parts. Part I is "Modeling and Control of Small and Mini Rotorcraft UAVs"; Part II is "Advanced Flight Control Systems for Rotorcraft UAVs and MAVs"; and Part III is "Guidance and Navigation of Short-Range UAVs."

Robotics and Systems Control Laboratory Kenzo Nonami, Professor
Chiba University

Acknowledgments

First I would like to express my gratitude to the contributors of some of the chapters of this book. Mr. Daisuke Iwakura, who is an excellent master's degree student in the MAVs research area, in particular, contributed Chapter 13. Also, I express my appreciation to Mr. Shyaril Azrad and Ms. Dwi Pebrianti for their contributions on vision-based flight control. They are Ph.D. students in MAV research in my laboratory. Dr. Jinok Shin, Dr. Daigo Fujiwara, Dr. Kensaku Hazawa, and Dr. Zhenyo Yu made a large contribution in UAV research, which is included in this book. I also appreciate the contributions of Dr. Shin, Dr. Fujiwara, Dr. Hazawa, and Dr. Yu. During the past 10 years, we at Chiba University have been engaged in joint research with Hirobo Co., Ltd., on unmanned fully autonomous helicopters; with Seiko Epson on autonomous micro flying robots; and with GH Craft on QTW-UAV. I thank all concerned for their cooperation and support in carrying out the joint research. I am especially grateful to Hirobo for providing my laboratory with technical support for 10 years. I appreciate the teamwork in the UAV group. I will always remember our field experiments and the support of each member. I thank you all for your help and support. Lastly, I would like to thank my family for their constant support and encouragement, for which I owe a lot to my wife.

Contents

Chapter 1
Introduction

Abstract This chapter contains a non-technical and general discussion about unmanned aerial vehicles (UAVs) and micro aerial vehicles (MAVs). This chapter presents some fundamental definitions related to UAVs and MAVs for clarification, and discusses the contents of this monograph. The goal of this chapter is to help the reader to become familiar with the contents of the monograph and understand what to expect from each chapter.

Video Links:

Auto-take off and Landing
http://mec2.tm.chiba-u.jp/monograph/Videos/Chapter1/1.avi
Cooperation between UAV, MAV and UGV
http://mec2.tm.chiba-u.jp/monograph/Videos/Chapter1/2.mpg
Formation flight control of two XRBs
http://mec2.tm.chiba-u.jp/monograph/Videos/Chapter1/3.mpg
Fully autonomous flight control of QTW-UAV
http://mec2.tm.chiba-u.jp/monograph/Videos/Chapter1/4.wmv
Fully autonomous hovering of micro-flying robot by vision
http://mec2.tm.chiba-u.jp/monograph/Videos/Chapter1/5.mpg
GPS_INS fusion flight control
http://mec2.tm.chiba-u.jp/monograph/Videos/Chapter1/6.avi
Operator assistance flight control
http://mec2.tm.chiba-u.jp/monograph/Videos/Chapter1/7.avi
Promotion video of UAVs and IMU sensor
http://mec2.tm.chiba-u.jp/monograph/Videos/Chapter1/8.avi
Power line inspection by Skysurveyor
http://mec2.tm.chiba-u.jp/monograph/Videos/Chapter1/9.avi
Rotation control of Eagle; onboard camera & ground camera views
http://mec2.tm.chiba-u.jp/monograph/Videos/Chapter1/10.mpg
UAV application by Skysurveyor
http://mec2.tm.chiba-u.jp/monograph/Videos/Chapter1/11.avi
Vision based auto-take off, hovering and auto-landing
http://mec2.tm.chiba-u.jp/monograph/Videos/Chapter1/12.avi

K. Nonami et al., *Autonomous Flying Robots: Unmanned Aerial Vehicles and Micro Aerial Vehicles*, DOI 10.1007/978-4-431-53856-1_1, © Springer 2010

1.1 What are Unmanned Aerial Vehicles (UAVs) and Micro Aerial Vehicles (MAVs)?

In recent years, there has been rapid development of autonomous unmanned aircraft equipped with autonomous control devices called unmanned aerial vehicles (UAVs) and micro aerial vehicles (MAVs). These have become known as "robotic aircraft," and their use has become wide spread. They can be classified according to their application for military or civil use. There has been remarkable development of UAVs and MAVs for military use. However, it can be said that the infinite possibilities of utilizing their outstanding characteristics for civil applications remain hidden. Figure 1.1 shows that there was a large number of registered UAVs in Japan in 2002. This was because of the many unmanned helicopters used for agricultural–chemical spraying, as can be seen in Table 1.1. Figure 1.2 shows the country-wise R&D expenditure and Fig. 1.3 indicates the application of UAVs for civil and military purposes.

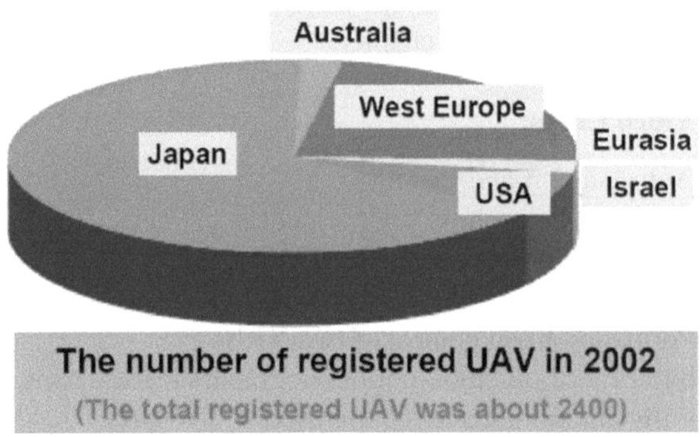

Fig. 1.1 Registered UAVs

Table 1.1 Number of registered UAVs in Japan

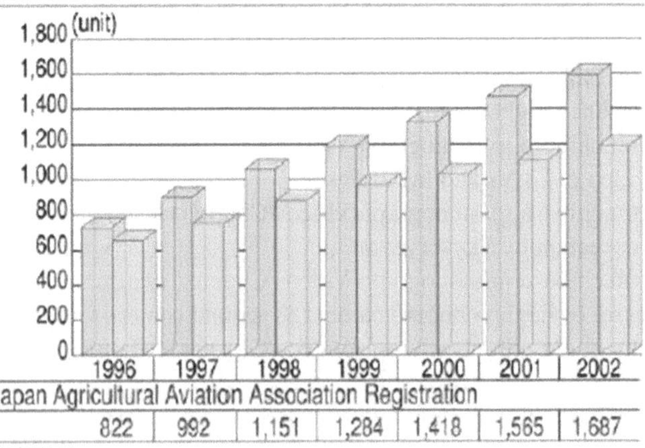

	1996	1997	1998	1999	2000	2001	2002
Japan Agricultural Aviation Association Registration	822	992	1,151	1,284	1,418	1,565	1,687
YAMAHA	748	844	973	1,067	1,121	1,202	1,281

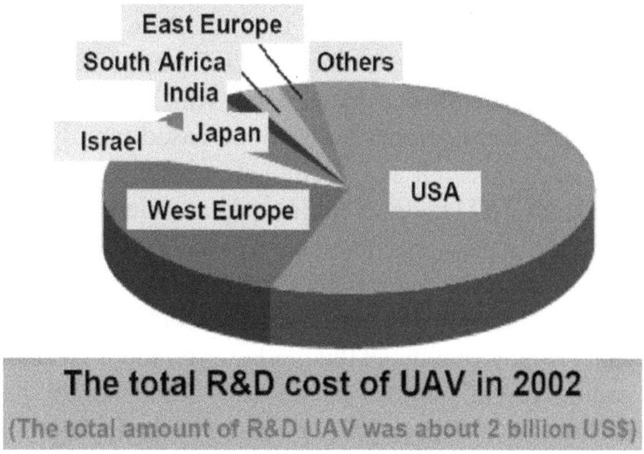

Fig. 1.2 Country-wise R&D expenditure on UAVs

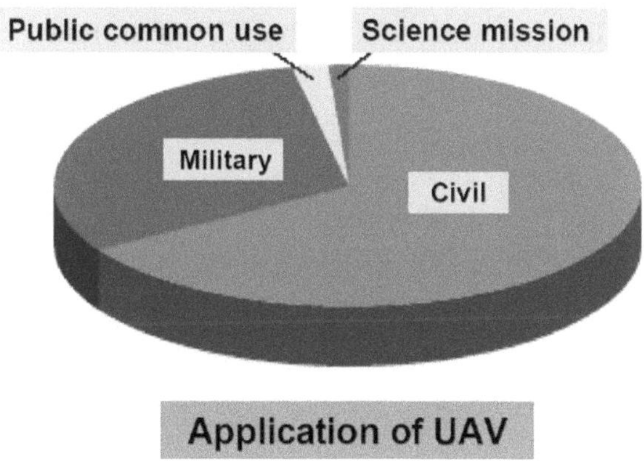

Fig. 1.3 Application of UAVs for civil and for military use in 2002

UAVs offer major advantages when used for aerial surveillance, reconnaissance, and inspection in complex and dangerous environments. Indeed, UAVs are better suited for dull, dirty, or dangerous missions than manned aircraft. The low downside risk and higher confidence in mission success are two strong motivators for the continued expansion of the use of unmanned aircraft systems. Furthermore, many other technological, economic, and political factors have encouraged the development and operation of UAVs. First, technological advances provide significant leverage. The newest sensors, microprocessors, and propulsion systems are smaller, lighter, and more capable than ever before, leading to levels of endurance, efficiency, and autonomy that exceed human capabilities. Second, UAVs have been used successfully in the battlefield, being deployed successfully in many missions. These factors have resulted in more funding and a large number of production orders. Third, UAVs can operate in dangerous and contaminated environments, and can also operate in other environments denied to manned systems, such as altitudes that are both lower and higher than those typically traversed by manned aircraft. Several market studies [1–3] have predicted that the worldwide UAV market will expand significantly in the next decade. These studies also estimated that UAV spending will more than triple over the next decade, totaling close to \$55 billion in the next 10 years [3]. As stated in [2, 4], over the next 5–7 years, the UAV market in the U.S. will reach \$16 billion, followed by Europe, which is spending about \$3 billion. In US for example, development budgets increased rapidly after 2001, as shown in Fig. 1.4, and UAV research and development was given a powerful push [5]. On the other hand, the R&D budgets in Europe have increased slowly, as seen in Fig. 1.5. Today, there are several companies developing and producing hundreds of UAV designs. Indeed, major defense contractors are involved in developing and producing UAVs. At the same time, newer or smaller companies have also emerged with innovative technologies that make the market even more vibrant, as seen in Fig. 1.6. U.S. companies currently hold about 63–64% of the market share, while European companies account for less than 7% [2]. As shown in Table 1.2, in 2005, some 32 nations were developing or manufacturing more than 250 models of UAVs, and about 41 countries were operating more than 80 types of UAVs, primary for reconnaissance in military applications [5]. Table 1.2 lists the results of an investigation that tracked

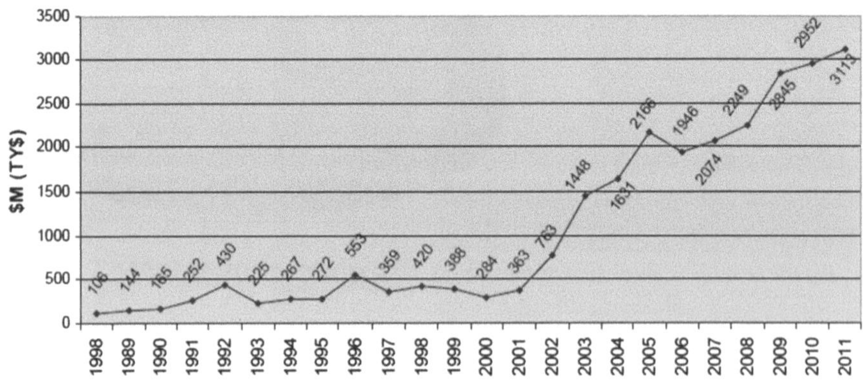

Fig. 1.4 Annual funding profile of the U.S. Department of Defense [5]

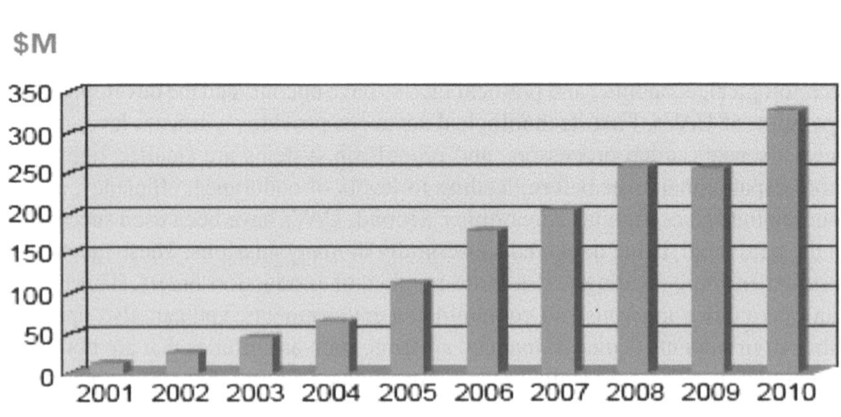

Fig. 1.5 Annual funding profile in Europe

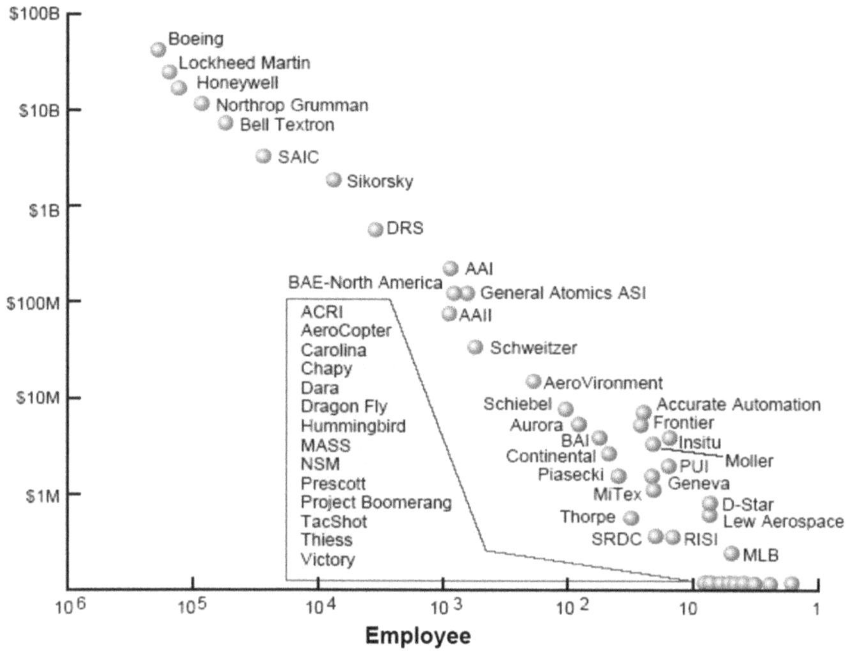

Fig. 1.6 The scale of the U.S. companies developing and manufacturing UAVs

and recorded the exporters, users, manufacturers, and developers of UAVs around the world. In some countries, including the group of seven (G7) industrialized countries and Russia, every category has a "Yes." Although their use varies, except for Japan and some other countries, the majority of the research and development is supported by defense expenditures. However, the civil UAV market is predicted to emerge over the next decade, starting first with government organizations requiring surveillance systems, such as coast guards, border patrol organizations, rescue teams, police, etc. Although armed forces around the world continue to strongly invest in researching and developing technologies with the potential to advance the capabilities of UAVs, commercial applications now drive many unmanned technologies. Among these technologies, some apply equally to manned aircraft like platform technologies (airframe, materials, propulsion systems, aerodynamics, etc.) and payload technologies (mission sensors, weapons, etc.). Other technologies are specific to UAVs in the sense that they compensate for the absence of an onboard pilot and thus enable unmanned flight and autonomous behavior. Indeed, UAVs rely predominantly on

- *Navigation sensors and microprocessors*: Sensors now represent one of the single largest cost items in an unmanned aircraft and are necessary for navigation and mission achievement. Processors allow UAVs to fly entire missions autonomously with little or no human intervention.

Table 1.2 Current exporters, operators, manufacturers, and developers of UAVs [5]

MTCR member	UA exporter	UA operator	UA manufacturer	UA developer
Argentina	No	Yes	Yes	Yes
Australia	Yes	Yes	Yes	Yes
Austria	Yes	No	Yes	Yes
Belgium	No	Yes	Yes	Yes
Brazil	No	No	No	No
Canada	Yes	No	Yes	Yes
Czech Republic	No	Yes	Yes	Yes
Denmark	No	Yes	No	No
Finland	No	Yes	No	No
France	Yes	Yes	Yes	Yes
Germany	Yes	Yes	Yes	Yes
Greece	No	No	No	Yes
Hungary	No	No	No	Yes
Iceland	No	No	No	No
Ireland	No	No	No	No
Italy	Yes	Yes	Yes	Yes
Japan	Yes	Yes	Yes	Yes
Luxembourg	No	No	No	No
The Netherlands	No	Yes	No	No
New Zealand	No	No	No	No
Norway	No	No	No	Yes
Poland	No	No	No	No
Portugal	No	No	No	Yes
Russia	Yes	Yes	Yes	Yes
South Africa	Yes	Yes	Yes	Yes
South Korea	No	Yes	Yes	Yes
Spain	No	No	Yes	Yes
Sweden	No	Yes	Yes	Yes
Switzerland	Yes	Yes	Yes	Yes
Turkey	Yes	Yes	Yes	Yes
Ukraine	Yes	Yes	Yes	Yes
United Kingdom	Yes	Yes	Yes	Yes
United States	Yes	Yes	Yes	Yes

- *Communication systems (data link)*: The principal issues for communication technologies are flexibility, adaptability, security, and cognitive controllability of the bandwidth, frequency, and information/data flows.
- *Ground Station Command, Control, and Communications (C3)*: There are several key aspects of the off-board C3 infrastructure that are being addressed, such as man–machine interfaces, multi-aircraft C3, target identification, downsizing ground equipment, voice control, etc. Advancing the state of the art in all of the areas discussed above will allow a single person to control multiple aircraft.

- *Aircraft onboard intelligence (guidance, navigation, and control)*: The intelligence that can be "packed" into a UAV is directly related to how complicated a task that it can handle, and inversely related to the amount of oversight required by human operators. More work needs to be done to mature these technologies in the near term to show their utility and reliability. The reader can refer to [5] for more details on forecasting trends in these technologies over the coming decades.

1.2 Unmanned Aerial Vehicles and Micro Aerial Vehicles: Definitions, History, Classification, and Applications

Before any discussion on UAV technologies, it is necessary to provide clarifications related to the terminology, classification, and potential applications of UAVs.

1.2.1 Definition

An uninhabited aircraft is defined using the general terms UAV (uninhabited aerial vehicle or unmanned aerial vehicle), ROA (remotely operated aircraft), and RPV (remotely piloted vehicle) [4]. A pilot is not carried by an uninhabited aerial vehicle, but the power source, which provides dynamic lift and thrust based on aerodynamics, is controlled by autonomous navigation or remote-control navigation. Therefore, neither a rocket, which flies in a ballistic orbit, nor a cruise missile, shell, etc. belong in this category. An unmanned airship that flies in the air with a help of gas is also not included in this category.

On the other hand, the AIAA defines a UAV as "an aircraft which is designed or modified, not to carry a human pilot and is operated through electronic input initiated by the flight controller or by an onboard autonomous flight management control system that does not require flight controller intervention." Although there is no strict definition of the difference between a UAV and MAV, according to a definition by DARPA (Defense Advanced Research Projects Agency) of the U.S. Department of Defense, an MAV has dimensions (length, width, or height) of 15 cm or less.

1.2.2 Brief History of UAVs

The first UAV was manufactured by the Americans Lawrence and Sperry in 1916. It is shown in Fig. 1.7. They developed a gyroscope to stabilize the body, in order to manufacture an auto pilot. This is known as the beginning of "attitude control," which came to be used for the automatic steering of an aircraft. They called their

Fig. 1.7 First UAV in the world, 1916

Fig. 1.8 UAVs in the 1960s and 1970s (Firebee)

device the "aviation torpedo" and Lawrence and Sperry actually flew it a distance that exceeded 30 miles. However, because of their practical technical immaturity, it seems that UAVs were not used in World War I or World War II.

The development of UAVs began in earnest at the end of the 1950s, taking advantage of the Vietnam War or the cold war, with full-scale research and development continuing into the 1970s. Figure 1.8 shows a UAV called Firebee. After the Vietnam War, the U.S. and Israel began to develop smaller and cheaper UAVs. These were small aircraft that adopted small engines such as those used in motorcycles or snow mobiles. They carried video cameras and transmitted images to the operator's location. It seems that the prototype of the present UAV can be found in this period.

Fig. 1.9 Predator in military use

Fig. 1.10 Civil use UAV by NASA (Helios)

The U.S. put UAVs into practical use in the Gulf War in 1991, and UAVs for military applications developed quickly after this. The most famous UAV for military use is the Predator, which is shown in Fig. 1.9. On the other hand, NASA was at the center of the research for civil use during this period. The most typical example from this time was the ERAST (Environmental Research Aircraft and Sensor Technology) project. It started in the 1990s, and was a synthetic research endeavor for a UAV that included the development of the technology needed to fly at high altitudes of up to 30,000 m, along with a prolonged flight technology, engine, sensor, etc. The aircraft that were developed in this project included Helios, Proteus, Altus, Pathfinder, etc., which are shown in Figs. 1.10–1.12. These were designed to carry out environmental measurements.

Fig. 1.11 Civil use UAV by NASA (Proteus)

Fig. 1.12 Civil use UAV by NASA (Altus)

1.2.3 Classification of UAV Platforms

During recent decades, significant efforts have been devoted to increasing the flight endurance and payload of UAVs, resulting in various UAV configurations with different sizes, endurance levels, and capabilities. Here, we attempt to classify UAVs according to their characteristics (aerodynamic configuration, size, etc.). UAV platforms typically fall into one of the following four categories:

- Fixed-wing UAVs, which refer to unmanned airplanes (with wings) that require a runway to take-off and land, or catapult launching. These generally have long endurance and can fly at high cruising speeds, (see Fig. 1.13 for some examples).
- Rotary-wing UAVs, also called rotorcraft UAVs or vertical take-off and landing (VTOL) UAVs, which have the advantages of hovering capability and high maneuverability. These capabilities are useful for many robotic missions, especially

in civilian applications. A rotorcraft UAV may have different configurations, with main and tail rotors (conventional helicopter), coaxial rotors, tandem rotors, multi-rotors, etc. (see Fig. 1.14 for some examples).

- Blimps such as balloons and airships, which are lighter than air and have long endurance, fly at low speeds, and generally are large sized (see Fig. 1.15 for some examples).

Fig. 1.13 Some configurations of fixed-wing UAVs

Fig. 1.14 Examples of rotary-wing UAVs

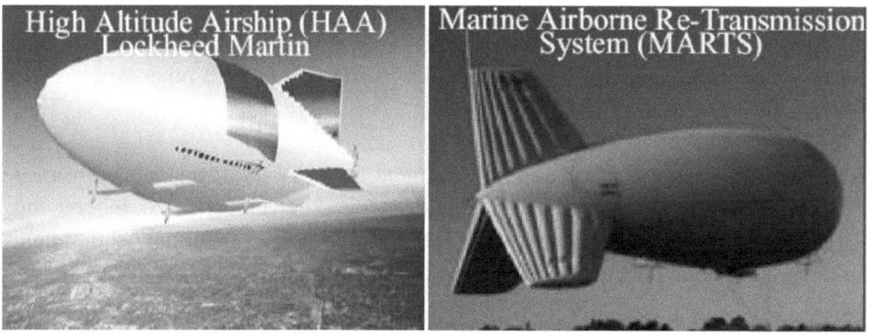

Fig. 1.15 Examples of airship-based UAVs

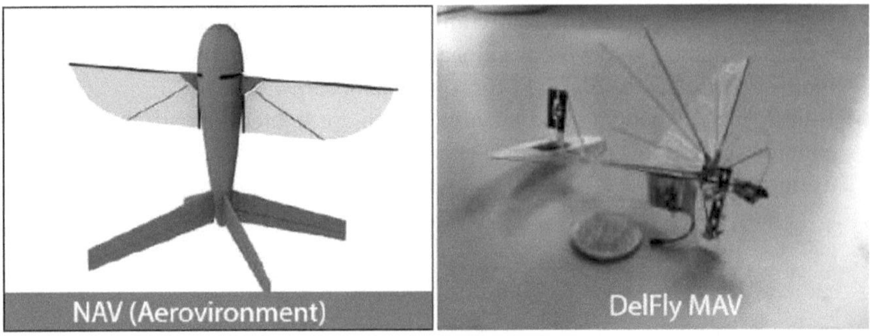

Fig. 1.16 Micro flapping-wing UAVs

- Flapping-wing UAVs, which have flexible and/or morphing small wings inspired by birds and flying insects, see Fig. 1.16.

There are also some other *hybrid* configurations or *convertible* configurations, which can take-off vertically and tilt their rotors or body and fly like airplanes, such as the Bell Eagle Eye UAV. Another criterion used at present to differentiate between aircraft is size and endurance [5]:

- High Altitude Long Endurance (HALE) UAVs, as for example, the *Northrop-Grumman Ryan's Global Hawks* (65,000 ft altitude, 35 h flight time, and 1,900 lb payload).
- Medium Altitude Long Endurance (MALE) UAVs, as for example *General Atomics's Predator* (27,000 ft altitude, 30/40 h flight time, and 450 lb payload).
- Tactical UAVs such as the *Hunter*, *Shadow 200*, and *Pioneer* (15,000 ft altitude, 5–6 h flight time, and 25 kg payload).
- Small and Mini man-portable UAVs such as the *Pointer/Raven (AeroVironment), Javelin* (*BAI*), or *Black Pack Mini* (*Mission Technologies*).
- Micro aerial vehicles (MAV): In the last few years, micro aerial vehicles, with dimensions smaller than 15 cm, have gained a lot of attention. These include the *Black Widow* manufactured by *AeroVironment*, the *MicroStar* from *BAE*, and many new designs and concepts presented by several universities, such as the *Entomopter* (*Georgia Institute of Technology*), *Micro Bat* (*California Institute of Technology*), and *MFI* (*Berkeley University*), along with other designs from European research centers (Fig. 1.17).

Currently, the main research and development for UAV platforms aims at pushing the limits/boundaries of the flight envelope and also the vehicle's size. Indeed, most ongoing ambitious projects (or prototypes in development) are about (1) unmanned combat air vehicles (UCAV) with high speed and high maneuverability or (2) micro aerial vehicles (MAVs) with insect-like size and performance.

Fig. 1.17 Unmanned aerial vehicles, from big platforms to micro flying robots

1.2.4 Applications

Currently, the main UAV applications are defense related and the main investments are driven by future military scenarios. Most military unmanned aircraft systems are primarily used for intelligence, surveillance, reconnaissance (ISR), and strikes. The next generation of UAVs will execute more complex missions such as air combat; target detection, recognition, and destruction; strike/suppression of an enemy's air defense; electronic attack; network node/communications relay; aerial delivery/resupply; anti-surface ship warfare; anti-submarine warfare; mine warfare; ship to objective maneuvers; offensive and defensive counter air; and airlift.

Today, the civilian markets for UAVs are still emerging. However, the expectations for the market growth of civil and commercial UAVs are very high for the next decade (Fig. 1.18). Potential civil applications of UAVs are

- Inspection of terrain, pipelines, utilities, buildings, etc.
- Law enforcement and security applications.

Fig. 1.18 Unmanned combat air vehicles (UCAV) and micro aerial vehicles (MAVs) as trends in UAV platform research and development

- Surveillance of coastal borders, road traffic, etc.
- Disaster and crisis management, search and rescue.
- Environmental monitoring.
- Agriculture and forestry.
- Fire fighting.
- Communications relay and remote sensing.
- Aerial mapping and meteorology.
- Research by university laboratories.
- And many other applications.

1.3 Recent Research and Development of Civil Use Autonomous UAVs in Japan

We next give a brief overview of the recent research and development in Japan of civil use autonomous UAVs. JAXA has been developing a multiple-purpose small unmanned aircraft made of CFRP since 2002, which is shown in Fig. 1.19. The goals

Fig. 1.19 UAV Prototype II by JAXA

Fig. 1.20 Noppi-III by NIPPI Corp

of this project are as follows: a gross load of about 20 kg (which includes a payload), cruising altitude of about 3,000 m, cruising time of 30 h, and the use of a UHF radio modem. The UAV made by Japan Aircraft Co., Ltd., which is shown in Fig. 1.20, has a full length of 3 m, 5 kg payload, and flying speed of about 90–120 km/h. The kite aircraft developed by the Sky Remote Company is shown in Fig. 1.21. It is being used to conduct desert investigations around Dunhuang in inland China, coastline investigations, etc. It has an airspeed of 36–54 km/h, cruising time of 2 h, gross weight of 27 kg, and payload of 6 kg. In Gifu prefecture, Kawajyu Gifu Engineering and Furuno Electric Co., Ltd. spent three fiscal years, starting in 2003, developing a new UAV supported by the "fire-fighting science-of-disaster-prevention technical research promotion system" of the Fire Defense Agency. As shown in Fig. 1.22 their UAV can be used to monitor a forest fire, seismic hazard situation, etc. Its payload is about 500 g and its cruising time is 30 min. It has a gross weight of 4–7 kg and can carry a digital camera, infrared camera, etc., according to application. They have developed from No. 1 to No. 4. No. 4 adopted an electric motor as the power plant and incorporated a U.S. autopilot.

Fig. 1.21 Kite plane by Sky Remote Co

Fig. 1.22 UAV by Gifu Industrial Association

Next, rotor-wing UAVs are described. Yamaha Motor began developing an unmanned helicopter for agricultural–chemical spraying in 1983, using a motorcycle engine. It succeeded in the use of an unmanned helicopter for the application of fertilizer for the first time in the world in 1989, and by the end of December, 2002, 1,281 of their UAVs were in use for agricultural–chemical spraying in Japan, as seen in Table 1.1. Using this unmanned helicopter as a base, the development of an autonomous type of UAV using a GPS sensor began in 1998, and succeeded in the observation of the volcanic activity at Usu-zan in April, 2000. Yamaha's RMAX, which has a flight distance of about 10 km, is shown in Fig. 1.23. Its gross-load is about 90 kg, and it has a payload of about 30 kg and a flight time of 90 min. The use of the unmanned helicopters in the agricultural field has increased yearly, with 1,687 registered in Japan at the end of December, 2002, as seen in Table 1.1. In this area, the unmanned helicopter of Yamaha has secured about 80% of the market share, based on the above-mentioned numbers. Moreover, in connection with this, 10% or more of the paddy fields in Japan are sprayed by unmanned helicopters. Moreover, about 8,000 individuals are licensed to operate an unmanned helicopter.

Fig. 1.23 Yamaha RMAX helicopter

Fig. 1.24 Unmanned RPH2 helicopter

This number is encouraging and it should be noted that of all the advanced nations, the rotor-wing UAV, i.e., unmanned helicopter, is an invention, not of the U.S., but of Japan.

At the same time, Fuji Heavy Industries is developing RPH2 for fully autonomous chemical spraying, as seen in Fig. 1.24. An automated take-off-and-landing technology has been developed for RPH2, which can carry a payload of 100 kg with a gross weight of about 300 kg. It has a spraying altitude of 5 ± 1 m, spraying width of 10 ± 1 m, and spraying speed 8 ± 1 m/s. Moreover, a recording of volcanic activity in Miyake Island has also been carried out.

In addition, polar zone observations for science missions have been performed at both the North and South Poles. Although Cessna planes, artificial satellites, etc. are presently used, more general-purpose UAV applications are being considered. In particular, ecosystem investigations of rock exposure, ice sheets, sea ice distribution, vegetation, penguins, seals, snow coverage states, crevasses, etc. are the main survey areas for earth science investigations.

Other applications include command transmissions from high altitudes, relaying communications, the acquisition of three-dimensional data for a region, etc. In addition, in February, 2005, the 46th observation party attempted the first flight of a small fixed wing UAV on Onguru Island. Moreover, sea geomagnetism observations by UAV have also been considered.

Chiba University's UAV group started research on autonomous control in 1998, advanced joint research with Hirobo, Ltd. from 2001, and realized a fully autonomously controlled helicopter for small-scale hobbyists [6–12]. A power line monitoring [12] application uses a UAV called SKYSURVEYOR, as seen in Fig. 1.25. This UAV has a gross load of 48 kg and a cruising time of 1 h, and regardless of the movements of the helicopter, various onboard cameras are capable of continuously monitoring a power line. In addition, it has a payload of about 20 kg. Although Fig. 1.26 shows a helicopter (SST-eagle2-EX) for hobbyists, with a gross load of 5–7 kg, autonomous control of this vehicle has already been accomplished. The cruising time is about 20 min and the payload is about 1 kg. This body was used for the automated commercial radio controlled helicopter for hobbyists, since it is capable of autonomous flight and can be freely flown by one person, is a cheap and simple system, and can apply chemical sprays, such as for an orchard, field, small garden, etc. In the future, it will also be used for aerial photography, various surveillance applications, and disaster prevention and rescue.

Fig. 1.25 SKYSURVEYOR by Chiba Univ., Hirobo Co., and Chugoku Electric Power Co

Fig. 1.26 Hobby class autonomous helicopter by Chiba Univ and Hirobo Co

1.4 Subjects and Prospects for Control and Operation Systems of Civil Use Autonomous UAVs

GH Craft and Chiba University are furthering the research and development of an autonomous control system for the four rotor-tilt-wing aircraft, seen in Fig. 1.27. This QTW (quad tilt wing) UAV has a gross load of about 30 kg. Take-offs and landings are performed in helicopter mode, while high-speed flight is carried out in airplane mode. Although the Bell Company in the U.S. developed a QTR (quad tilt rotor)-UAV and its first flight [13] occurred in January, 2006, this QTW-UAV is not yet available. Moreover, Seiko Epson and Chiba University tackled the autonomous control of the smallest micro flying robot in the world (12.3 g), shown in Fig. 1.28, and succeeded in perfecting autonomous control inside a room using image

Fig. 1.27 QTW-UAV by GH Craft and Chiba Univ

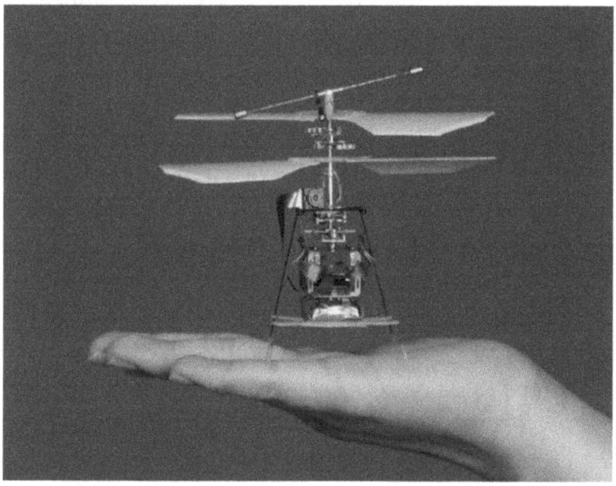

Fig. 1.28 Micro flying robot by Seiko-Epson and Chiba Univ

Fig. 1.29 Autonomous landing of MAV

Fig. 1.30 Autonomous hovering

processing and a camera [14]. At about 170 g, the larger XRB by Hirobo, Ltd. also demonstrated autonomous control at Chiba University. Autonomous control for free flight is currently being tested in interior locations.

In addition, we have been aggressively studying and developing our own advanced flight control algorithm by means of a quad-rotor MAV provided by a German company (Ascending Technologies GmbH) as a hobby helicopter. Figures 1.29 and 1.30 show two photos of autonomous landing and hovering as examples. We chose this platform because it presents good performance in terms of weight and payload. The original X-3D-BL Kit consists of a solid airframe; brushless motors and the associated motor drivers; X-base, which is an electronic card that decodes the receiver outputs and sends commands to the motors; and an X-3D board that incorporates three gyroscopes for stabilization. The total weight of the original platform is about 400 g, including batteries, and it has a payload of about 200 g. The flight time is up to 20 min without a payload and about 10 min with a payload of 200 g. The X-3D-BL helicopter can fly at high speeds of close to 8 m/s. These good characteristics are due to its powerful brushless motors, which can rotate at a very high speed. Furthermore, the propellers are directly mounted on the motors without using mechanical gears, thereby reducing vibration and noise.

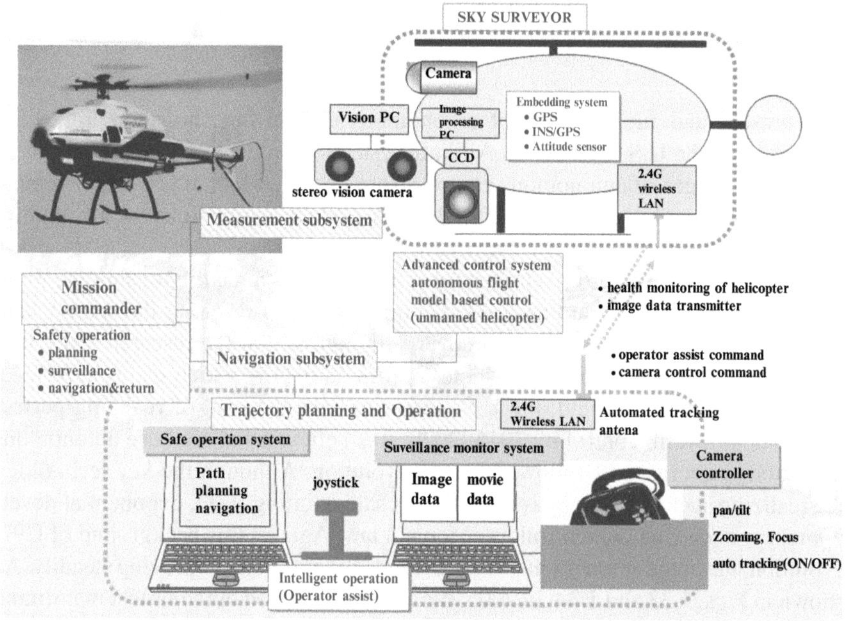

Fig. 1.31 System overview of SKY SURVEYOR for power transmission line monitoring by Hirobo Co., Chiba Univ., and Chugoku Electric Power Co

The configuration of the autonomous control system in the power line monitoring helicopter of Fig. 1.25 is shown in Fig. 1.31. Generally, the configuration of a civil use autonomous UAV for a specific mission is as shown in Fig. 1.31. We could roughly divide system configurations into the power line monitoring helicopter SKY SURVEYOR, which uses a ground station, and autonomous UAVs. The power line monitoring devices or the various devices used by the autonomous control system such as sensors and a computer are carried in the body. The sensors used for autonomous control include a GPS receiver, attitude sensor, and compass. Flight using GPS/INS for compound inertial navigation or a three-dimensional stereo vision based method is also possible if needed. A ground station can monitor the flight and surveillance operation, and can cancel it if needed. As for attitude control, an operator performs only position control of a helicopter with autonomous control, although the so-called operator assisted flight can also be performed. In addition, although a power line surveillance image is recorded on the video camera of a UAV utilizing automatic capture mode and is simultaneously transmitted to a ground station, an operator can also interrupt and control the direction and zoom of a power line monitoring camera at any time.

1.5 Future Research and Development of Autonomous UAVs and MAVs

The present and future levels of autonomous control are shown in Fig. 1.32. According to the U.S. Unmanned Aircraft Systems Roadmap 2005–2030 [5], there are various stages of autonomous control, from level 1, which refers to the remote control of one vehicle, to level 10, which is perfect autonomous swarm control similar to the formation flight of insects or birds.

The present level performs trajectory re-planning during a flight using the flight program, vision sensor, and embedded computer, and is reaching the stage where obstacle avoidance is possible. Moreover, although still at the research level, it is now possible to fly two or more vehicles in formation [15], which seems to be level 4 or 5. In the military field, the U.S. seems to have the goal of realizing perfect autonomous swarm control by 2015–2020. It is believed that civil use autonomous uninhabited aircraft will follow the same evolution. Although the key technology for realizing such technology is the CPU, as shown in Fig. 1.33, exponential development is occurring, which follows Moore's law. Against this background of CPU evolution, the autonomous control of a UAV also seems to be improving steadily. As shown in Figs. 1.33 and 1.34, in 2005, the computing speed of the fastest mainframe (CRAY supercomputer) was nearly equal to the human brain. Furthermore, Moore's law predicts that the performance of the microprocessor for a personal computer will be equal to that of the human brain by around 2015, and will be equal to the brain's

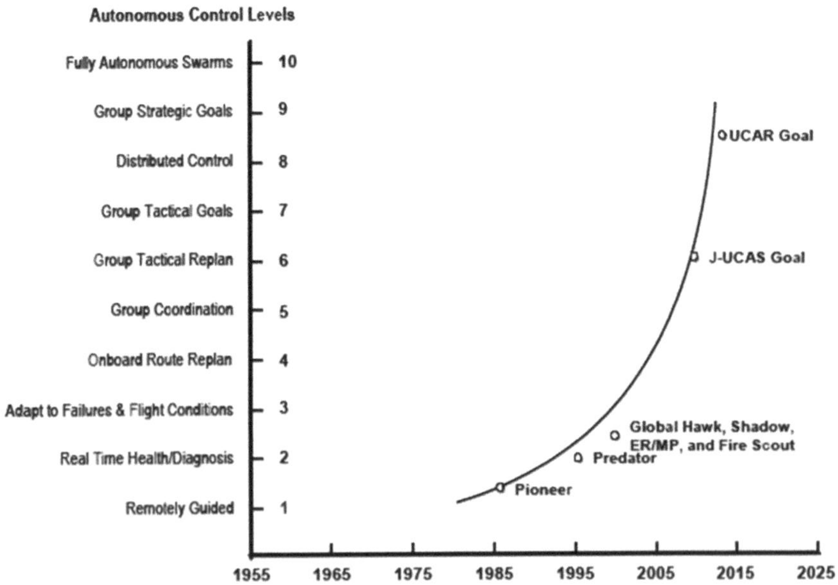

Fig. 1.32 Trend in UAV autonomy [5]

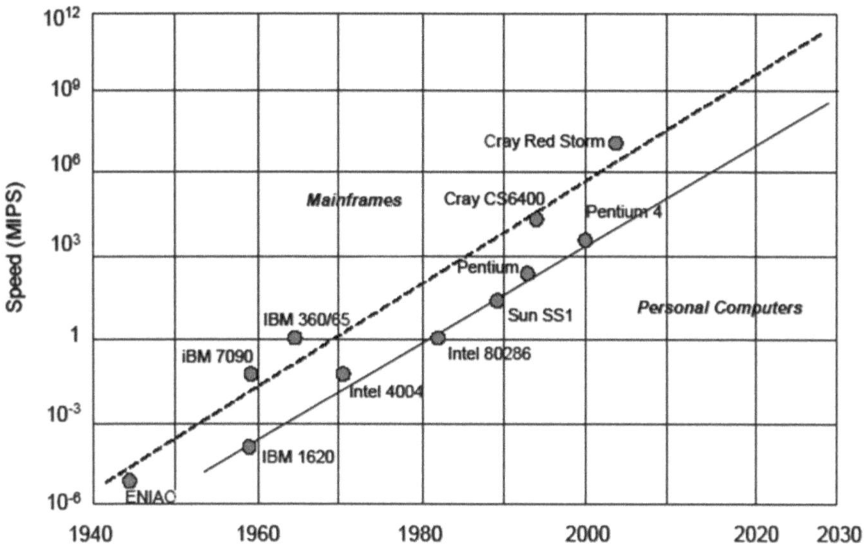

Fig. 1.33 Trend in processor speed [5]

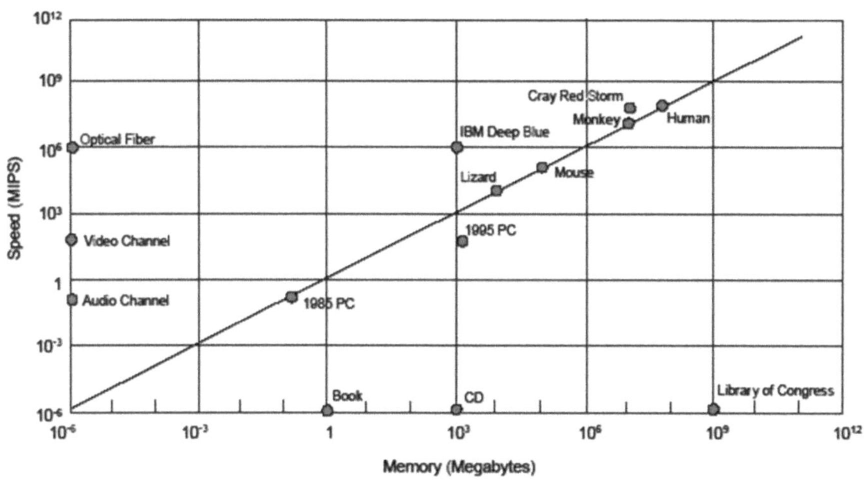

Fig. 1.34 Relationship between processor speed and memory [5]

storage capacity by around 2030. However, if there is no evolution to this level, it will be difficult for autonomous uninhabited aircraft to carry out formation flight like birds.

Moreover, this is also important from the viewpoint of body design, including the loading and reliability of a data link and advanced sensors, the design of a more lightweight body, high propulsion per unit of weight, body structure with high stability, and body specifically suitable for autonomous control.

The following are important subjects for the sake of increasing the efficiency of inspection and surveillance work, data relay, refueling in the air, etc.

1. Formation-flight control: noncommercial use as a future research task, with an accuracy of several cm depending on the case.
2. Integrated hierarchical control of UAVs to MAVs: the ability to fly various classes simultaneously.
3. For example, high precise missions could be performed by controlling several vehicles simultaneously, from big UAVs to small MAVs.
4. Super-high-altitude flight: since a UAV does not carry people, flying into the stratosphere, etc. is also attainable.
5. Consequently, prolonged flights suitable for science observation missions can also be attained.
6. High precision orbital flight: this is a technology that will be needed in the future.
7. All weather flights.
8. Radar payloads for impact prevention, etc.
9. An intelligent flight system and operation management.
10. Advanced reliability, etc.

There are an infinite number of public welfare applications for UAVs. They could be used for detailed perpendicular direction weather surveys, ozone layer observations, air pollution observations, coastline observations, fire detection activities, vegetation growth observations and chemical spraying, glacier and snow coverage investigations, three-dimensional mapping, gravity surveys, magnetic field measurements, polar zone observations, river surveillance, observations of typhoon and hurricane generation processes, tornado observations and predictions, forest surveillance, ecosystem surveillance, the inspection of large-scale national parks, traffic surveillance, disaster prevention and rescue operation support, power line surveillance, the surveillance of industrial complexes or pipelines, next-generation logistics distribution systems, etc. Their applications will be endless. Such research and development of a civil use UAV should place our country in a powerful position as a world leader. On the other hand, because this represents ultramodern technology, it will become very important who uses it, and for what purpose. Although human beings are capable of abusing technology, if it is used correctly, history will show its contribution to mankind's happiness. In parallel with the development of such ultramodern technology, it is necessary to develop a mechanism to prevent its abuse. The author will be pleased if this book serves as an aid to researchers and engineers in this field.

1.6 Objectives and Outline of the Book

The main objective of this book is the systematic description of the current research and development of small or miniature unmanned aerial vehicles and micro aerial vehicles, with a focus on rotary wing vehicles and the integrated prototypes developed by the Nonami research laboratory at Chiba University. In particular, this book might be used as a comprehensive overview for beginners. Most chapters have some demonstration videos. Readers can use these videos to verify the performance discussed. This book is divided into three parts.

Part I consists of four chapters, Chaps. 2–5, which discuss the modeling and control of small and mini rotorcraft UAVs and MAVs. These chapters provide an introduction designed to motivate and guide readers gradually into the field of the fundamental modeling and control of UAVs and MAVs.

Chapter 2, which is entitled "Fundamental Modeling and Control of Small and Miniature Unmanned Helicopters," focuses on the modeling and control system design of unmanned helicopters. Two types of unmanned helicopters exist today – a single-rotor helicopter and a coaxial-rotor helicopter. In general, small helicopters with weights from 1–50 kg use the single-rotor mechanism, while miniature helicopters with weights of less than 500 g use the coaxial-rotor mechanism. In this chapter, small and miniature unmanned helicopters are discussed. First, their mathematical models are obtained as a transfer function or state space equation. Subsequently, the optimal control design method for both small and miniature unmanned helicopters is introduced.

Chapter 3, which is entitled "Autonomous Control of a Mini Quadrotor Vehicle Using LQG Controllers," presents techniques for modeling and designing the attitude controller for a quad-rotor MAV. Compared with a small helicopter that uses a single rotor or contra-rotating propellers, the advantages of quad-rotor MAVs are: the ability to carry larger payloads, they're more powerful and can more easily handle turbulence such as wind, and they're easier to design using a compact airframe. At present, the autonomous control of quad-rotor MAVs is a very active area of research. A key characteristic of quad-rotor MAVs is that all the degrees of freedom of the airframe are controlled by tuning the rotational speeds of the four motors. Moreover, because their internal controller calculates the angular velocity feedback by using a gyro sensor, the nonlinearity of the airframe becomes weaker and a linear model is more appropriate. Therefore in this chapter, we introduce the technique of linear modeling and model based controller design for quad-rotor MAVs, along with the performance of the designed controllers.

Chapter 4, which is entitled "Development of Autonomous Quad-Tilt-Wing (QTW) Unmanned Aerial Vehicle: Design, Modeling, and Control," provides an autonomous attitude control for a quad-tilt-wing unmanned aerial vehicle (QTW-UAV). A QTW-UAV can achieve vertical takeoff and landing; further, hovering flight, which is a characteristic of fixed-wing helicopters, and high cruising speeds, which are a characteristic of fixed-wing aircraft, can be achieved by changing the angle of the rotors and wings by a tilt mechanism. First, we construct an attitude model of the QTW-UAV by using the identification method. We then design the attitude control system with a Kalman filter-based linear quadratic integral (LQI) control method. Experimental results show that a model-based control design is very useful for the autonomous control of a QTW-UAV.

Chapter 5, which is entitled "Linearization and Identification of Helicopter Model for Hierarchical Control Design," presents an analytical modeling and model-based controller design for a small unmanned helicopter. Generally, it can be said that helicopter dynamics are nonlinear, with the coupling of each axis. However, for low speed flights, i.e., speeds less than 5 m/s, the dynamics can be expressed by a set of linear equations of motion as a SISO (single input single output) system. The dynamics of the helicopter are divided into several components. We derive a model for each component from either the geometric relation or equation of motion. By combining all of the components, we derive two linear state equations that describe the helicopter's lateral and longitudinal motions. The parameters of the model are determined by the helicopter's specs. Based on the derived models, we design a control system by using the linear quadratic integral (LQI). The validity of these approaches is then verified by flight tests.

Part II consists of four chapters, Chaps. 6–9, which discuss advanced flight control systems for rotorcraft UAVs and MAVs. These chapters cover relatively advanced modeling and control methods, like the nonlinear modeling and control of UAVs and MAVs.

Chapter 6, which is entitled "Analysis of the Autorotation Maneuver in Small-Scale Helicopters and Applications for Emergency Landing," shows linear and nonlinear models of a helicopter autorotation maneuver, focusing on the main rotor revolution and descent rate dynamics. The nonlinear model for the descent rate dynamics is derived from the Blade Element Theory combined with an induced velocity linear approximation proposed in this paper. The linear model is derived by implementing a neural network optimization method based on experimental data. The derived nonlinear and linear models are verified by comparing the simulated data of the model with experimental data. We also carry out a preliminary experiment for autorotation landing with a simple PI controller. The rotor revolution and descent rate are found to be well controlled by the designed controller.

Chapter 7, which is entitled "Autonomous Acrobatic Flight of Small Robotic Helicopter Using Feedforward Sequence Control," focuses on a simple logic that enables a small unmanned helicopter to perform an autonomous axial roll using a feedforward sequence control method. To develop the feedforward input sequence, we study the maneuvering logic and strategies developed by expert human pilots to achieve acrobatic maneuvers. Here, we first explain the hardware structure of our SF 40 helicopter and the avionics system used to overcome hardware limitations in achieving aggressive maneuvers and then explain the simple feedforward control logic used based on recorded pilot control command sequences. This approach may be useful in achieving smooth, less time consuming, high-speed maneuvers, as this method does not require complicated nonlinear control methods.

Chapter 8, which is entitled "Mathematical Modeling and Nonlinear Control of VTOL Aerial Vehicles," contains the following contents. In an effort to make autonomous flight behaviors available to mini and micro rotorcraft, an embedded and inexpensive autopilot was developed. In this chapter, we present the main steps for designing a nonlinear flight controller for mini rotorcraft unmanned aerial vehicles (UAVs). The proposed control system is based on a nonlinear model of rotorcraft UAVs and uses the inner and outer-loop control scheme. It considers the system's nonlinearities and coupling and results in a practical controller that is easy to implement and tune. The asymptotic stability of the complete closed-loop system was proven by exploiting the theories of systems in cascade. In addition to a controller design and stability analysis, the chapter provides information about the air vehicle, sensor integration, and real-time implementation of guidance, navigation, and control algorithms. A mini quad-rotor UAV, equipped with an embedded autopilot, has undergone an extensive program of flight tests, resulting in various flight behaviors under autonomous control from takeoff to landing. Experimental results demonstrating the capabilities of our autonomous UAV are presented.

Chapter 9, which is entitled "Formation Flight Control of Multiple Small Autonomous Helicopters Using Predictive Control," focuses on the model-based formation flight control of multiple small unmanned helicopters as an example of the advanced control of unmanned aerial vehicles (UAVs). We design an autonomous formation flight control system using a "leader-following" configuration. In order to achieve good control performance under the system constraints, the "model predictive control" is used for the translational position control of the follower helicopters. Position constraints such as the moving range and collision avoidance problem are considered in the real-time optimal control calculations. To achieve robustness against disturbance, a minimal-order disturbance observer is used to estimate the unmeasurable state variables and disturbance. Simulation results are presented to show the feasibility of the control strategy. A formation flight control experiment is performed using two helicopters. The experimental results demonstrate an accurate control performance. The position constraint capability is confirmed through experiments with a single helicopter. Finally, the robustness against wind is verified by a windy condition experiment.

Part III consists of four chapters, Chaps. 10–13, which discuss the guidance and navigation of short-range UAVs and MAVs. These chapters include the development of low-cost attitude sensors and the vision based flight control of UAVs and MAVs.

Chapter 10, which is entitled "Guidance and Navigation Systems for Small Aerial Robots," contains the following contents. As the capabilities of unmanned aerial vehicles (UAVs) expand, increasing demands are being placed on the hardware and software that comprise their guidance and navigation systems. Guidance, navigation, and control algorithms are the core of a UAV's flight software, which is used to successfully complete the assigned mission through autonomous flight. This chapter describes some guidance and navigation systems that we have designed and successfully applied to the autonomous flight of a mini rotorcraft UAV that weighs less than 0.7 kg. The real-time flight test results show that the vehicle can perform autonomous flight reliably in indoor and outdoor environments.

Chapter 11, which is entitled "Design and Implementation of Low-Cost Attitude Quaternion Sensor," describes the development of a low-cost attitude sensor. In the previous chapters, the control system design method for several UAVs/MAVs with rotary wings was shown, and several kinds of controllers were designed. The most important controller is the attitude controller because if attitude control is not achieved, no other control such as velocity and position control can be achieved. To achieve attitude control, it is necessary to measure the attitude of a UAV/MAV. Hence, we require an attitude sensor. However, conventional attitude sensors are somewhat expensive and heavy, and thus cannot be used for the attitude control of small UAVs and MAVs. Therefore, the design of an attitude estimation algorithm using low-cost sensors, accelerometers, gyro sensors, and magnetic sensors is introduced. Finally, a low-cost attitude sensor is developed and evaluated by comparing it with a conventional high-accuracy sensor.

Chapter 12, which is entitled "Vision Based Navigation and Visual Servoing of Mini Flying Machines," focuses on the following contents. The design of reliable navigation and control systems for unmanned aerial vehicles (UAVs) based only on visual cues and inertial data has many unsolved and challenging problems, ranging from hardware and software development to pure control–theoretical issues. This chapter addresses these issues by developing and implementing an adaptive vision-based autopilot for the navigation and control of small and mini rotorcraft UAVs. The proposed autopilot includes a visual odometer (VO) for navigation in GPS denied environments and a nonlinear control system for flight control and target tracking. The VO estimates the rotorcraft ego-motion by identifying and tracking visual features in the environment, using a single camera mounted onboard the vehicle. The VO has been augmented by an adaptive mechanism that fuses optic flow and inertial measurements to determine the range and recover the three-dimensional position and velocity of the vehicle. The adaptive VO pose estimates are then exploited by a nonlinear hierarchical controller to achieve various navigational tasks, including take-off, landing, hovering, trajectory tracking, target tracking, etc. Furthermore, the asymptotic stability of the entire closed-loop system has been established using systems in cascade and adaptive control theories. Experimental flight test data over various ranges of the flight envelope illustrate that the proposed vision-based autopilot performs well and allows a mini rotorcraft UAV to achieve autonomously advanced flight behaviors by using vision.

Chapter 13, which is entitled "Autonomous Indoor Flight and Precise Automated-Landing Using Infrared and Ultrasonic Sensors," presents a method for making a safe landing using low cost sensors. The accuracy of the data provided by a small and low cost GPS system is insufficient for precisely landing micro aerial vehicles (MAVs). This study shows how an MAV can land on a small targeted landing site by using rangefinders rather than imprecise GPS data. This chapter describes a proposed movable range finding sensor system for measuring the environment and an algorithm for position measurement. This range finding system consists of four infrared (IR) rangefinders, four servo motors, and one ultrasonic rangefinder. In order to measure the MAV's position, the sensor system vertically swings each IR rangefinder using the servo motors, and these IR sensors detect the edge of the landing target. This sensor system calculates the position from the measured edge direction and the ultrasonic altitude. Additionally, autonomous hovering experiments over a 52 $\times$ 52 cm table, as well as autonomous landings, were carried out indoors using the proposed sensor system. Our experiments succeeded, and as a result, the MAV kept flying horizontally within a 17.9-cm radius circle, and then landed on the table from a height of 50 cm. The IR rangefinders were selected because of the payload limitations of small MAVs. If the MAV's payload capacity were higher, then a laser rangefinder would be used, since lasers can operate better in sunlight than IR sensors and also tend to have longer scanning ranges and higher accuracy.

References

1. The unmanned aerial vehicles (UAV) market 2009–2019 (2009) Visiongain
2. Dickerson L (2007) UAV on the rise. Aviat Week Space Technol, Aerospace Source Book 2007 166(3)
3. http://www.roboticstrends.com/displayarticle880.html (2006) Robotics Trends
4. Unmanned aerial vehicles roadmap 2002–2027 (2002) Office of the Secretary of Defense
5. Unmanned aircraft systems roadmap 2005–2030 (2005) Office of the Secretary of Defense
6. Fujiwara D, Shin J, Hazawa K, Igarashi K, Fernando D, Nonami K (2003) Autonomous flight control of small hobby-class unmanned helicopter, report 1: hardware development and verification experiments of autonomous flight control system. J Robot Mechatron 15(5):537–545
7. Fujiwara D, Shin J, Hazawa K, Nonami K (2004) H∞ hovering and guidance control for autonomous small-scale unmanned helicopter. Trans Jpn Soc Mech Eng C 70(694):1708–1714 (in Japanese)
8. Shin J, Fujiwara D, Hazawa K, Nonami K (2002) Attitude control and hovering control of radio-controlled helicopter. Trans Jpn Soc Mech Eng C 68(675):3284–3291 (in Japanese)
9. Hazawa K, Shin J, Fujiwara D, Igarashi K, Fernando D, Nonami K (2003) Autonomous flight control of small hobby-class unmanned helicopter, report 2: modeling based on experimental identification and autonomous flight control experiments. J Robot Mechatron 15(5):546–554
10. Shin J, Fujiwara D, Hazawa K, Nonami K (2004) Model based optimal attitude and positioning control of small-scale unmanned helicopter. Trans Jpn Soc Mech Eng C 70(697):2631–2637
11. Hazawa K, Shin J, Fujiwara D, Igarashi K, Fernando D, Nonami K (2004) Autonomous autorotation landing of small unmanned helicopter. Trans Jpn Soc Mech Eng C 70(698):2862–2869 (in Japanese)
12. Shimo Y, Nishioka T (2006) Small unmanned helicopter and development for actual use of power transmission line monitoring system. Trans Jpn Soc Aeronaut Space Sci 54(627):22–25 (in Japanese)
13. Hirschberg M (2006) Bell makes first flight with new eagle eye demonstrator. Vertiflight, The American Helicopter Society 52(1):14–19
14. Wang W, Hirata M, Nonami K, Miyazawa O Modeling and H∞ hovering control of micro-miniature ultra-light weight micro flying robot, CD-8
15. Masuko Y (2005) Toward cooperative aircraft swarm control. Mitsubishi Heavy Industries Report 42(1):20–21 (in Japanese)
16. Castillo P, Lozano R, Dzul A (2005) Modelling and control of mini-flying machines. Springer-Verlag, New York
17. http://www.vectorsite.net/twuav.html. Unmanned aerial vehicles
18. Valvanis KP (ed) (2007) Advances in unmanned aerial vehicles. Intelligent systems, control, and automation: science and engineering, vol 33
19. Cox TH et al (2004) Civil UAV capability assessment -draft version-, NASA Report
20. Aviation Week and Space Technology (2005) "2005 Aerospace source book"
21. Presentation materials of the 2nd workshop on trends of small unmanned aircraft and application for scientific observation, 2005 (in Japanese)
22. Proceedings of the 47th joint conference on automatic control, 2004 (in Japanese)
23. Valavanis KP, Oh P, Piegl LA (2008) Unmanned aircraft systems. Springer, Berlin
24. Ollero A, Maza I (eds) (2007) Multiple heterogeneous unmanned aerial vehicles, Springer, Berlin

Part I
Modeling and Control of Small and Mini Rotorcraft UAVs

Chapter 2
Fundamental Modeling and Control of Small and Miniature Unmanned Helicopters

Abstract In this chapter, the modeling and control system design of unmanned helicopters are introduced. Two types of unmanned helicopters exist today – a single-rotor helicopter and a coaxial-rotor helicopter. In general, small helicopters with weights from 1 to 50 kg use the single-rotor mechanism, while miniature helicopters with weights less than 500 g use the coaxial-rotor mechanism. In this chapter, small and miniature unmanned helicopters are discussed. First, their mathematical models are obtained as a transfer function or state space equation. Subsequently, the optimal control design for the control method for both small and miniature unmanned helicopters is introduced.

Video Links:

Fully autonomous hovering control
http://mec2.tm.chiba-u.jp/monograph/Videos/Chapter2/1.avi
Square trajectory following control
http://mec2.tm.chiba-u.jp/monograph/Videos/Chapter2/2.wmv
Circle trajectory following control
http://mec2.tm.chiba-u.jp/monograph/Videos/Chapter2/3.avi
S character trajectory following control by preview optimal control
http://mec2.tm.chiba-u.jp/monograph/Videos/Chapter2/4.wmv

2.1 Introduction

In recent times, unmanned/Micro aerial vehicles (UAVs/MAVs) have been operated across the world; they have also been the subject of considerable research. In particular, UAVs/MAVs with rotary wings have been expected to perform various tasks such as monitoring at fixed points and surveillance from the sky since they can perform not only perform static flights by hovering but also achieve vertical takeoffs and landing. Helicopters have been used for personnel transport, carrying goods, spreading information, and performing monitoring duties for long periods. A manned helicopter has to be used for all these duties. On the other hand, unmanned

K. Nonami et al., *Autonomous Flying Robots: Unmanned Aerial Vehicles and Micro Aerial Vehicles*, DOI 10.1007/978-4-431-53856-1_2, © Springer 2010

helicopters that can be operated by radio control have been developed as a hobby. Since unmanned helicopters are often superior to manned helicopters in terms of cost and safety, in recent years, accomplishing tasks using unmanned helicopters has become popular. Actually, chemical spraying of paddy fields has been performed for a long time by using manned helicopters. but this task has become to be performed by unmanned helicopter named YAMAHA-RMAX. However, unmanned helicopters have a few disadvantages. First, considerable expertise is required to operate unmanned helicopters by radio control, and hence, vast labor resources are employed to train operators. Moreover, it is impossible to operate unmanned helicopters outside visual areas because of lack of radio control, and the working area is hence limited remarkably. For solving the above problems, it is necessary to realize autonomous control of unmanned helicopters. Hence, several researches on the autonomous control of unmanned helicopters have been performed across the world. Some representative studies include [1–14] and [17–31].

However, no general method for designing the control system of small unmanned helicopters has been developed yet – today, various control system design techniques by different study groups using different helicopters exist. In this chapter, a model-based control system design will be introduced as a general control method for small unmanned helicopters. Further, a mathematical model for the model-based control system design and a design of the optimal control system by using the derived model will be introduced.

2.2 Fundamental Modeling of Small and Miniature Helicopters

2.2.1 Small and Miniature Unmanned Helicopters

In this section, single-rotor and coaxial-rotor helicopters are introduced. A single-rotor helicopters consist of four parts – fuselage, tail system, main rotor, and stabilizer (Fig. 2.1). Coaxial-rotor helicopters also consist of four parts – fuselage, lower rotor, upper rotor, and stabilizer (Fig. 2.2). Each helicopter has a special actuator mechanism called a swash plate. The swash plate is mounted at the main mast of the helicopter to actuate the rotor blade pitch angle. An overview of the swash plate is shown in Figs. 2.3 and 2.4. In the following sections, the modeling of the single-rotor and coaxial-rotor helicopters is shown.

2.2.2 Modeling of Single-Rotor Helicopter

In this section, a mathematical model of the single-rotor helicopter is derived. The single-rotor helicopter comprises a large number of rigid bodies; hence, we must first derive the model of each rigid body. We then combine the models of all rigid

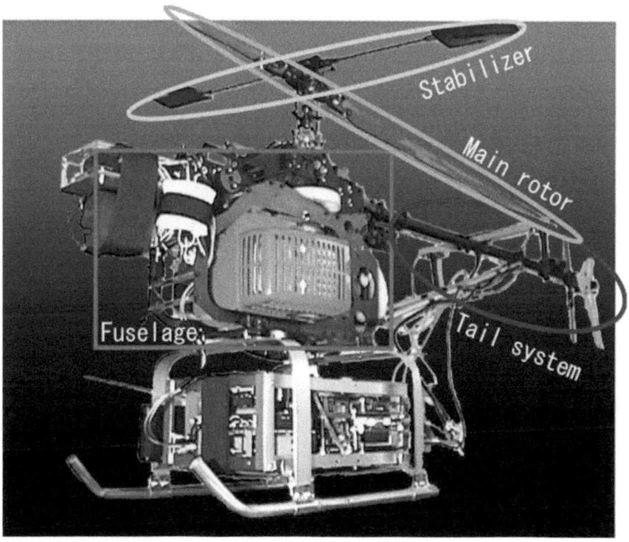

Fig. 2.1 Overview of single-rotor helicopter

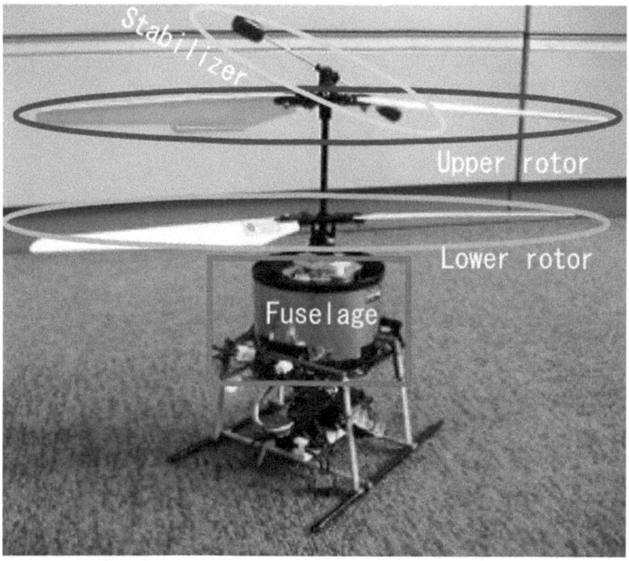

Fig. 2.2 Overview of coaxial-rotor helicopter

bodies, and derive the entire mathematical model of a single-rotor helicopter. In the following section, we introduce the modeling of the dynamics of servomotors, dynamics of rolling and pitching motions, dynamics of yawing motions, dynamics of

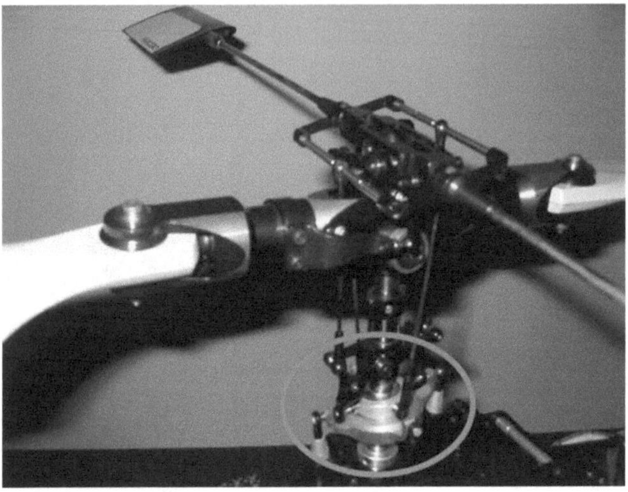

Fig. 2.3 Swash plate of single-rotor helicopter

Fig. 2.4 Swash plate of coaxial-rotor helicopter

heaving motions, and dynamics of horizontal velocities and positions. It is assumed that the coupling terms about each axis can be neglected. Hence, the derived models are a simple single input single output (SISO) model.

2.2.2.1 Dynamics of Servomotors

Hobby-class helicopters typically use an actuator with servomotors. In single-rotor helicopters, there are five servomotors for actuation – aileron servomotors, elevator

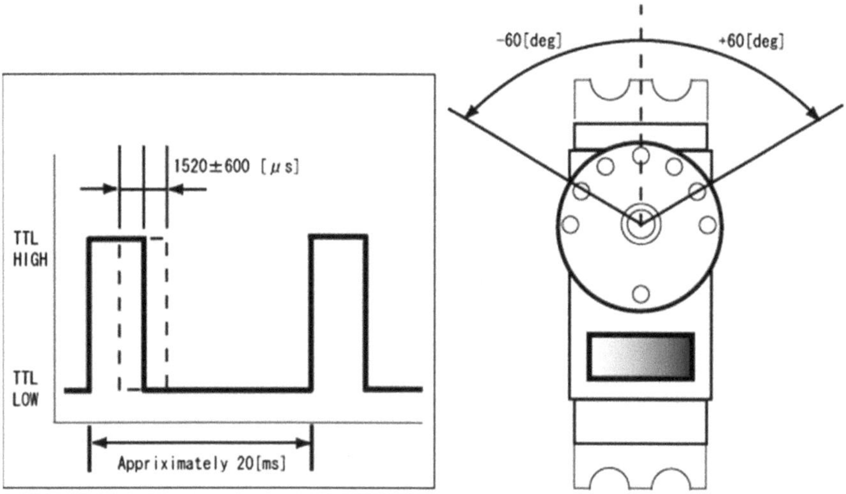

Fig. 2.5 Servomotor actuator

servomotors, collective pitch servomotors, rudder servomotors, and throttle servomotors. The input pulse widths of the servomotor are proportional to the rotation angle of an arm (Fig. 2.5). We assume that the transfer function of the servomotor is the following equation:

$$G_s(s) = \frac{\omega_{ns}^2}{s^2 + 2\zeta_s \omega_{ns} s + \omega_{ns}^2} \tag{2.1}$$

Here the parameter ζ_s is a damping coefficient and the parameter ω_{ns} is a natural angular frequency. These parameters can be determined using an identification method. The frequency response of this model is shown in Fig. 2.6. The cut-off frequency is approximately 5 Hz.

2.2.2.2 Pitching and Rolling Motions

When the rotation angles of the aileron and elevator servomotors change, the cyclic pitch angle of the main blade also changes. Accordingly, the rotor disc rotates and gyro moments are generated. Pitching and rolling moments are produced by these gyro moments of the main rotor blades. We assume that the effect of the gyro moments on pitching and rolling directions are identical; therefore, differences in the model gain exist due to the difference in the inertia moments of fuselage. In such a case, the transfer function from the rotation angle of the servomotor to the rotation angle of the fuselage is defined as follows:

$$G(s) = \frac{K}{(Ts + 1)s} \tag{2.2}$$

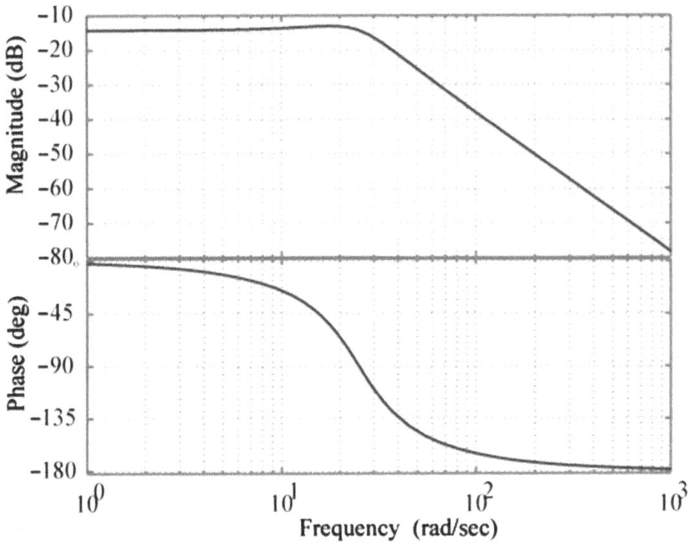

Fig. 2.6 Bode diagram of servomotor model

Here T is the time constant of the system and K is the model gain. We connect the servomotor model derived in the previous section and (2.2) serially. The transfer function from the pulse input to the rotation angle of the fuselage is obtained as follows:

$$
\begin{aligned}
G_\theta(s) &= \frac{\omega_{ns}^2 K_\theta}{(s^2 + 2\zeta_s\omega_{ns}s + \omega_{ns}^2)(T_\theta s + 1)s} \\
G_\phi(s) &= \frac{\omega_{ns}^2 K_\varphi}{(s^2 + 2\zeta_s\omega_{ns}s + \omega_{ns}^2)(T_\varphi s + 1)s}
\end{aligned}
\tag{2.3}
$$

Here the suffixes θ and φ imply pitching and rolling directions, respectively. Our control system loops have time delays – time delay of sensor output data and time delay of data transfer via a wireless modem. The total time delay is approximately thrice the data sampling time. We consider the time delay when we design controllers. The transfer functions of pitching and rolling motions including time delays are obtained as follows:

$$
\begin{aligned}
G_\theta(s) &= e^{-Ls}\frac{\omega_{ns}^2 K_\theta}{(s^2 + 2\zeta_s\omega_{ns}s + \omega_{ns}^2)(T_\theta s + 1)s} \\
G_\phi(s) &= e^{-Ls}\frac{\omega_{ns}^2 K_\varphi}{(s^2 + 2\zeta_s\omega_{ns}s + \omega_{ns}^2)(T_\varphi s + 1)s}
\end{aligned}
\tag{2.4}
$$

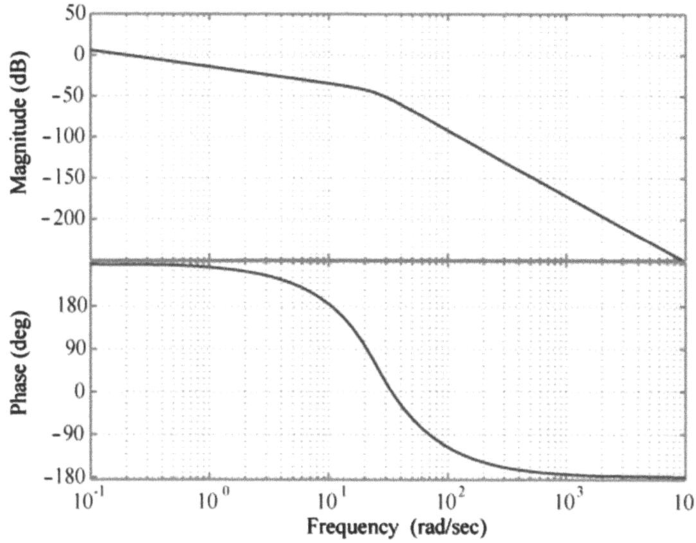

Fig. 2.7 Bode diagram of pitch model

We determined the value of the parameters T_θ, T_ϕ, K_θ, and K_ϕ, from repeated data tuning using simulations and experiments. We then set the time delay e^{-Ls} to the first-order Pade approximation. The first-order Pade approximation is as follows:

$$e^{-Ls} \approx \frac{1 - \frac{L}{2}S}{1 + \frac{L}{2}S} \tag{2.5}$$

As a result, the order of the model is 5. We show the bode diagram of the pitching model in Fig. 2.7.

2.2.2.3 Dynamics of Yawing Motion

When the rotation angle of a rudder servomotor is changed, the tail rotor blade pitch angle changes. Accordingly, the thrust of the tail rotor also changes. Yawing moments are generated by the tail rotor thrust, while the helicopter rotates in the yawing direction. It should be noted that the system of the yawing motion of the small helicopter is not a simple open-loop system. If the yawing system was an open-loop system, the helicopter will be rotated by the counter moment of the main rotor. However, in such a case, it is extremely difficult to control the heading of the helicopter. For the abovementioned reasons, all small helicopters have a local feedback system called the rate gyro. The rate gyro feedbacks the yawing rate, which is measured by a gyro sensor, to counteract the counter moments of the main rotor. In our model, we use an active velocity control system (AVCS) gyro in which a proportional integral (PI) control system is implemented. The block diagram of the

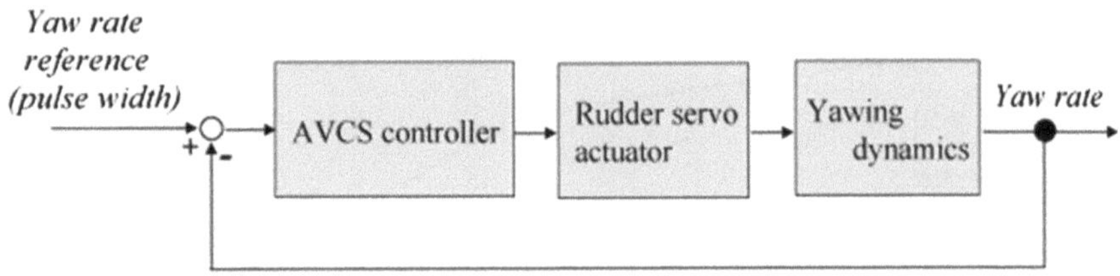

Fig. 2.8 Yawing dynamics with AVCS control system

AVCS gyro system is shown in Fig. 2.8. We assume that the closed-loop system with the AVCS gyro, actuator, and yaw dynamics is a second-order system. Further, we add the integral elements and delays mentioned in the previous section. The transfer function from the rudder pulse input to the rotation angle of the yawing direction is obtained as follows:

$$G_\psi(s) = e^{-Ls}\frac{\omega_{n\psi}^2 K_\psi}{\left(s^2 + 2\zeta_s\omega_{n\psi}s + \omega_{n\psi}^2\right)s} \tag{2.6}$$

Here the suffix ψ implies the yawing direction.

2.2.2.4 Dynamics of Heave Direction

The helicopter moves up and down as a result of changes in the thrust of the main rotor. According to the blade element theory, the thrust generated by the main rotor is calculated as follows:

$$T = \frac{b}{4}\rho a c\Omega^2 R^3(\theta_t + \phi_t) \tag{2.7}$$

Here T is the thrust of the main rotor; b, the number of rotor blades; ρ, the air density; a, a 2-dimensional lift curve slope; c, the blade chord length; Ω, the rotational speed of the main rotor; R, the radius of the main rotor; and θ_t, the collective pitch angle of the main rotor. In addition, φ_t is the inflow angle and defined as follows.

$$\phi_t = \frac{v_d + V_z}{2R\Omega} \tag{2.8}$$

v_d is the induced velocity and V_z is the upward velocity of the helicopter. When the helicopter is in the sky, most of the abovementioned parameters are constant, and there are only three variable parameters – Ω, θ_t, and φ_t. In particular, in hovering flight, V_z is zero and v_d is constant; hence, φ_t changes with only Ω. The changes

Fig. 2.9 Electric governor

in Ω and θ_t are proportional to the rotational angles of the throttle servomotor and collective servomotor, respectively. At this point, we used an engine speed controller called a Governor. The Governor maintains a constant rotor speed Ω. A typical Governor is shown in Fig. 2.9. Additionally, we assume that $v_d \ll \Omega$, and hence, $\varphi_t \cong 0$. Under these assumptions, it can be said that the main-rotor thrust T changes only with the collective pitch angle θ_t. We can rewrite (2.7) into a simpler equation as follows:

$$T = K_T \theta_t \tag{2.9}$$

Here K_T is the gain that depends on any constant parameter. We consider M to be the mass of the helicopter, and the equation of motion of the heave direction is obtained as follows:

$$M\dot{V}_z = -K_v V_z + K_T \theta_t \tag{2.10}$$

The first term on the right-hand side is air drag and K_v is the mean coefficient of the air drag. We add an integral element, and the transfer function from the collective pulse input to the height of the helicopter is obtained as follows:

$$G_{ud}(s) = \frac{K_T K_{\theta_t}}{(Ms + K_v)s} \tag{2.11}$$

K_{θ_t} is the ratio of the collective pulse input to the collective pitch angle of the main-rotor blade in hovering flight.

2.2.2.5 Dynamics of Horizontal Velocity and Position

The main-rotor disc is tilted by the rotational motion in the pitching direction. Accordingly, the main-rotor thrust is decomposed into a horizontal direction (forward) and a vertical direction. Further, the horizontal component of the main-rotor thrust makes the helicopter move in the forward direction (Fig. 2.10). We now assume that the helicopter remains stationary or moves with a constant speed in an up-down direction. The vertical component of the main-rotor thrust is equal to the weight of the helicopter. Under this assumption, we consider M to be the mass of the helicopter; g, the acceleration due to gravity; θ, the pitching rotation angle of the helicopter; and V_x, the forward velocity of the helicopter. We assume that θ is sufficiently small, and obtain the equation of motion of the forward direction as follows.

$$M\dot{V}_x = -Mg\tan\theta \cong -Mg\theta \tag{2.12}$$

In a similar manner, we consider ϕ as the rolling rotation angle of the helicopter and V_y as the rightward velocity of the helicopter. We assume that ϕ is sufficiently small, and the equation of motion of the rightward direction is obtained as follows:

$$M\dot{V}_y = Mg\tan\varphi \cong Mg\varphi \tag{2.13}$$

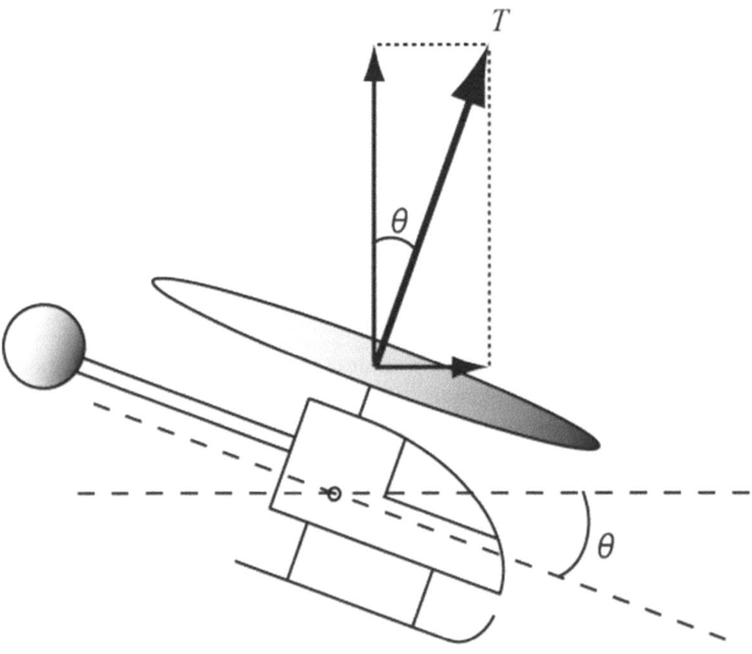

Fig. 2.10 Decomposition of thrust

We apply the Laplace transform to (2.12) and (2.13). The transfer function from the rotation angle of the helicopter to its horizontal velocity is obtained as follows:

$$\begin{aligned} G_{vx}(s) &= \frac{V_x}{\theta} = -\frac{g}{s} \\ G_{vy}(s) &= \frac{V_y}{\varphi} = \frac{g}{s} \end{aligned} \tag{2.14}$$

However, in the case of the simple model shown by (2.14), the model output does not agree with the helicopter velocity collected from flight experiments. Hence, we introduced an unstable pole instead of the integrator in (2.14). The modified model is shown as follows:

$$\begin{aligned} G_{vx}(s) &= -\frac{g}{s-a} \\ G_{vy}(s) &= \frac{g}{s-a} \end{aligned} \tag{2.15}$$

a is a positive constant, which, in this study, is considered to be $a = 0.2$. This unstable pole was introduced because the stability of the control system designed based on the model with an unstable pole might not degrade (only an unstable tendency will be present in the experimental data). In fact, it is well-known that the control system design makes the closed-loop system become strongly stable by adding unstable poles to the model). Additionally, it is clear that some dynamics between the rotation angles and horizontal accelerations are present, when we compare the acceleration data from the differential calculus of the horizontal velocity data with the rotation angle data measured at the time of the flight experiment. We assumed that this dynamics is first-order and added it to (2.15) [15]. The modified velocity model is obtained as follows:

$$\begin{aligned} G_{vx}(s) &= -g\frac{b}{s+b}\frac{1}{s-a} \\ G_{vy}(s) &= g\frac{b}{s+b}\frac{1}{s-a} \end{aligned} \tag{2.16}$$

b is an arbitrary positive constant. It is difficult to conceptualize such a first-order dynamics and unstable pole being included in the dynamics from the rotation angle of a helicopter to the velocity of the helicopter; however, it is clearly caused by the inside dynamics of the attitude sensor that are used for rotation-angle measurements during the experiment (as per analysis after) [16].

We add an integrator to the model of the horizontal velocity of the helicopter mentioned above, and the transfer function from the rotation angle of the helicopter to its horizontal position is obtained as follows:

$$\begin{aligned} G_{vx}(s) &= -g\frac{b}{s+b}\frac{1}{s-a}\frac{1}{s} \\ G_{vy}(s) &= g\frac{b}{s+b}\frac{1}{s-a}\frac{1}{s} \end{aligned} \tag{2.17}$$

2.2.3 Modeling of Coaxial-Rotor Helicopter

In this section, the mathematical model of a coaxial-rotor helicopter is derived. It is assumed that the coupling terms about each axis can be neglected (identical to the case of the single-rotor helicopter). Further, the dynamics of heaving and heading motions are similar to that in a single-rotor helicopter; hence, only the dynamics of a servomotor, dynamics of rolling and pitching motions, and dynamics of horizontal motion are introduced.

2.2.3.1 Dynamics of Servomotors

In the case of a coaxial-rotor helicopter, which is addressed in this book, it is not required to actuate the collective pitch angle and tail-rotor pitch angle. Hence, only servomotors are used to actuate the pitching and rolling motions. The servomotor itself is small, and its range of movement is not significantly large; hence, highorder dynamics can be neglected. We assume that the dynamics of a servomotor can be an approximated first-order equation.

$$G_m(s) = \frac{K_m}{T_m s + 1} \tag{2.18}$$

Here T_m represents the time constant of servomotors and K_m, the appropriate scalar gain.

2.2.3.2 Dynamics of Rolling and Pitching Motions

In the case of coaxial-rotor helicopters, when the rotation angles of aileron and elevator servomotors are changed, the swash plate moves; the cyclic pitch angle of the lower rotor changes. The rotation plane of the lower rotor then tilts and the flapping direction is obtained. Here, the system between the rotation angle of the servomotor and the tilt angle of the lower rotor can be approximated as a first-order transfer function, as shown in (2.19).

$$G_{\varphi l}(s) = \frac{K_{\varphi l}}{T_{\varphi l} s + 1} \tag{2.19}$$

The suffix ϕl implies that this is the transfer function of the tilt angle of the lower rotor. The moment is then generated by the tilt of the lower rotor, and the rotation angle of the fuselage changes. However, the stabilizer bar mounted on top of the body maintains the rotational plane using the gyro effect. This implies that the rotational angle between the main mast and stabilizer bar changes, and this system can be approximated as the following transfer function:

$$G_{\varphi s}(s) = K_{\varphi s}\left(1 - \frac{1}{T_{\varphi s}s + 1}\right) \tag{2.20}$$

The suffix ϕs implies that this is the transfer function of the tilt angle of the stabilizer bar. Next, when the tilt angle of the stabilizer bar changes, the cyclic pitch angle of the upper rotor, which is connected to the stabilizer bar, changes, and the rotational plane of the upper rotor tilts. Here, the transfer function between the tilt angles of the stabilizer bar and upper rotor can be approximated as follows:

$$G_{\varphi u}(s) = \frac{K_{\varphi u}}{T_{\varphi u}s + 1} \tag{2.21}$$

The suffix ϕu implies that this is the transfer function of the tilt angle of the upper rotor. We now describe the transfer function that represents the dynamics of rolling and pitching motions by a combination of the abovementioned three transfer functions and the transfer function of the servomotor. The external moment, which has significant influence on the changes in the attitude of the fuselage, is generated by changes in the tilt angles of the upper and lower rotors. Now, let us consider T_u and T_l as the thrusts generated by the upper and lower rotors, respectively; L_u and L_l, the distance between the center of gravities of the fuselage and the upper and lower rotors, respectively; and ϕ_u and ϕ_l, the tilt angles of the upper and lower rotors, respectively. Assuming that ϕ_u and ϕ_l are sufficiently small, the moments M_u and M_l, which are generated by the upper and lower rotors, respectively, around the center of the gravity of the fuselage are as follows:

$$\begin{aligned} M_u &= T_u L_u \sin\varphi_u \cong T_u L_u \varphi_u \\ M_l &= T_l L_l \sin\varphi_l \cong T_l L_l \varphi_l \end{aligned} \tag{2.22}$$

Here θ is the pitch or roll attitude angle, and J represents the inertia moment of each axis; the general form of the equation of motion about the attitude of the fuselage is given as follows.

$$J\ddot{\theta} = M_u + M_l \tag{2.23}$$

Figure 2.11 combines (2.18)–(2.21) and (2.23) and shows the block diagram of the system between the input and attitude angles of the fuselage. In Fig. 2.11, D_u and D_l imply a damping factor that is based on air resistance, and both are scalar values. Here, ϕ_m represents the rotational angle of the servomotor; ϕ_s, the tilt angle of the stabilizer bar; and $x_a = \begin{bmatrix} \theta & \dot{\theta} & \phi_l & \phi_u & \phi_s & \phi_m \end{bmatrix}^T$, the state vector of the system. According to the block diagram, the general form of the state space equation, which describes the dynamics of the system between the input and attitude angles of the fuselage, can be represented as follows.

$$\dot{x}_a = A_a x_a + B_a u \tag{2.24}$$

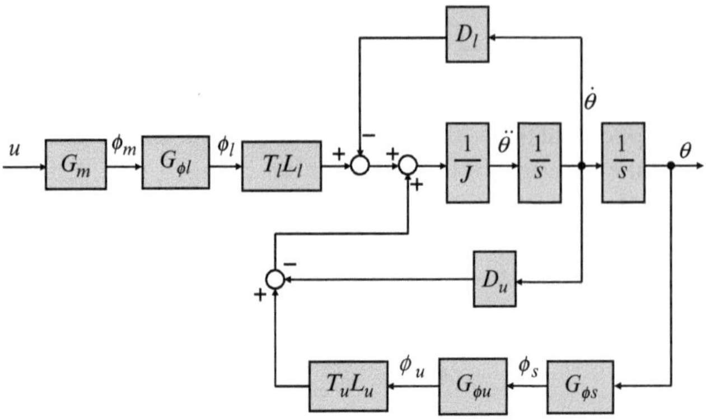

Fig. 2.11 Block diagram of attitude dynamics of coaxial-rotor helicopter

Each matrix in the equation is defined as follows:

$$A_a = \begin{bmatrix} 0 & 1 & 0 & 0 & 0 & 0 \\ 0 & -\frac{D_u + D_l}{J} & \frac{T_l L_l}{J} & \frac{T_u L_u}{J} & 0 & 0 \\ 0 & 0 & -\frac{1}{T_{\varphi l}} & 0 & 0 & \frac{K_{\varphi l}}{T_{\varphi l}} \\ 0 & 0 & 0 & -\frac{1}{T_{\varphi u}} & \frac{K_{\varphi u}}{T_{\varphi u}} & 0 \\ 0 & -K_{\varphi s} & 0 & 0 & -\frac{1}{T_s} & 0 \\ 0 & 0 & 0 & 0 & 0 & -\frac{1}{T_s} \end{bmatrix} \tag{2.25}$$

$$B_a = \begin{bmatrix} 0\,0\,0\,0\,0\,\frac{K_m}{T_m} \end{bmatrix}^T$$

The dynamics of the servomotor, upper and lower rotors, and the stabilizer bar are symmetrical about the roll and pitch directions. Hence, we can obtain the attitude model of the roll and pitch directions by substituting the value of the inertia moment of the roll and pitch axis, respectively, into J in (2.25).

2.2.3.3 Dynamics of Horizontal Velocity and Position

The horizontal component of the thrust generated by the rotor primarily affects the dynamics of the horizontal motion. Now, consider the horizontal components of the thrust generated by the upper and lower rotors as F_{uh} and F_{lh} respectively; the relation between these forces and the tilt angles of the upper and lower rotors can be represented as (2.26).

$$\begin{aligned} F_{uh} &= T_u \sin(\theta + \varphi_u) \cong T_u(\theta + \varphi_u) \\ F_{lh} &= T_l \sin(\theta + \varphi_l) \cong T_l(\theta + \varphi_l) \end{aligned} \tag{2.26}$$

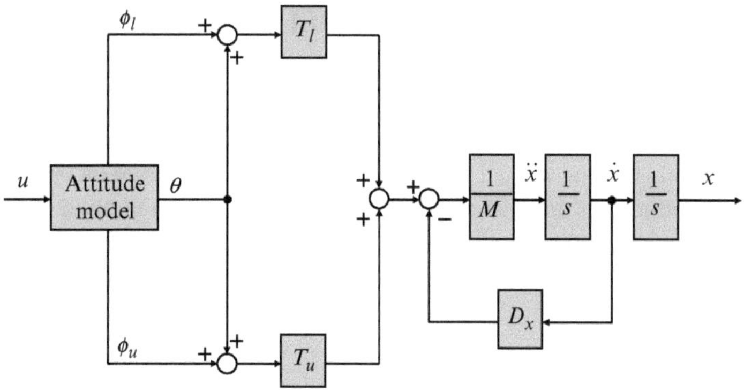

Fig. 2.12 Block diagram of the horizontal dynamics of coaxial-rotor helicopter

Here we assume that the attitude angle and tilt angle of each rotor are sufficiently small. x is considered to be the displacement in the X-axis and Y-axis directions and M as the mass of the body, and the general form of the equation of motion is represented as (2.27).

$$M\ddot{x} = -D_x\dot{x} + F_{uh} + F_{lh} \tag{2.27}$$

The dynamics of the horizontal motion, including the dynamics of the attitude of fuselage, is described in Fig. 2.12. In this block diagram, the state space equation that expresses the dynamics between the input and displacement in the horizontal direction is represented by (2.28).

$$\dot{x}_h = A_h x_h + B_h u \tag{2.28}$$

Each matrix in this equation is defined as follows:

$$A_h = \begin{bmatrix} 0 & 1 & 0 & 0 & 0 & 0 & 0 & 0 \\ 0 & -\frac{D_u+D_l}{J} & \frac{T_lL_l}{J} & \frac{T_uL_u}{J} & 0 & 0 & 0 & 0 \\ 0 & 0 & -\frac{1}{T_{\varphi l}} & 0 & 0 & 0 & 0 & \frac{K_{\varphi l}}{T_{\varphi l}} \\ 0 & 0 & 0 & -\frac{1}{T_{\varphi u}} & \frac{K_{\varphi u}}{T_{\varphi u}} & 0 & 0 & 0 \\ 0 & -K_{\varphi s} & 0 & 0 & -\frac{1}{T_s} & 0 & 0 & 0 \\ 0 & 0 & 0 & 0 & 0 & 0 & 1 & 0 \\ \frac{F_{lh}+F_{uh}}{M} & 0 & \frac{F_{lh}}{M} & \frac{F_{uh}}{M} & 0 & 0 & -D_x & 0 \\ 0 & 0 & 0 & 0 & 0 & 0 & 0 & -\frac{1}{T_s} \end{bmatrix} \tag{2.29}$$

$$B_a = \begin{bmatrix} 0\,0\,0\,0\,0\,0\,0\,\frac{K_m}{T_m} \end{bmatrix}^T$$

$$x_h = \begin{bmatrix} \theta\ \dot{\theta}\ \varphi_l\ \varphi_u\ \varphi_s\ x\ \dot{x}\ \varphi_m \end{bmatrix}^T$$

2.3 Control System Design of Small Unmanned Helicopter

In the previous section, the mathematical model that expresses the dynamics of small and miniature helicopters was derived. In this section, a control systetm design based on this mathematical model is shown. A method based on optimal control theory is introduced as an example of a representative design method for stationary flight control such as hovering. Moreover, for trajectory control, the optimal preview control is efficiency; hence, it is introduced in a subsequent section.

2.3.1 Optimal Control

Before performing control system design, it is neccesary to transform the transfer function that was derived in the previous section into a state space equation because the state space equation is required for optimal control design. Transformation of the transfer function into a state space equation is called realization. In general, it is well known that multiple realizations exist for a transfer function. In the following discussion, it is desirable that the system is controllable and observable; hence, we assume such a system. The realization that is controllable and observable is called minimum realization. The realization of the transfer function that was derived in the previous section is shown as follows:

$$\begin{aligned} \dot{\mathbf{x}} &= \mathbf{A}\mathbf{x}(t) + \mathbf{B}u(t), \quad && \mathbf{x} \in R^n, \quad u \in R \\ y &= \mathbf{C}\mathbf{x}(t), && y \in R \end{aligned} \tag{2.30}$$

(***A***, ***B***) are controllable and (***A***, ***C***) is observable.

For the control of small unmanned helicopters, it is desired that the output of the system follows an appropriate value. For example, for attitude control, the attitude of the helicopter follows the reference attitude, while in the case of height control, the height of the helicopter follows the reference height. In these cases, the servo sytstem design is necessary for following the reference value without any steady-state error. Considering that r is a constant reference value, the integral value of the error between the reference and output is defined by the following equation.

$$e(t) = \int_0^t \{r - y(\tau)\} d\tau \tag{2.31}$$

Here $\boldsymbol{x}_a$ is the state value of the augmented system and is defined as $\boldsymbol{x}_a = [\boldsymbol{x}^T, e]^T$, and the state space equation of the augmented system is defined as follows:

$$\dot{\mathbf{x}}_a = \mathbf{A}_a \mathbf{x}_a + \mathbf{B}_a u + \mathbf{E}r \tag{2.32}$$

The matrices in this equation are defined as follows:

$$\mathbf{A}_a = \begin{bmatrix} \mathbf{A} & \mathbf{0} \\ -\mathbf{C} & 0 \end{bmatrix}, \quad \mathbf{B}_a = \begin{bmatrix} \mathbf{B} \\ 0 \end{bmatrix}, \quad \mathbf{E} = \begin{bmatrix} \mathbf{0} \\ 1 \end{bmatrix} \tag{2.33}$$

The optimal controller will now be designed on the basis of the servo augmented system represented by (2.32).

Optimal control is an objective of the control system design method. In this method, the optimal control inputs are calculated by using state feedback that can minimize certain criteria. Let us now consider the quadratic criteria shown as follows:

$$J = \int_0^\infty \left[\mathbf{x}_a^T(t)\mathbf{Q}\mathbf{x}_a(t) + Ru^2(t)\right]dt \tag{2.34}$$

The weighting matrix Q for the control objective is a non-negative definite symmetric matrix. Further, the weight R for the control input is a positive scalar. The optimal control input that minimizes (2.34) is obtained as (2.35).

$$u(t) = -\mathbf{F}\mathbf{x}_a(t), \quad \mathbf{F} = R^{-1}\mathbf{B}_a^T\mathbf{P} \tag{2.35}$$

The matrix $\boldsymbol{P}$ is the solution of the Riccati equation represented by (2.36), and it is a positive definete symmetric matrix.

$$\mathbf{P}\mathbf{A}_a + \mathbf{A}_a^T\mathbf{P} - \mathbf{P}\mathbf{B}_a R^{-1}\mathbf{B}_a^T\mathbf{P} + \mathbf{Q} = \mathbf{0} \tag{2.36}$$

In the case of $\boldsymbol{Q}$ and R, the control performance increases if we increase $\boldsymbol{Q}$, and we can decrease the control input by increasing R. An example of the design guidance is shown below. For servo system design, we first set a value for $\boldsymbol{Q}$ depending on the integral calculus level of the error in order to improve the control performance. We then suppress any overshoots by regulating the value of $\boldsymbol{Q}$ without depending on the integral calculus level of the error. Finally, we maintain the control input to the desired size by coordinating R.

For calculating the optimal control input represented by (2.35), it is necessary that the state vector $\boldsymbol{x}_a$ is fed back. However, all the elements of the state vector may not be observed. In particular, in the case of a state equation that is converted from a transfer functiton, these state variable often do not have physical meaning, and it is difficult to observe them directly. In such cases, we must design the observer. We can observe the integral calculus level of the error in an element of $\boldsymbol{x}_a$ clearly; hence, in this case, the observer for the state vector $\boldsymbol{x}$ is designed. Considering $\hat{\boldsymbol{x}}$ to be the estimate of the state vector $\boldsymbol{x}$, the observer can calculate $\hat{\boldsymbol{x}}$ using the following equation

$$\dot{\hat{\mathbf{x}}}(t) = \mathbf{A}\hat{\mathbf{x}}(t) + \mathbf{B}u(t) + \mathbf{L}(y(t) - \mathbf{C}\hat{\mathbf{x}}(t)) \tag{2.37}$$

The matrix $\boldsymbol{L}$ is the observer gain. The observer with these configurations is called an identity observer. Various methods have been developed for designing an observer gain L; here, the configuration of (2.37) is similar to the configuration of the steady-state Kalman filter. Hence, the design method using the theory of the Kalman filter has been introduced.

First, the following system is assumed for the Kalman filter.

$$\begin{aligned} \dot{\mathbf{x}} &= \mathbf{A}\mathbf{x}(t) + \mathbf{B}u(t) + \mathbf{G}\mathbf{w}(t) \\ y &= \mathbf{C}\mathbf{x}(t) + \mathbf{v}(t) \end{aligned} \tag{2.38}$$

Here $\boldsymbol{w}$(t) is the system noise and $\boldsymbol{v}$(t) is the measurement noise. It can be assumed that the coefficient matrix $\boldsymbol{G}$, which depends on the system noise, is equal to the input matrix $\boldsymbol{B}$; hence, we assume that $\boldsymbol{G} = \boldsymbol{B}$. The filter equation of the Karman filter is expressed using (2.37), and the filter gain is calculated using the following equation.

$$\mathbf{L} = \overline{\mathbf{X}}\mathbf{C}^T\mathbf{V}^{-1} \tag{2.39}$$

The matrix $\overline{\boldsymbol{X}}$ is the solution of the Riccati equation represented by (2.40), and it is a positive definete symmetric matrix.

$$\mathbf{A}\overline{\mathbf{X}} + \overline{\mathbf{X}}\mathbf{A}^T - \overline{\mathbf{X}}\mathbf{C}^T\mathbf{V}^{-1}\mathbf{C}\overline{\mathbf{X}} + \mathbf{G}\mathbf{W}\mathbf{G}^T = \mathbf{0} \tag{2.40}$$

The matrices $\boldsymbol{V}$ and $\boldsymbol{W}$ are covariance matrices depending on the measurement noise and system noise, respectively. Moreover, they can be obtained as follows:

$$\begin{aligned} E[\mathbf{w}(t)\mathbf{w}^T(\tau)] &= \mathbf{W}(t)\delta(t-\tau) \\ E[\mathbf{v}(t)\mathbf{v}^T(\tau)] &= \mathbf{V}(t)\delta(t-\tau) \end{aligned} \tag{2.41}$$

Here, $\delta(t)$ implies Dirac's delta function and $E[\cdot]$ implies an expectation value.

We can calculate the control input by feeding back the estimated state vector $\hat{x}_a = [\hat{x}^T, e]^T$, which is estimated by the abovementioned observer design.

The controller is obtained by combining the observer with the state feedback in the manner shown by the following equation.

$$\begin{aligned} \dot{\hat{\mathbf{x}}}_a &= (\mathbf{A}_a - \mathbf{B}_a\mathbf{F} - \mathbf{L}\mathbf{C})\hat{\mathbf{x}}_a + \mathbf{L}\mathbf{y} \\ u &= -\mathbf{F}\hat{\mathbf{x}}_a \end{aligned} \tag{2.42}$$

2.3.2 Optimal Preview Control

In this section, optimal preview control is introduced as a method for trajectory control. First, the controlled object is described as the following discrete linear state space.

$$\begin{aligned} \mathbf{x}(k+1) &= \mathbf{A}\mathbf{x}(k) + \mathbf{B}u(k) + \mathbf{E}d(k) \\ y(k) &= \mathbf{C}\mathbf{x}(k) \end{aligned} \tag{2.43}$$

Here x, y, u, and d are the state variable, system output, system input, and disturbance, respectively. $R(k)$ is used as the reference signal, and the tracking error is defined as follows:

$$e(k) = R(k) - y(k) \tag{2.44}$$

The error system can be derived as follows [32]:

$$\begin{aligned} \begin{bmatrix} e(k+1) \\ \Delta \mathbf{x}(k+1) \end{bmatrix} &= \begin{bmatrix} \mathbf{I}_m & -\mathbf{CA} \\ 0 & \mathbf{A} \end{bmatrix} \begin{bmatrix} e(k) \\ \Delta \mathbf{x}(k) \end{bmatrix} + \begin{bmatrix} -\mathbf{CB} \\ \mathbf{B} \end{bmatrix} \Delta u(k) \\ &\quad + \begin{bmatrix} \mathbf{I}_m \\ 0 \end{bmatrix} \Delta R(k+1) + \begin{bmatrix} -\mathbf{CE} \\ \mathbf{E} \end{bmatrix} \Delta d(k) \end{aligned} \tag{2.45}$$

or,

$$\mathbf{X}_0(k+1) = \Phi \mathbf{X}_0(k) + \mathbf{G}\Delta u(k) + \mathbf{G}_R \Delta R(k+1) + \mathbf{G}_d \Delta d(k) \tag{2.46}$$

The criterion function is defined as follows:

$$J = \sum_{k=-M_R+1}^{\infty} \left[\mathbf{X}_0^T(k) \mathbf{Q} \mathbf{X}_0(k) + \Delta u^T(k) H \Delta u(k) \right] \tag{2.47}$$

Q is the positive definite, and H is the positive semi-definite. The control input that minimizes the criterion function is as follows:

$$\Delta u(k) = \mathbf{F}_0 \mathbf{X}_0(k) + \sum_{j=1}^{M_R} F_R(j) \Delta R(k+j) + \sum_{j=0}^{M_d} F_d(j) \Delta d(k+j) \tag{2.48}$$

Here

$$\begin{aligned} \mathbf{F}_0 &= -\left[H + G^T P G \right]^{-1} G^T P \Phi \\ F_R(j) &= -\left[H + G^T P G \right]^{-1} G^T (\xi^T)^{j-1} P G_R \\ F_d(j) &= -\left[H + G^T P G \right]^{-1} G^T (\xi^T)^{j} P G_d \\ \xi &= \Phi + G\mathbf{F}_0 \\ P &= Q + \Phi^T P \Phi - \Phi^T P G \left[H + G^T P G \right]^{-1} G^T P \Phi \end{aligned} \tag{2.49}$$

M_R and M_d are the parameters that indicate the preview steps, i.e., how many future reference or disturbance steps are used for the control calculation. In our study, it is difficult to preview future disturbance, and hence, we use only the reference preview. Figure 2.13 shows the block diagram of the control structure, including the preview feed-forward. By eliminating the preview feed-forward, the controller becomes a type 1 servo feedback controller that maintains the stability of the closed loop.

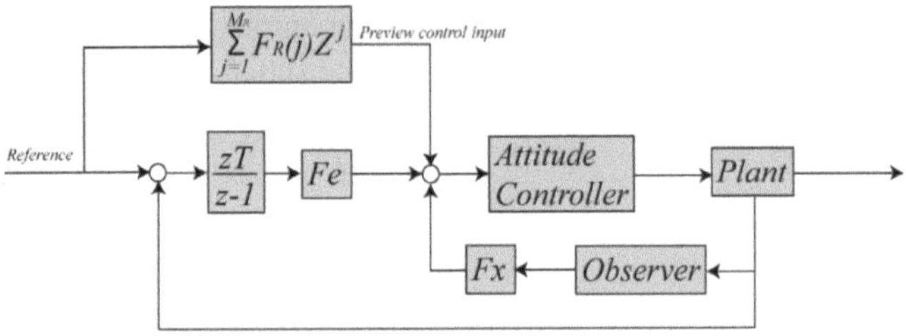

Fig. 2.13 Block diagram of optimal preview control

2.4 Experiment

Flight experiments were performed by using the controller designed in the previous sections. We now show the experimental setup for a single-rotor helicopter and coaxial-rotor helicopter for outdoor and indoor operations. Experimental results of static flight control such as hovering by using a optimal controller and trajectory flight control by using optimal preview control are described.

2.4.1 Experimental Setup for Single-Rotor Helicopter

In this section, an overview of the experimental setup for a single-rotor helicopter in an outdoor environment is introduced. In this case, we use a control system called a host-based control system. In the following discussion, the merits and demerits of the host-based control system are described in detail.

Figure 2.14 shows the configuration and signal flow of the host-based control system. According to this figure, in the host-based control system, a ground station and a control device that are equipped on a helicopter are combined; this constitutes a closed loop. First, a microcomputer collects sensor data that are mounted on a helicopter; the data is sent to a ground station by wireless communication. The control calculations are performed by a computer on the ground station by using data sent from the helicopter. The control input is then converted to a servo pulse by a pulse generator and sent to servomotors through a radio transmitter and receiver. By using this control system, we can use a high-performance computer for the control calculations, and hence, it is possible to calculate any control calculations even if it is large and complicated. This is the merit of the host-based control system. However, there is some delay because of the size of the radio transmission section, and it is impossible to perform control operations over the entire distance that a radio touches by using this control system. This is the demerit of this system.

Next, we will introduce a control device mounted on the helicopter. Figure 2.15 shows an overview of our control device, while Fig. 2.16 shows its configuration. The control device consists of a micro-computer board, GPS sensor, attitude sensor,

Fig. 2.14 Host-based control system

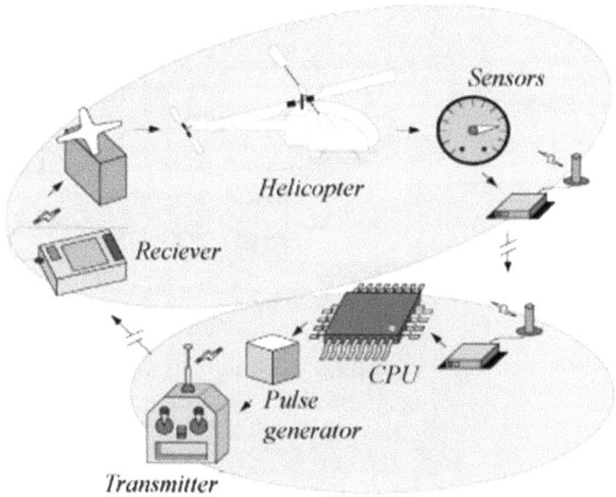

Fig. 2.15 Overview of control device

and wireless module. The total weight of this device including the electric battery is approximately 2 kg. This is sufficiently light to mount it on a small helicopter.

2.4.2 Experimental Setup of Coaxial-Rotor Helicopter

In this section, an overview of the experimental setup for a coaxial-rotor miniature helicopter operated in indoor environments is described. Figure 2.17 shows an overview of the experimental system. In the case of indoor flight, a GPS sensor

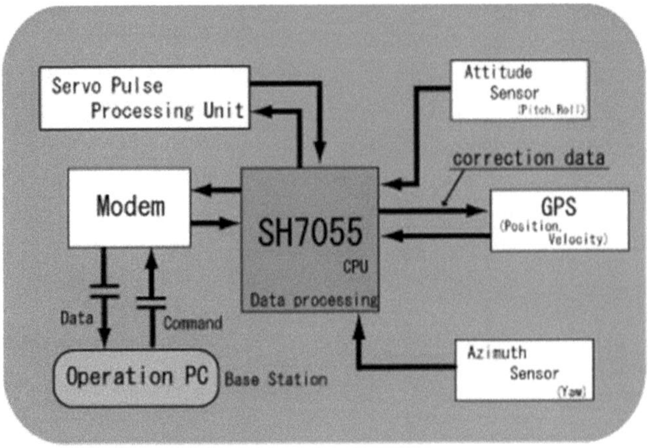

Fig. 2.16 Block diagram of control device

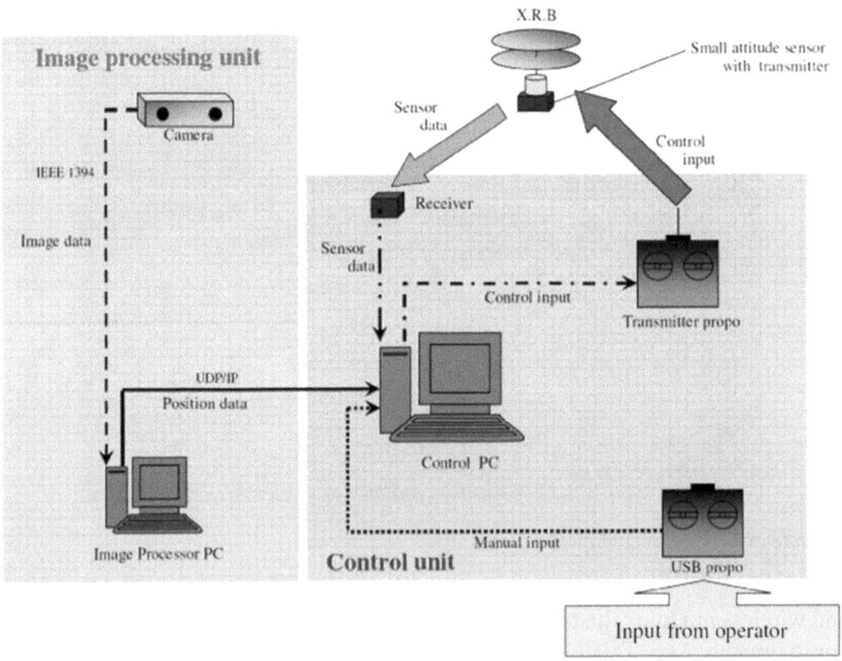

Fig. 2.17 Experimental setup of miniature helicopter

cannot be used. Hence, we use a stereo vision camera to measure the position of the helicopter. First, the image data captured by the stereo vision camera is sent to an image-processing computer. In the image-processing computer, the position of the helicopter is calculated by using image data and sent to a control computer. In contrast, the attitude data of a helicopter is measured by a small-attitude sensor and also

Fig. 2.18 Small-attitude sensor

sent to control computer. Control calculations are performed by a control computer by using the position and attitude data of the helicopter. The control input is sent to a helicopter through a transmitter that is connected to the control computer. An overview of the small-attitude sensor is shown in Fig. 2.18. The weight of this senor is only 20 g, and it can be easily mounted on miniature helicopters.

2.4.3 Static Flight Control

Static flight experiments were performed. The results of the experiments are shown below. Figure 2.19 shows the waypoint experiment results of a single-rotor helicopter. The position reference is expressed in steps. From this fugure, it is clear that a very stable flight was achieved. Experiment results of the miniture helicopter hovering in indoor environment are shown in Fig. 2.20. The miniture helicopter flies to maintain stability using an optimal controller.

2.4.4 Trajectory-Following Control

In this section, the trajectory-following control results when an optimal preview controller was used are shown. Figures 2.21–2.24 show the experimental results. Figures 2.21 and 2.22 show the experimental results when we use a reference trajectory that goes around on a circular trajectory (radius = 10 m) within 30 s. In this experiment, the yaw angle was fixed at north. The phase delay due to preview control improved considerably as compared with the simulated results without preview control, as shown in Fig. 2.22. The feedback controller of these two results is

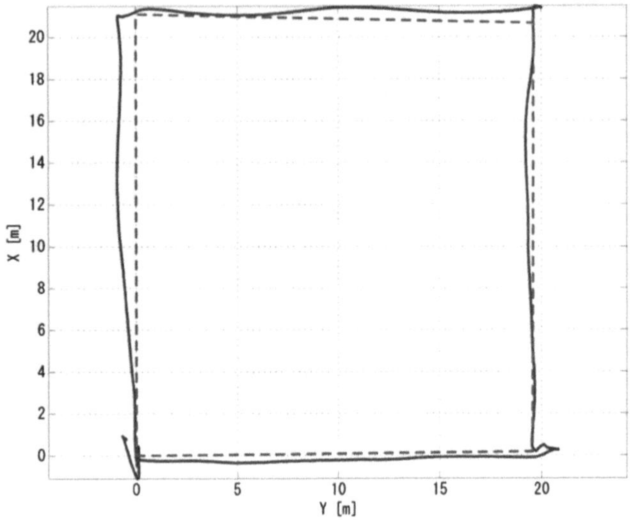

Fig. 2.19 Waypoint experiment result for single-rotor helicopter

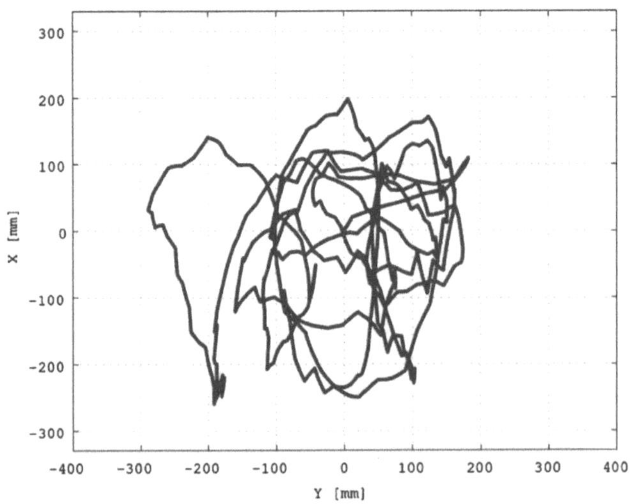

Fig. 2.20 Hovering experiment result for miniature helicopter

identical. Therefore, the difference is due to the existence of the preview loop. Figures 2.23 and 2.24 show the experimental results of the "S" character-like trajectory. In this experiment, we control the yaw angle heading in the moving direction. In Fig. 2.23, the sharpest edge of the isosceles triangle indicates the heading direction.

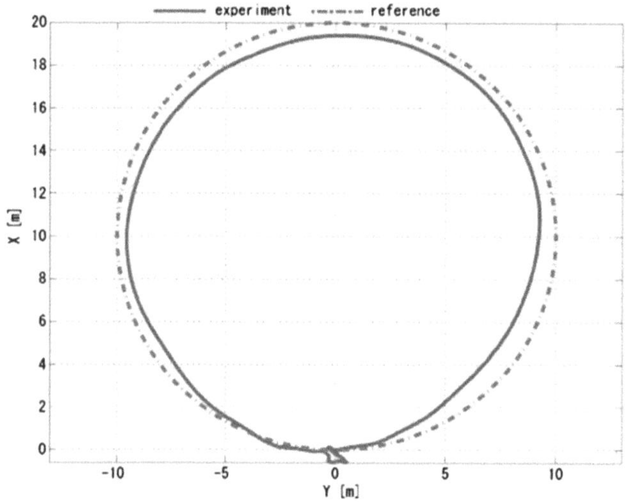

Fig. 2.21 Experimental results of preview control (X-Y plot, circular trajectory)

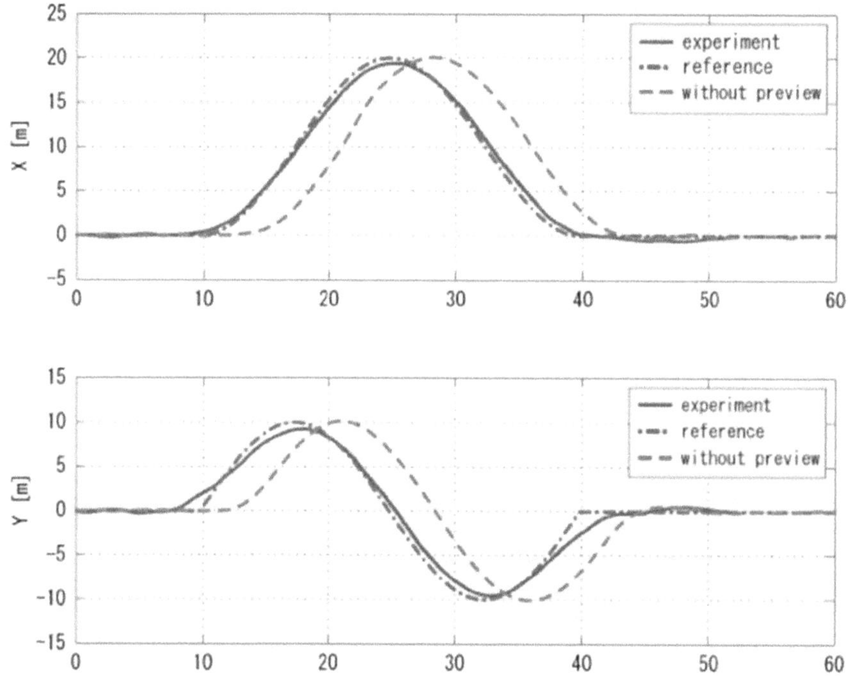

Fig. 2.22 Experimental results of preview control (time domain response, circular trajectory)

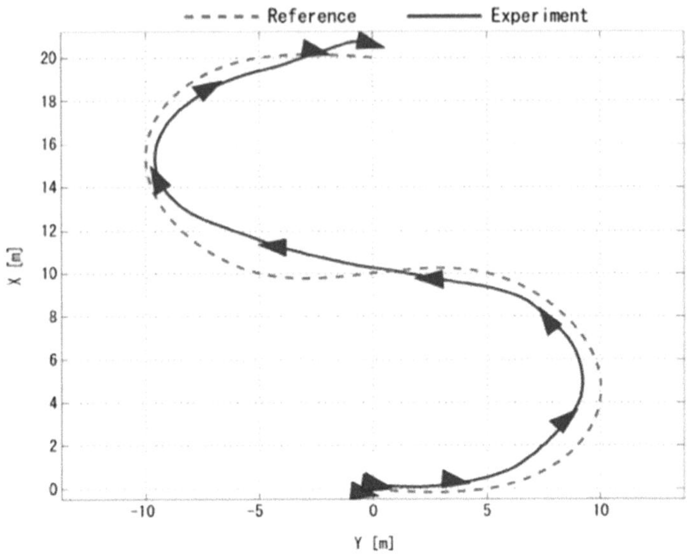

Fig. 2.23 Experimental results of preview control (X-Y plot, S trajectory)

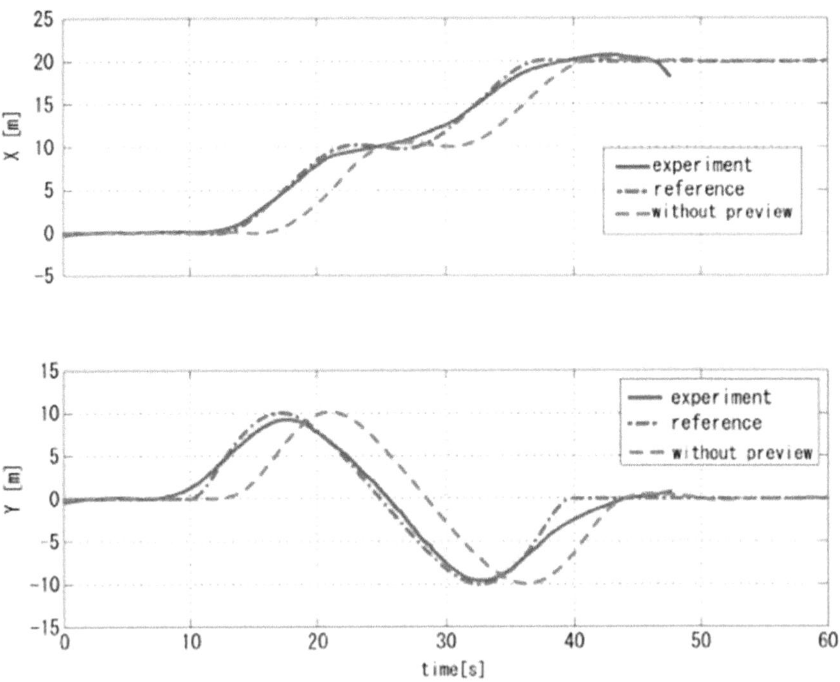

Fig. 2.24 Experimental results of preview control (time domain response, S trajectory)

2.5 Summary

In this chapter, control system design by using a mathematical model was introduced an efficient method of control system design for small unmanned helicopters. We chose single-rotor type and coaxial-rotor type small helicopters as the control object. The mathematical model that expresses the dynamics of both types of helicopters was derived. Moreover, optimal control system design based on the derived mathematical model was conducted.

References

1. Watanabe K, Iwatani Y, Nonaka K, Hashimoto K (2008) A visual-servo-based assistant system for unmanned helicopter control. In: Proceedings of the IEEE/RSJ international conference on intelligent robots and systems (IROS) 2008, pp 822–827
2. Ping S, Guangping Q, Kejie L (2009) The flight control system based on multivariable PID neural network for small-scale unmanned helicopter. In: Proceedings of the international conference on information technology and computer science (ITCS) 2009, vol 1, pp 538–541
3. Jiade W, Guoyong H, Yugang F (2008) An attitude control method of unmanned helicopter based on adaptive output feedback. In: Proceedings of the 3rd international conference on intelligent system and knowledge engineering (ISKE) 2008, vol 1, pp 748–753
4. Zhou F, Jiande W, Ping L (2008) Control system design and flight testing for a miniature unmanned helicopter. In: Proceedings of the 7th World Congress on intelligent control and automation (WCICA) 2008, pp 2315–2319
5. Mettler B, Tischler MB, Kanade K (2002) System identification modeling of a small-scale unmanned rotorcraft for flight control design. J Am Helicopter Soc 47(1):50–63
6. Esmailifar SM, Saghafi F (2009) Autonomous unmanned helicopter landing system design for safe touchdown on 6-DOF moving platform. In: Proceedings of the 5th international conference on autonomic and autonomous systems (ICAS) 2009, pp 245–250
7. Oh SR, Pathak K, Agrawal SK, Roy Pota H, Garrett M (2006) Approaches for a tether-guided landing of an autonomous helicopter. IEEE Trans Robot 22(3):536–544
8. Oh CK, Barlow GJ (2004) Autonomous controller design for unmanned aerial vehicles using multi-objective generic programming. In: Proceedings of congress on evolutional computation 2004, vol 2, pp 1538–1545
9. David HS, Jin K, Shankar S (2002) A flight control system for aerial robots: algorithm and experiments. In: 15th IFAC World Congress on automatic control. Control engineering practice, vol 11(12), pp 1389–1400
10. Jinok S, Daigo F, Kensaku H, Kenzo N (2002) Attitude control and hovering control of radio controlled helicopter. Trans Jpn Soc Mech Eng Ser C 68(675):3284–3291
11. Satoshi S, Jinok S, Makoto T, Yoshiro K, Daisuke N, Kenzo N (2007) Analytical modeling and autonomous hovering control of hobby class 5 kg unmanned helicopter. Trans Jpn Soc Mech Eng Ser C 73(726):562–569
12. Ishidori A, Marconi L, Serrani A (2003) Robust nonlinear motion control of a helicopter. IEEE Trans Automat Contr 48(3):413–426
13. Sakamoto T, Katayama H, Ichikawa I (2006) Attitude control of a helicopter model by robust PID controllers. In: Proceedings of the 2006 IEEE international conference on control applications, pp 1971–1976
14. Daigo F, Jinok S, Kensaku H, Nonami K (2004) H ∞ hovering and guidance control for autonomous small-scale unmanned helicopter. Trans Jpn Soc Mech Eng Ser C 70(694): 1708–1714

15. Kensaku H, Jinok S, Daigo F, Kazuhiro I, Delshan F, Kenzo N (2004) Autonomous flight control of hobby-class small unmanned helicopter (modeling based on experimental identification and autonomous flight control experiment). Trans Jpn Soc Mech Eng Ser C 70(691):720–727
16. Kensaku H, Kenzo N (2006) Analysis of transition model of hobby-class small unmanned helicopter. Trans Jpn Soc Mech Eng Ser C 72(723):3540–3547
17. Bauchau OA, Wang J (2007) Efficient and robust approaches for rotorcraft stability analysis. In: Proceedings of American Helicopter Society 63rd annual forum, Virginia Beach, USA, 2007
18. Benedict M, Jarugumilli T, Chopra I (2009) Design and development of a hover-capable cyclocopter micro air vehicle. In: American Helicopter Society 65th annual forum, Grapevine, USA, 2009
19. Bermes C, Schafroth D, Bouabdallah S, Siegwart R (2009) Design parameters for coaxial rotary wing MAVs with passive roll and pitch stability. In: American Helicopter Society 65th annual forum, Grapevine, USA, 2009
20. Bhandari S, Colgren R (2006) 6-DoF dynamic model for a raptor 50 UAV helicopter including stabilizer bar dynamics. In: AIAA modeling and simulation technologies conference and exhibit, Keystone, USA, 2006
21. Bohorquez F, Rankinsy F, Baederz J, Pines D (2003) Hover performance of rotor blades at low reynolds numbers for rotary wing micro air vehicles. An experimental and CFD study. In: AIAA applied aerodynamics conference, Orlando, USA, 2003
22. Chriette A, Cheviron T, Plestan F (2009) Generic nonlinear model of reduced scale UAVs. In: IEEE international conference on robotics and automation (ICRA), Kobe, Japan, 2009
23. Fusato D, Celi R (2003) Design sensitivity analysis for helicopter flight dynamic and aeromechanic stability. J Guid Control Dyn 26(6):918–927
24. Gress GR (2007) Lift fans as gyroscopes for controlling compact VTOL air vehicles: overview and development status of oblique active tilting. In: American Helicopter Society 63rd annual forum, Virginia Beach, USA, 2007
25. Haosheng C, Darong C (2005) Identification of a model helicopters yaw dynamics. J Dyn Syst Meas Control 127(1):140–145
26. Hirosue W, Ookura A, Sunada S (2003) A study of a coaxial helicopter (ii): Analysis on effects of a stabilizer bar on fuselage motion. In: 41st aircraft symposium of the Japan Society for Aeronautical and Space Sciences, 2003
27. Mettler B, Dever C, Feron E (2004) Scaling effects and dynamic characteristics of miniature rotorcraft. J Guid Control Dyn 27:466–478
28. Samuel P, Sirohi J, Bohorquez F, Couch R (2005) Design and testing of a rotary wing MAV with an active structure for stability and control. In: Amercian Helicopter Society 61st annual forum, Grapevine, USA, 2005
29. Marco LC, George P, William CM, Kanade T (2002) Design and flight testing of a high-bandwidth H ∞ loop shaping controller for a robotic helicopter. In: Proceedings of the AIAA guidance, navigation, and control conference, Monterey, CA, Aug 2002
30. YAMAHA Motors Corp. (2000) Flight control system for unmanned helicopter, Japan Patent 2000–118498
31. Harbick K, Montgomery JF, Sukhatme GS (2001) Planar spline trajectory following for an autonomous helicopter. 2001 IEEE international symposium on computational intelligence in robotics and automation
32. Tsuchiya T, Egami M (1992) Digital preview control, Sangyo-Tosyo, Tokyo

Chapter 3
Autonomous Control of a Mini Quadrotor Vehicle Using LQG Controllers

Abstract This chapter describes the techniques of modeling and controller design of attitude for Quad-Rotor MAVs Compared with a single rotor or a contra-rotating propeller small helicopter, the advantages of Quad-Rotor MAVs are that: they have larger payloads, they are more powerful and can more easily handle turbulence such as wind and they are easier to design using a compact airframe. The research about autonomous control for Quad-Rotor MAVs is very active now. A key characteristic of Quad-Rotor MAVs is that all the degrees of freedom of the airframe are controlled by tuning the rotational speed of four motors. Moreover, because their internal controller calculates angular velocity feed back by using a gyro sensor, the nonlinearity of the airframe becomes weaker and a linear model is more appropriate. Therefore in this chapter, we introduce the technique of linear modeling and model based controller design for Quad-Rotor MAVs and the performance of the designed controller.

Video Link:

Fully Autonomous Waypoint Flight Control of Quad-Rotor MAV
http://mec2.tm.chiba-u.jp/monograph/Videos/Chapter3/1.mpg

3.1 Introduction

If a natural disaster such as an earthquake or an accident occurs and a poisonous substance (gas, radiation, and medicine, etc.) exists, it is difficult to access the site since the area is dangerous. Also, there can be narrow, difficult to access spaces. Moreover, in buildings, high place etc., often it can be difficult for a person to easily enter. As these examples illustrate it is often very useful if a flying robot can monitor dangerous or difficult to access areas with a camera and various sensors. It could enter tight spaces and reach the targeted position. In addition, these types of robots could also be used for rescue operations and preventing secondary disasters. Research on Unmanned Air Vehicles (UAVs) for surveillance and gathering information has been conducted recently and some results are being reported. Research on the micro UAVs (Micro Air Vehicles: MAVs) is also being carried out. MAVs have many advantages such as flying even in narrow areas due to their vertical take-off and landing capabilities.

K. Nonami et al., *Autonomous Flying Robots: Unmanned Aerial Vehicles and Micro Aerial Vehicles*, DOI 10.1007/978-4-431-53856-1_3, © Springer 2010

Since the 1990s, the United States Defense Advanced Research Projects Agency (DARPA) has paid particular attention to Micro Electro Mechanical Systems (MEMS) technology, and has aggressively engaged in joint research with research laboratories and universities for the development of MAVs. MAVs started as unmanned aircraft developed for advancing nanotechnology and as small, low-cost vehicles for short-distance search operations during warfare. Other possible uses include scouting areas contaminated by nuclear, biological, and chemical (NBC) weapons, radio relaying in places where communication is difficult, and following rescue signals and finding the descent point when an aircraft pilot ejects. Normally, MAVs are not flown by a radio controller. Instead, they can autonomously fly and find their target. However, given their small size there are many technical problems, especially regarding the installation of their power, communications apparatus and sensors.

Recently, the research of MAVs is increasing due to the limitations of fixed-wing aircraft. While fixed-wing aircraft can fly fast and for relatively long distances, their main disadvantages are that they cannot hover and or fly at low speeds to precisely provide aerial photography and reconnaissance. Also, it is difficult for fixed-wing aircraft to automatically takeoff and land. On the other hand, rotating-wing MAVs can hover and fly at low speeds and they are ideal for precise aerial photography, for short range surveillance or for reconnaissance in cluttered urban areas such as a large city. And rotating-wing MAVs can vertically takeoff and land (VTOL). If you classify MAVs by the number of propellers, there are two kinds, single and multi-rotor vehicles. Compared with the single-rotor vehicle, multi-rotor units obtain their lift using two or more propellers; and a more compact airframe can be designed. Therefore, in the research of autonomous VTOL MAVs, the coaxial reversing type and Quad-Rotor MAVs are the main areas of current interest [1–19]. While the small double reversing MAVs have good stability, their mobility is poor; therefore they are unsuitable for autonomous outdoor flight. Compared with double reversing type MAVs, the same size Quad-Rotor type MAVs have more payload and better mobility, and they can autonomously fly in winds of 5 m/s. In addition, because they are lightweight, even if they crash the damage to the vehicle is relatively small. Therefore Quad-Rotor MAVs can be ideal for reconnaissance or aerial photography in the large cities. And in this chapter, we focus on Quad-Rotor type MAVs and review the composition of our platform and experiment system, the technique of attitude modeling and attitude control system design.

3.2 Description of the Experimental Platform

In this research, a versatile hobby Quad-Rotor, the X-3D-BL produced by Ascending Technologies GmbH in Germany is adopted as our controlled object [7]. X-3D-BL is four-channel radio controlled type, powered by a 2,100 mAh 3-cell lithium polymer battery. Its hovering flight time is about 20 min without payload. Using three Gyro sensors and two micro controllers that are installed in the main

Fig. 3.1 Photo of autonomous controlled Quad-Rotor MAV

Table 3.1 Specifications of the X-3D-BL

Parameter	Value
Maximum diameter	540 mm
Total mass (with battery)	420 g
Payload	400 g
Motor	Brushless, 29 g, (hacker)
Motor driver	Order made (I2C)
Receiver	4-Top (PWM, Serial output)
Gyro	ENC-03 × 3
Micro controller	ATMEGA8–16AU×1
	ATMEGA88–20AU×1

body, it offers very steady flight even when it is controlled manually. The vehicle design consists of a carbon fiber airframe, so it has a light and strong body. The lift results from four brushless motor, and by using a special driver, the maximum speed reaches about 8,000 rpm. The complete vehicle is shown in Fig. 3.1. And the specification of X-3D-BL is shown in Table 3.1.

The following explains the flight principles of a four rotor type MAV. The motors in Fig. 3.2a are numbered as No.1 at the front motor and then clockwise to No. 4 (also shown in Fig. 3.2c). Lift is obtained from the total force by all motors. The moment around X axis is generated by the rotational speed difference between No. 2 and No. 4 motors (Fig. 3.2a), so as the attitude angle around X axis of the airframe changes, the thrust is converted into the component force on Y direction. Using the same principle, by using the rotational speed difference between No. 1 and No. 3 motors, it is possible to control the X direction of the airframe (Fig. 3.2b). Moreover, the No. 1 and No. 3 rotors are rotating clockwise while the No. 2 and No. 4 rotors are rotating in the opposite direction (counter-clockwise) (Fig. 3.2c). The rotation around Z axis (Yaw) of the airframe is controlled by counterbalancing the moment (Fig. 3.2d). The internal PID controller of the airframe does the stabilization control

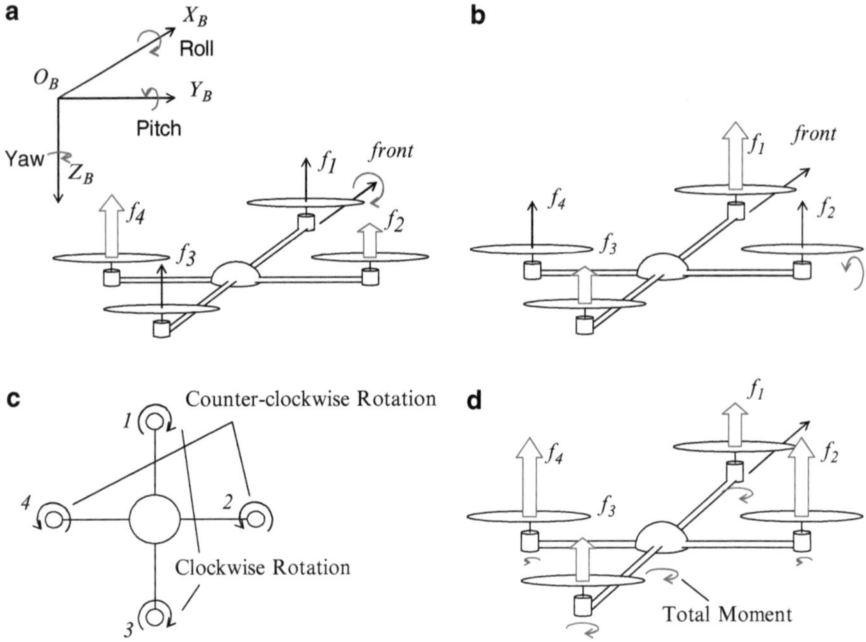

Fig. 3.2 Image of principles of Quad-Rotor MAVs flight. (**a**) Principle of roll control, (**b**) Principle of pitch control, (**c**) Motor rotation direction, (**d**) Principle of yaw control

by using three gyro sensors. In the manual control mode, the X-3D-BL behaves just like a radio controlled helicopter and it can be operated using a four channel Propo (RC controller). With attitude control mode, signals of the joystick are transmitted to the airframe as control instruction values of Altitude, Roll, Pitch, and Yaw angle. After the airframe receives the instructions, it calculates the appropriate response using its internal control system, and then sends the rotational speed instructions to the driver of each motor. Each degree of freedom of the airframe can be controlled by setting the rotational speeds of the motors.

3.3 Experimental Setup

Experiment setup is composed of Ground Control Station (GCS) and embedded control system which is installed to the airframe.

3.3.1 Embedded Control System

To decrease the delay due to radio communications and other factors, we installed all sensors and the micro controller on the airframe and used an onboard embedded controlled system. This system is composed of a sensor unit MNAV, an SH2A micro controller board, a wireless modem, a wireless camera and other parts.

3.3.1.1 Sensor Unit MNAV

The MNAV100CA is a calibrated digital sensor and servo control system designed for use in Radio Control (R/C) vehicles. The onboard sensor package includes accelerometers, angular rate sensors, and magnetometers for use in inner loop control applications as well as static pressure (altitude) and dynamic pressure (airspeed) sensors for use in airborne robotics. A GPS sensor is also included for both path planning and navigation.

The MNAV100CA's comprehensive onboard servo control solution includes both R/C servo control hardware and an R/C receiver Pulse Position Modulation (PPM) interface. R/C servo hardware provides users with software-based control of up to NAVIGATION & SERVO CONTROL BOARD nine separate servos while the PPM interface enables software interpretation of R/C receiver commands thereby offering users both automated software control as well as manual "takeover" capability. The highest update rate of the IMU sensor is 120 Hz, and the GPS sensor (u-blox Corp.) is 4 Hz. The photo of MNAV is shown in Fig. 3.3.

3.3.1.2 Micro Controller Board AP-SH2A-0A

Because all the calculations are made onboard, a high calculation relatively fast microcomputer AP-SH2A-0A was installed in the vehicle in this research. AP-SH2A-0A is general-purpose CPU board that adopted SH7211F (SH-2A) for CPU core. All signals necessary for an external enhancing are drawn out to external connected connector. The vehicle does the data processing of sensors, the operations of the controller and communication processing with the earth station by using AP-SH2A-0A.

Fig. 3.3 Photo of sensor unit MNAV

Fig. 3.4 Photo of wireless modem XBee-pro

3.3.1.3 Wireless Modem XBee-Pro

Communicates between the vehicle and the Ground Control Station is done using a wireless modem XBee-pro which is shown in Fig. 3.4. While the XBee-pro can communicate with up to a range of 1.6 Km by using its maximum power of 63 mW, due to Japanese regulations, we cannot use the maximum output setting. And due to these restrictions, the maximum range in Japan was limited to about 400 m. When it needs to fly a longer distance, we can record the flight route in the micro controller beforehand. While there is no communications between the vehicle and the Ground Control Station at longer ranges, the mission can still be accomplished by using the autonomous flight controls.

3.3.1.4 Wireless Camera

To collect external information, a wireless camera system (MST-BB050M: Fig. 3.5) was also installed. To avoid interference from the wireless 2.4 GHz frequency modem, the transmitter of the wireless camera was set at 1.2 GHz frequency. And with this system, it is possible to send a clear picture to the Ground Control Station in real time.

A photo during flight from the Quad-Rotor type MAV equipped with the embedded control system is shown in Fig. 3.6.

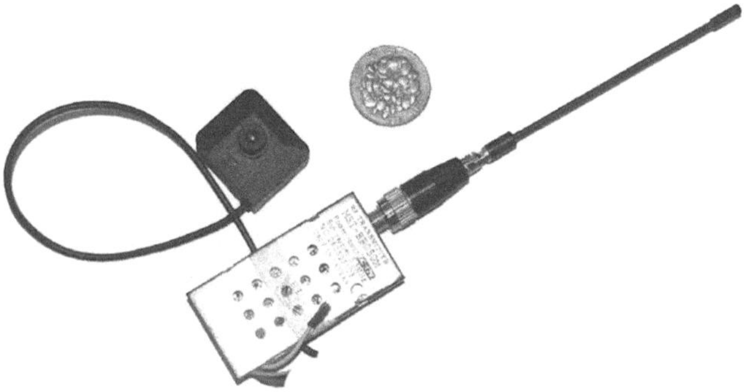

Fig. 3.5 Wireless camera system

Fig. 3.6 Image from wireless camera system

3.3.2 *Ground Control Station: GCS*

The Ground Control Station (Fig. 3.7) is composed of a mobile computer, a USB controller, a wireless modem an XBee, and the wireless camera receiver, etc. Instructions for the USB controller by the manual flight, the flight mode and the reference data by autonomous flight are sent to the vehicle by XBee, and necessary

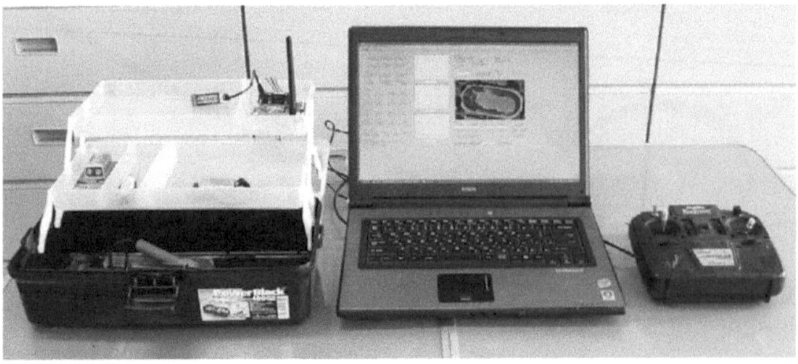

Fig. 3.7 Photo of Ground Control Station (GCS)

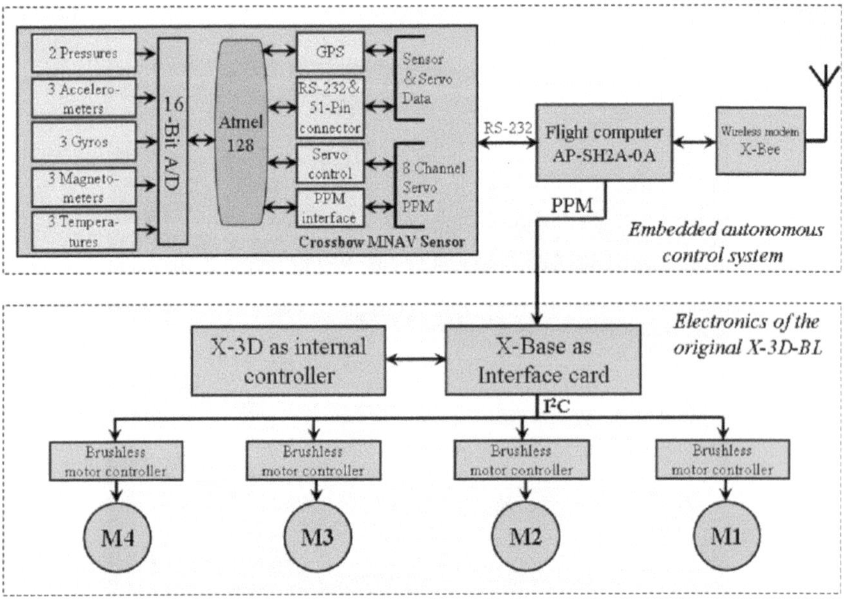

Fig. 3.8 Architecture of embedded control system

information is received from the vehicle. The flight situation data of attitude, speed present location of the vehicle and the image from camera are displayed on the monitor, the service watch and the information gathering, etc. are done.

Figure 3.8 shows the architecture of embedded control system, and Fig. 3.9 shows the image of the entire experiment system. The USB controller is used control the MAV manually. The instruction value from USB controller is taken into the onboard computer by XBee and converted into PPM signal then forwarded to the X-UFO directly.

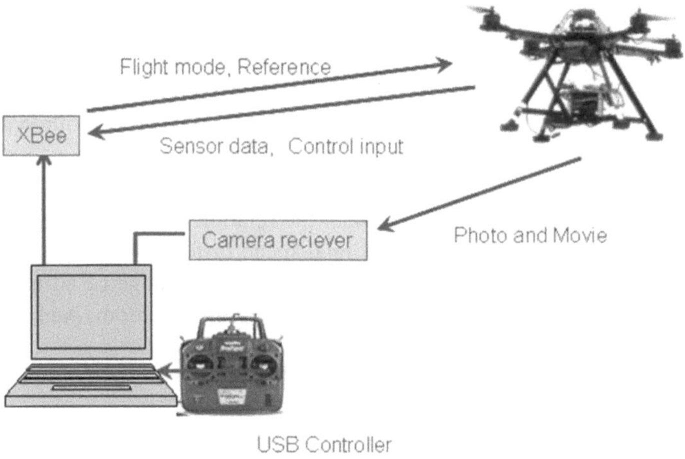

Fig. 3.9 Image of the data transmission

3.4 Modeling and Controller Design

In order to direct the flight of a small unmanned helicopter, the control system is separated into attitude, velocity, and position control loop to achieve autonomous control. Because the manual control of Quad-Rotor type MAV is similar to a helicopter, we designed the control system to mimic the autonomous control techniques of small unmanned helicopter. Since each degree of freedom of the Quad-Rotor MAV is controlled by tuning rotational speed of every rotor, the motion appears coupled. However, because it is not easy to model such a multiple-input, multiple-output (MIMO) system, in this research we disregard the coupled motion and consider each degree of freedom as an independent single-input, single-output (SISO) system and designed the control system accordingly. For the X and Y directions, we divided the control loop into three control loops of attitude (Roll, Pitch), velocity, and position. For Yaw angle, there is only one control loop. The controller design technique of vertical direction usually separates into two loops of velocity control loop and position control loop by using GPS. However, this research uses inexpensive GPS (u-blox). And while the control of vertical direction is possible when the GPS signal is good, it is not easy to control when the signal is poor. Therefore in this situation, the vertical direction position of the vehicle is measured using the atmospheric pressure sensor instead of the GPS sensor. However, since the velocity cannot be detected by atmospheric pressure sensor, it differs from the advanced control system that uses usual GPS. So in this case, the velocity control loop is pulled out and only the positional control loop is used. At first modeling is done for each loop, and then the model based controller is designed. For the multi-loop control system, the inner loop controller is designed at first and then the outer loop controller is adjusted. This chapter introduces the technique of modeling and the control system design of the attitude control loop of Roll, Pitch, and Yaw angle [9, 10, 20, 21].

3.4.1 Modeling

It is thought that the attitude model of Roll is same as Pitch because the symmetry of Quad-Rotor type MAV. Therefore, modeling is done to roll attitude angle and Yaw angle.

Because the attitude of Quad-Rotor type MAV is controlled according to the rotational speed of the motor, at first consider the relational expression of the attitude and the torque generated from the rotational speed difference of the motor around the center of the vehicle. As a precondition, the change range of attitude angle is minute and there is no coupling between each degree of freedom. And, the expression below is obtained [3, 5, 6].

$$\tau = J\ddot{\eta} \tag{3.1}$$

Here, τ is a torque vector, J is an inertia matrix, and η is attitude Euler angle. Because the torque was obtained according to the rotational speed difference of the motor, make Laplace transforms of expression (3.1), and assume the transfer function between the instruction value of the rotational speed and the rotational speed actually obtained to be the first order delay system. The transfer function from the motor rotational speed instruction value to the attitude angle becomes expression (3.2).

$$G_{\eta n}(s) = \frac{\eta(s)}{n(s)} = \frac{k}{J(1+Ts)s^2} \tag{3.2}$$

For the usual small unmanned radio controlled helicopter, for Roll and Pitch angle stability mechanism a stabilizer bar is installed, and yaw angle is stabilized by using the rate gyro. For a Quad-Rotor type MAV, since the Roll and the Pitch angle are controlled by the rotor rotational speed, a stabilizer cannot be used. To stabilize the vehicle, the gyro sensor of Roll angle and Pitch angle are used to do angular velocity feedback control. Here, assume the feedback control of angular velocity is P control, and then the transfer function from the input to the angular velocity can be obtained.

$$G_{\eta u}(s) = \frac{\eta(s)}{u(s)} = \frac{k}{(JTs^2 + Js + kk_p)s} \tag{3.3}$$

To request the unknown parameters, set input as PPM signal and angular velocity as output, then collect experiment data and each parameter is identified excluding the integration of expression (3.3). As an example, Fig. 3.10 shows the frequency response of the transfer function of the obtained Roll angle. Moreover, Fig. 3.11 and Fig. 3.12 show the verification result in the time area by using the experiment data. The output of Fig. 3.11 is angular velocity, and Fig. 3.12 is attitude angle. It can be confirmed that the simulation result almost corresponds to the experimental data.

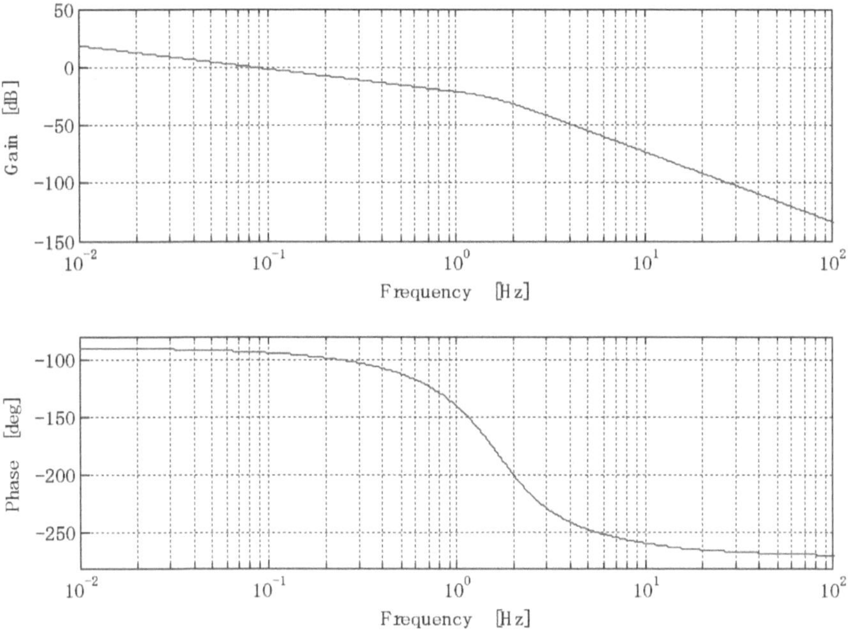

Fig. 3.10 Frequency response of attitude model

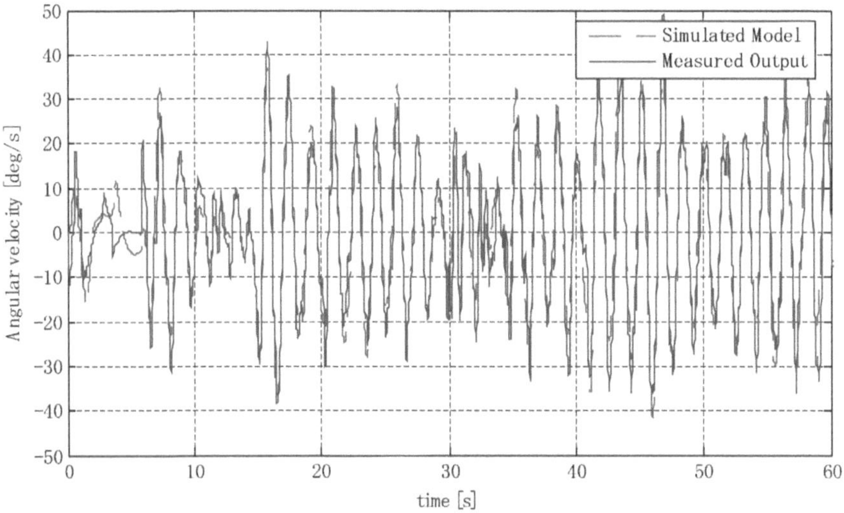

Fig. 3.11 Cross validation result of angular velocity output

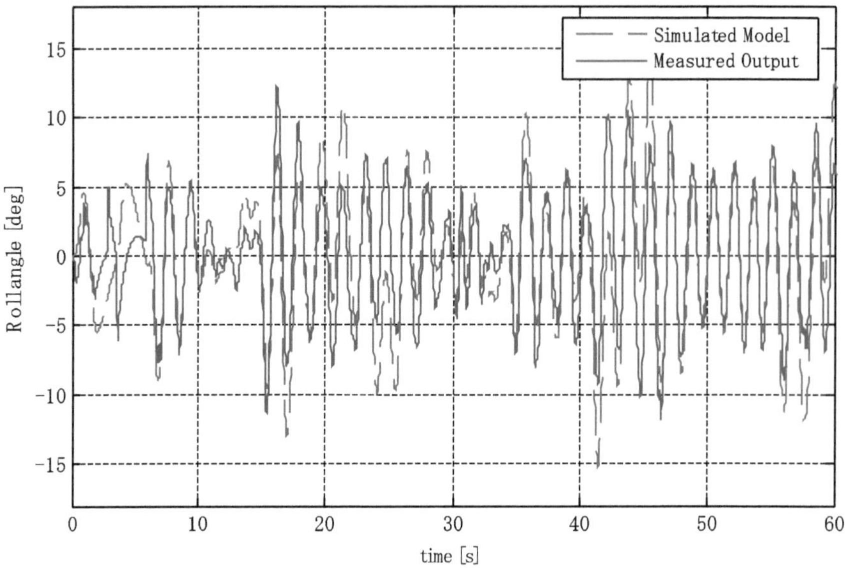

Fig. 3.12 Cross validation result of Roll angle output

3.4.2 Controller Design

In the controller design, we adopt the LQI control method for attitude control. LQI is one of the most popular design methods of servo controllers based on the modern control theory [20].

When transfer function equation (3.3) is converted into the state equation, it becomes equation (3.4).

$$\begin{aligned} \dot{x}(t) &= Ax(t) + Bu(t) \\ y(t) &= Cx(t) \\ A &= \begin{bmatrix} -\frac{1}{T} & -\frac{kk_p}{JT} & 0 \\ 1 & 0 & 0 \\ 0 & 1 & 0 \end{bmatrix} \qquad B = \begin{bmatrix} kk_p \\ 0 \\ 0 \end{bmatrix} \\ C &= \begin{bmatrix} 0 & 0 & 1 \end{bmatrix} \end{aligned} \tag{3.4}$$

To design the LQI servo control system, a novel state variable $x_r(t)$ is introduced as the integral value of the displacement error, and the following expansion systems are established.

$$\frac{d}{dt}\begin{bmatrix} x(t) \\ x_r(t) \end{bmatrix} = \begin{bmatrix} A & 0 \\ -C & 0 \end{bmatrix}\begin{bmatrix} x(t) \\ x_r(t) \end{bmatrix} + \begin{bmatrix} B \\ 0 \end{bmatrix} u(t) + \begin{bmatrix} 0 \\ I \end{bmatrix} r(t) \tag{3.5}$$

The LQI control corresponds to the optimal control problem for the system (3.5) based on the quadratic form criterion function

$$J = \int_0^\infty \left[x_a^T Q x_a(t) + u^T(t) R u(t) \right] dt \tag{3.6}$$

Its control input is obtained by the state feedback algorithm using the state feedback gains as

$$\begin{aligned} u(t) &= -F_1 x(t) - F_2 x_r(t) = -F_1 x(t) - F_2 \int_0^t e(t)dt \\ F &= \begin{bmatrix} F_1 & F_2 \end{bmatrix} = R^{-1} B^{\#T} P \end{aligned} \tag{3.7}$$

where P is the solution of the following Riccati equation,

$$\begin{aligned} & PA^{\#} + A^{\#T} P - PB^{\#} R^{-1} B^{\#T} P + Q = 0 \\ & A^{\#} = \begin{bmatrix} A & 0 \\ -C & 0 \end{bmatrix}, \quad B^{\#} = \begin{bmatrix} B \\ 0 \end{bmatrix} \end{aligned} \tag{3.8}$$

The weights of the criterion function (3.6) are often tuned by the following procedure. The weight on the input R is set to a constant, e.g., 1.0. Simultaneously, the elements of the weighting matrix on the state variables Q are tuned such that the designed control system satisfies some control specifications. Especially, the LQI controller design pays attention to the tuning of the weight on $x_r(t)$. Other elements are zeros in many LQI designs.

Though the attitude angle and the angular velocity can be observed by IMU sensor, but there is still one state variable that cannot be observed. Then we design a Kalman filter to estimate the unknown state variable.

3.5 Experiment and Experimental Result

In order to verify the effectiveness of the technique introduced in this chapter, flight experiments were carried out. Because the drift of the sensor exists, the airframe cannot keep hovering by zero reference of attitude angle. To stabilize the airframe, it is necessary to always change the reference. Then, the reference is given to the airframe by the operator's using USB Propo. The performance of the attitude controller is verified by letting the airframe keeps hovering or does high-speed moving. Results from a flight test are shown in Fig. 3.13. One can see that the output angle tracked successfully the reference angle. Figures 3.14 and 3.15 shows the result of state variable estimate by Kalman filter. By Figs. 3.14 and 3.15, we can see both attitude angle and angular velocity obtained good estimate performance.

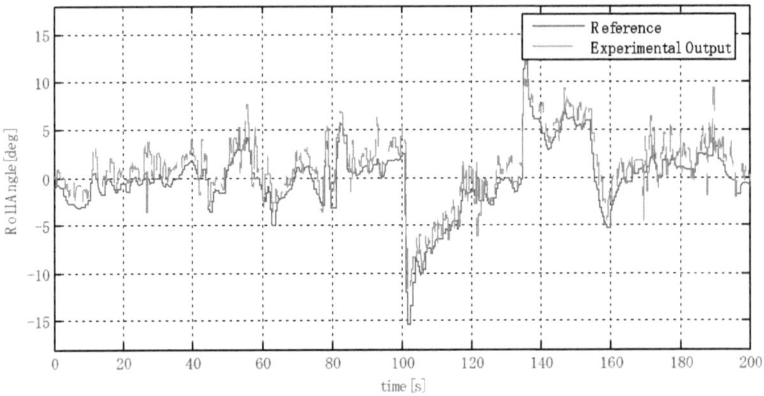

Fig. 3.13 Experimental result of Roll Angle

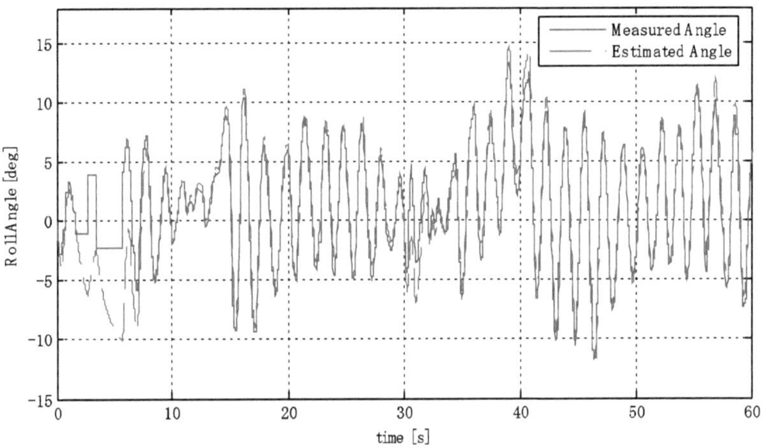

Fig. 3.14 Estimate result of Roll Angle

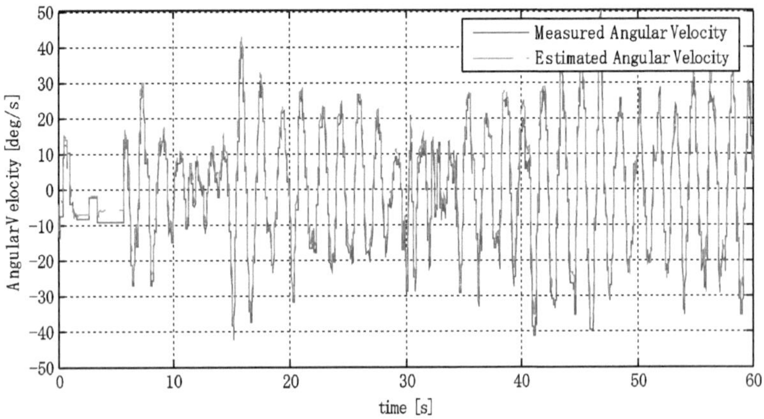

Fig. 3.15 Estimate result of Angular Velocity

3.6 Summary

This chapter introduced the technique of linear modeling and model based controller design for Quad-Rotor MAVs. Since Quad-Rotor MAVs have an internal angular velocity control loop the nonlinearity of the airframe may become weak, therefore we used linear control theory instead and obtained acceptable control performance. As the design of the attitude control loop is completed, the velocity control loop and the position control loop can adopt the same modeling techniques as a typical unmanned helicopter. However, the performance of the LQI control technique introduced in this chapter depends on the accuracy of model. When the model has uncertain parameters such as an increase or decrease the payload of the airframe, the control performance can deteriorate. For the designer who is must insure that the attitude control system can operate in such a situation, the recommended approach is to use sliding mode control theory and the H_∞ control theory which have good robustness in this situation.

References

1. McKerrow P (2004) Modeling the draganflyer four-rotor helicopter. In: Proceedings of the 2004 IEEE international conference on robotics and automation, New Orleans, LA, 2004, pp 3596–3601
2. Tayebi A, McGilvray S (2004) Attitude stabilization of a four-rotor aerial robot. In: *43rd IEEE conference on decision and control (CDC)*, 2, 2004
3. Tayebi A, McGilvray S (2006) Attitude stabilization of a VTOL quadrotor aircraft. *IEEE Trans Control Syst Technol* 14(3):562–571
4. Bouabdallah S, Siegwart R (2007) Full control of a quadrotor. In: *Proceedings of the IEEE/RSJ international conference on intelligent robots and systems (IROS)*, 2007
5. Kendoul F, Lara D, Fantoni I, Lozano R (2007) Real-time nonlinear embedded control for an autonomous quad-rotor helicopter. J Guid Control Dyn 30(4):1049–1061
6. Kendoul F, Fantoni I, Lozano R (2008) Asymptotic stability of hierarchical inner-outer loop-based flight controllers. In: Proceedings of the 17th IFAC World Congress, Seoul, Korea, 6–11 July 2008, pp 1741–1746
7. Ascending Technologies gmbh (2009) http://www.asctec.de
8. Microdrones gmbh (2009) http://www.microdrones.com
9. Wang W, Nonami K, Hirata M (2008) Model reference sliding mode control of small helicopter X.R.B based on vision. Int J Adv Robot Syst 5(3):235–242
10. Wang W, Hirata M, Nonami K, Miyazawa O (2007) Modeling and autonomous control of microminiature ultra-small scale/light weight micro-flying robot. J Robot Soc Jpn 25(1):121–129
11. Nonami K (2007) Prospect and recent research and development for civil use autonomous unmanned aircraft as UAV and MAV. J Syst Des Dyn 1(2) Special issue on new trends of motion and vibration control:120–128
12. Hoffmann GM, Tomlin CJ (2008) Decentralized cooperative collision avoidance for acceleration constrained vehicles. In: Proceedings of the 47th IEEE conference on decision and control cancun, Mexico, Dec 9–11 2008
13. Hoffmann GM, Waslander SL, Tomlin CJ (2008) Quadrotor helicopter trajectory tracking control. In: Proceedings of AIAA guidance, navigation, and control conference, Honolulu, HI, August 2008
14. Pounds P, Mahony R, Corke P (2006) Modeling and control of a quad-rotor robot. In: Proceedings of the Australasian conference on robotics and automation, Auckland, New Zealand, 2006

15. Hoffmann GM, Huang H, Waslander SL, Tomlin CJ (2007) Quadrotor helicopter flight dynamics and control: theory and experiment. In: Proceedings of AIAA guidance, navigation, and control conference, Hilton Head, SC, August, 2007
16. Roberts JF, Stirling T, Zufferey JC, Floreano D (2007) Quadrotor using minimal sensing for autonomous indoor flight. European air vehicle conference and flight competition EMAV2007
17. Ahrens S, Levine D, Andrews G, Jonathan P (2009) How, vision-based guidance and control of a hovering vehicle in unknown, gps-denied environments. In: 2009 IEEE International conference on robotics and automation, Kobe International Conference Center, Kobe, Japan, 12–17 May, 2009
18. Gurdan D, Stumpf J, Achtelik M, Doth KM, Hirzinger G, Rus D (2007) Energy-efficient autonomous four-rotor flying robot controlled at 1 khz. In: IEEE international conference on robotics and automation, Roma, Italy, April, 2007
19. Wang W, Suzuki S, Nonami K, Iwakura D, Pebrianti D, Zhu LD, Kendoul F (2008) Fully autonomous quad-rotor MAV and flying performance with complete embedded system. The 9th international conference on motion and vibration control, Paper No.1201, 2008
20. Nonami K, Nishimura H, Hirata M (1998) Control system design by MATLAB. Tokyo Denki University Press, Tokyo. ISBN4–501–31940–2
21. Mahony R, Hamel T (2004) Robust trajectory tracking for a scale model autonomous helicopter. Int J Robust Nonlinear Control 14(12):1035–1059

Chapter 4
Development of Autonomous Quad-Tilt-Wing (QTW) Unmanned Aerial Vehicle: Design, Modeling, and Control

Abstract In this chapter, we propose an autonomous attitude control of a quad tilt wing-unmanned aerial vehicle (QTW-UAV). A QTW-UAV can achieve vertical takeoff and landing; further, hovering flight, which is characteristic of helicopters, and high cruising speeds, which are a characteristic of fixed-wing aircraft, can be achieved by changing the angle of the rotors and wings by a tilt mechanism. First, we construct an attitude model of the QTW-UAV by using the identification method. We then design the attitude control system with a Kalman filter-based linear quadratic integral (LQI) control method; the experiment results show that a model-based control design is very useful for the autonomous control of a QTW-UAV.

Video Link:

Principle of flight for QTW-UAV
http://mec2.tm.chiba-u.jp/monograph/Videos/Chapter4/1.avi

4.1 Introduction

In recent times, many unmanned aerial vehicles (UAVs) have been operated and studied [1–21]. These UAVs can be roughly divided into two groups – UAVs based on fixed-wing aircraft and UAVs based on rotary wings (e.g., helicopter). Both types have their advantages and disadvantages. UAVs based on fixed-wing aircraft can fly with a large cruising speed, but an extremely large site is required for takeoff and landing, and it is unsuitable for missions in urban regions. On the other hand, in the case of UAVs based on rotary wings, taking off and landing sites need not be chosen because they can takeoff and land vertically. In addition, they can also perform a steady flight operation called hovering; hence, they are advantageous for monitoring fixed points. However, their cruising speed is less than that of a fixed-wing aircraft, and hence, the area on which they can carry out their mission is small.

In recent years, tilt-rotor type UAVs that combine the advantages of both types of UAVs have been developed. The most famous tilt rotor type UAV is EagleEye manufactured by Bell [22]. An overview of EagleEye is shown in Fig. 4.1. EagleEye

K. Nonami et al., *Autonomous Flying Robots: Unmanned Aerial Vehicles and Micro Aerial Vehicles*, DOI 10.1007/978-4-431-53856-1_4, © Springer 2010

Fig. 4.1 Overview of EagleEye [22]

has rotors with a tilt mechanism on the tip of its wing; this is a key feature of tilt-rotor type UAVs. EagleEye can realize vertical takeoff and landing, hovering, and high cruising speed flight by changing the angle of the rotor side by a tilt mechanism.

However, the disadvantage of using two rotors, such as in EagleEye, is that it is impossible to fly when one rotor has a malfunction. Hence, it is desirable to have four rotors. Moreover, since only rotors are tilted, the efficiency of the slipstream effect decreases because the wings are not linked.

In this chapter, the development of a quad tilt wing (QTW)-UAV, which has four rotors and wings and a tilt mechanism for each, for resolving the abovementioned problems is described. We aim to realize autonomous control of a quad tilt wing-unmanned aerial vehicle (QTW-UAV). In this chapter, the design of an attitude control system for the helicopter mode of a QTW-UAV is introduced; this represents fundamental research on the autonomous control of QTW-UAV.

4.2 Quad Tilt Wing-Unmanned Aerial Vehicle

An overview of a QTW-UAV is shown in Fig. 4.2 and the specifications of QTW-UAV are shown in Table 4.1. This UAV was developed for various observation missions such as polar region observations. It has four rotors, four wings, and a tilt mechanism that is actuated by electric motors; the flight mode of this UAV can be changed by tilting the rotors and wings. The first flight mode is the helicopter mode, and in this mode, the QTW-UAV can takeoff and land vertically; and it can hover on a fixed point. The second flight mode is the airplane mode; in this mode, the QTW-UAV can fly an airplane, and it is possible to move at high speeds. In the following discussion, the steerage of the QTW-UAV in each mode is introduced

Fig. 4.2 Overview of QTW-UAV

Table 4.1 QTW-UAV specifications

Power source	Electrical motor
Body length	1,610 mm
Propeller diameter	508 mm
Wingspread	1,342 mm
Rotor speed	About 6,000 rpm
Weight	24 kg

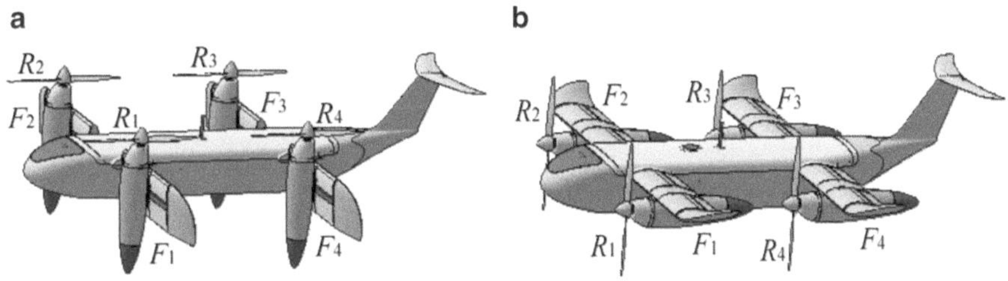

Fig. 4.3 Flight model of QTW-UAV. (**a**) Helicopter mode. (**b**) Airplane mode

by using Fig. 4.3. First, in the helicopter mode, the tilt angles of rotors and wings are almost 90° over the horizontal plane, as shown in Fig. 4.3a. In this state, the QTW-UAV rotation in the rolling direction by a moment is caused by the difference between the rotor thrusts R_1, R_4 and rotor thrusts R_2, R_3; and, the QTW-UAV rotation in the pitching direction by the moment is caused by the difference between the rotor thrusts R_1, R_2 and rotor thrusts R_3, R_4. Moreover, the QTW-UAV can rotate around the yawing direction by a yawing moment that is caused by a slipstream effect of each rotor and flaperon F_1, F_2, F_3, and F_4 (Fig. 4.4).

Next, the airplane mode is introduced. In the airplane mode, the tilt angles of the rotors and wings are almost 0° in the horizontal plane, as shown in Fig. 4.3b. In this state, the QTW-UAV rotates in the rolling direction by a moment caused by

Fig. 4.4 Slipstream effect

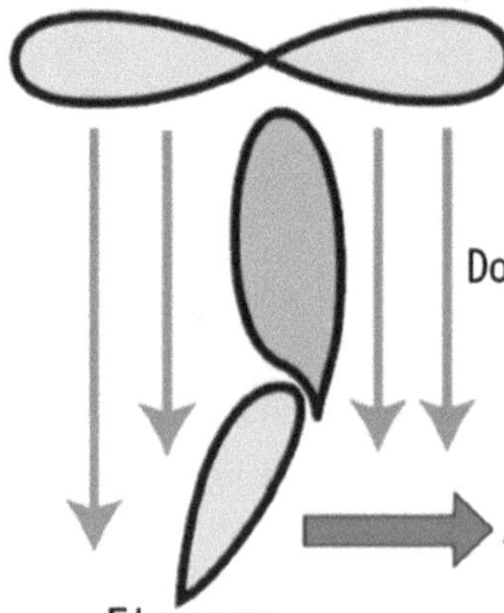

the difference between the lift forces of flaperons F_1, F_4 and those of flaperons F_2, F_3; further, the QTW-UAV rotates in the pitching direction by the moment caused by the difference between the lift forces of the flaperons F_1, F_2 and those of the flaperons F_3, F_4. Moreover, the QTW-UAV can rotate in the yawing direction by a yawing moment caused by the difference between the rotor thrusts R_1, R_4 and R_2, R_3.

4.3 Modeling of QTW-UAV

In this section, a mathematical model of the 3-axis attitude, roll, pitch, and yaw of the QTW-UAV in the helicopter mode is derived.

4.3.1 *Coordinate System*

The definitions of the coordinate systems used in this paper are shown in Fig. 4.5. The coordinate system F_i is a ground-fixed frame. The origin of this frame is fixed at an arbitrary point on the ground; X_i axis indicates north; Y_i axis, the east; and Z_i axis, the direction of gravity. The coordinate system F_b is a body-fixed frame. The origin of this frame is fixed on the center of gravity of QTW-UAV. X_b axis indicates the forward direction of the body; Y_b axis, the rightward direction of the body; and Z_b axis, the downward direction of the body.

In this chapter, we use Euler's angle for attitude expression. The roll angle ϕ, the pitch angle θ, and yaw angle ψ are defined as the attitude of the F_b frame for the F_i frame. Moreover, the 3-axis angular rate p, q, r is defined around all the three axes of the body-fixed frame, respectively.

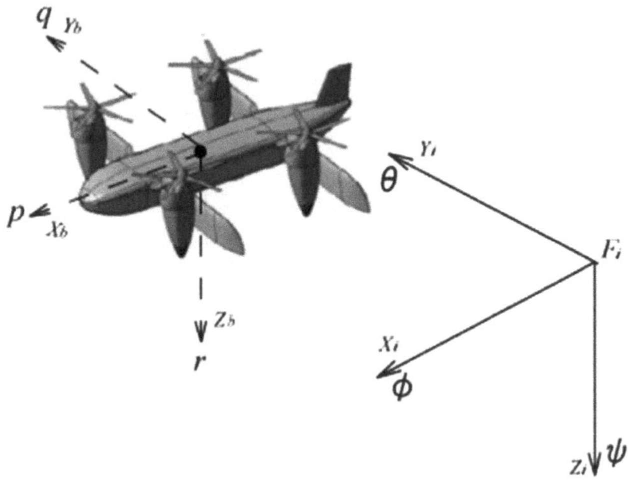

Fig. 4.5 Coordinate system

4.3.2 Yaw Model

In this section, the yaw attitude model of a QTW-UAV is derived by using the simple figure of the QTW-UAV shown in Fig. 4.6. Four rotors are located symmetrically around the center of gravity of the body. L is the length between the center of the rotor and center of gravity. Moreover, the X'_b axis and Y'_b axis are along the center of rotors R_1 and R_2, respectively, and are at right angles to each other. Therefore, Ψ in the figure is 45°. Now, let us consider that f_p is the force on each flaperon caused by the slip stream effect; the direction of f_p is defined as shown in Fig. 4.6. The moment around the center of the gravity caused by f_p is calculated as $f_p L \sin\Psi$, and the entire moment around the Z_b axis is calculated as follows:

$$M_z = 4 f_p L \sin \Psi \tag{4.1}$$

Here the transfer function between this moment and yaw rate r is assumed to be first order; further, $4L\sin\Psi$ in (4.1) is constant and determined by the geometrical arrangement of rotors.

Now, we consider this constant value to be K_1.

In such a case, the mathematical model between the force caused on the flaperon and yaw rate is obtained as follows:

$$R = \frac{T_r K_1}{s + T_r} F_p \tag{4.2}$$

In this equation, R and F_p are the Laplace transforms of r and f_p, respectively, and T_r is the time constant of the system. Moreover, it is believed that there is a certain

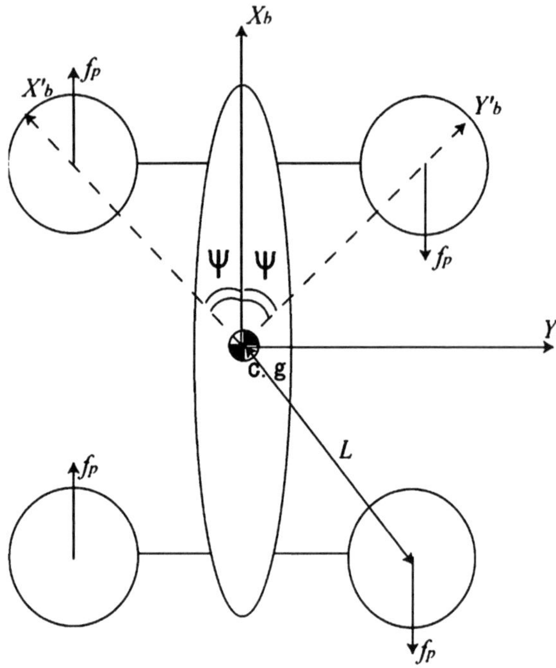

Fig. 4.6 Simplified model of QTW-UAV

Table 4.2 Parameters of yaw model

Parameter	Value	Parameter	Value
a_{r1}	15.826	a_{r2}	1.775
b_{r1}	2.13	–	–

delay between the angle of each flaperon and f_p, and we assume this delay to be first order. Finally, the mathematical model of the dynamics between Θ_{fp} and r is a second-order system, and the model of the yaw angle ψ is obtained as (4.4) after adding integral element to this second-order model.

$$R = \frac{b_{r1}}{s^2 + a_{r1}s + a_{r2}} \Theta_{fp} \tag{4.3}$$

$$\psi = \frac{b_{r1}}{s(s^2 + a_{r1}s + a_{r2})} \Theta_{fp} \tag{4.4}$$

a_{r1}, a_{r2}, and b_{r1} in this equation are model parameters. They were obtained by comparisons between experimental data, which is obtained from manually operated flight, and the output of the model. The obtained parameters are shown in Table 4.2, and the comparison between the experimental data and model response is shown in Figs. 4.7 and 4.8, respectively. From Figs. 4.7 and 4.8, we find that the model output matches the experimental data well, and thus, the model validity is demonstrated.

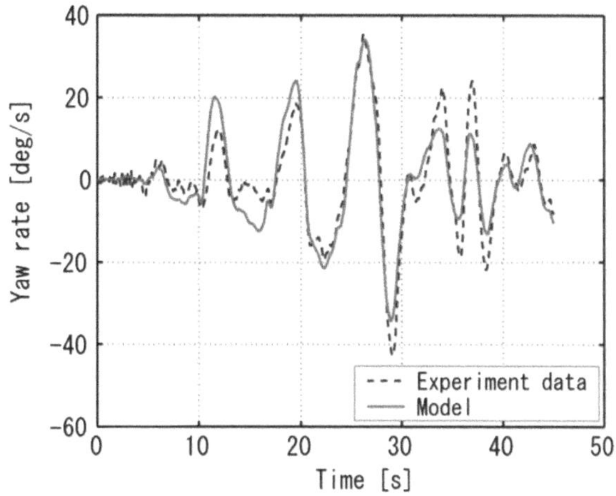

Fig. 4.7 Comparison of yaw rate

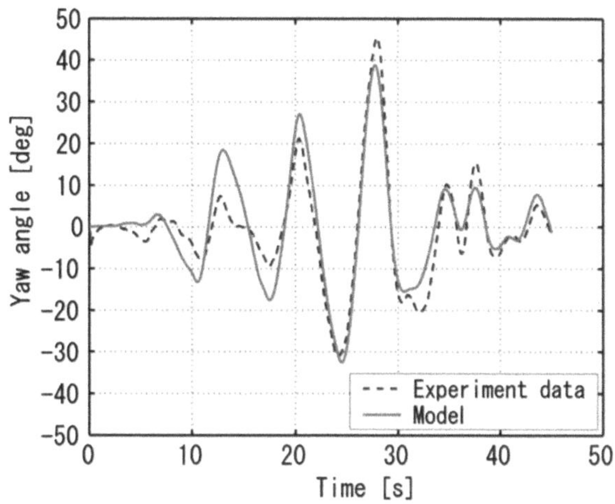

Fig. 4.8 Comparison of yaw angle

4.3.3 Roll and Pitch Attitude Model

In this section, the construction of a roll and pitch attitude model is introduced. The roll and pitch attitude angles of the QTW-UAV vary with the moments generated by the differences in the thrust of the four rotors. First, the relationship between the moment in the roll or pitch direction and the thrust of four rotors is introduced. The thrust generated by each rotor is expressed as the sum of DC elements and fluctuations, and shown in (4.5)–(4.8). Here, T_R is the thrust of each rotor; T_d, the DC element of thrust; and T_{xR}, the fluctuation in thrust. Further, the suffix implies the number of rotors.

$$T_{R_1} = T_d + T_{xR_1} \tag{4.5}$$

$$T_{R_2} = T_d + T_{xR_2} \tag{4.6}$$

$$T_{R_3} = T_d + T_{xR_3} \tag{4.7}$$

$$T_{R_4} = T_d + T_{xR_4} \tag{4.8}$$

Next, a simple figure of the QTW-UAV, which is shown in Fig. 4.6, is used in a similar manner as in the previous section. The moment around the center of gravity of the body, which is generated by the differences in the thrust of each rotor, is expressed by (4.9), (4.10).

$$M_{X'_b} = -L(T_{R_2} - T_{R_4}) = -L(T_{xR_2} - T_{xR_4}) \tag{4.9}$$

$$M_{Y'_b} = L(T_{R_1} - T_{R_3}) = L(T_{xR_1} - T_{xR_3}) \tag{4.10}$$

Here $M_{X'_b}$ and $M_{Y'_b}$ in the above equation are the X'_b axis and Y'_b axis components of the moment around the center of gravity of the body, respectively. The coordinate transformation of $M_{X'_b}$ and $M_{Y'_b}$ in the coordinate system F_b is then performed, and the next equation is obtained.

$$M_{X_b} = M_{X'_b} \cos\Psi + M_{Y'_b} \sin\Psi = K_2(T_{xR_4} - T_{xR_2}) + K_3(T_{xR_1} - T_{xR_3}) \tag{4.11}$$

$$M_{Y_b} = -M_{X'_b} \sin\Psi + M_{Y'_b} \sin\Psi = K_3(T_{xR_4} - T_{xR_2}) + K_2(T_{xR_1} - T_{xR_3}) \tag{4.12}$$

Here $K_2 = L\cos\Psi$ and $K_3 = L\sin\Psi$ are constants determined by the geometrical arrangement of rotors in a similar manner to K_1 in the previous section. We now assume a transfer function of the dynamics between the moment and angular rate to be of the first order in a similar manner to the case of the yaw model, and the mathematical model of the angular rate is obtained as (4.13) and (4.14). Here, P and Q are the Laplace transforms of the angular rates p and q, respectively. T_p and T_q are the time constants of the system.

$$P = \frac{T_p}{s + T_p}\{K_2(T_{xR_4} - T_{xR_2}\} + K_3(T_{xR_1} - T_{xR_3})\} \tag{4.13}$$

$$Q = \frac{T_q}{s + T_q}\{K_3(T_{xR_2} - T_{xR_4}\} + K_2(T_{xR_1} - T_{xR_3})\} \tag{4.14}$$

Moreover, considering the roll and pitch control inputs as δ_ϕ and δ_θ, the relationship between these inputs and the fluctuations in the thrust of each rotor are expressed as equations (4.15)–(4.18). Then, substituting these equations into (4.13) and (4.14), and converting the input of the model into δ_ϕ and δ_θ, the converted model is obtained as (4.19) and (4.20).

$$T_{xR_1} = K_{4R_1}\delta_\phi + K_{5R_1}\delta_\theta \tag{4.15}$$

$$T_{xR_2} = K_{4R_2}\delta_\phi + K_{5R_2}\delta_\theta \tag{4.16}$$

$$T_{xR_3} = K_{4R_3}\delta_\phi + K_{5R_3}\delta_\theta \tag{4.17}$$

$$T_{xR_4} = K_{4R_4}\delta_\phi + K_{5R_4}\delta_\theta \tag{4.18}$$

$$P = \frac{T_p}{s+T_p}\{K_2(K_{4R_4} - K_{4R_2}\} + K_3(K_{4R_1} - K_{4R_3})\}\delta_\phi$$
$$+\{K_2(K_{5R_4} - K_{5R_2}\} + K_3(K_{5R_1} - K_{5R_3})\}\delta_\theta \tag{4.19}$$

$$Q = \frac{T_q}{s+T_q}\{K_2(K_{4R_1} - K_{4R_3}\} + K_3(K_{4R_2} - K_{4R_4})\}\delta_\phi$$
$$+\{K_2(K_{5R_1} - K_{5R_3}\} + K_3(K_{5R_2} - K_{5R_4})\}\delta_\theta \tag{4.20}$$

Here $K_{4R_1} \sim K_{4R_4}$ and $K_{5R_1} \sim K_{5R_4}$ in this equation are constant; they are determined by the characteristics of each rotor. According to the above equation, in the angular velocity dynamics of the QTW-UAV, the axial inputs mix, and it is a difficult system to perform manual operations. Therefore, in this research, the angular rate feedback loop shown in Fig. 4.9 is constructed by using a rate gyro for the hobby helicopter shown in Fig. 4.10. In this manner, the improvement in the stability of the system and reduction in the influence of the coupling of each axis simplifies the QTW-UAV operation; K_g in this figure is a gyro gain. In this case, we used a 2-axial gyro GYA534 manufactured by the FUTABA electric engineering company. The influence of each axial coupling by using a rate gyro is insignificant, and it is possible to approximate the dynamics of the system as a single input single output (SISO) model. Finally, the angular rate model of the roll and pitch directions can be approximated as a second-order system, as shown in (4.21), (4.22).

$$P = \frac{b_{p1}}{s^2 + a_{p1}s + a_{p2}}\delta_\phi \tag{4.21}$$

$$Q = \frac{b_{q1}}{s^2 + a_{q1}s + a_{q2}}\delta_\theta \tag{4.22}$$

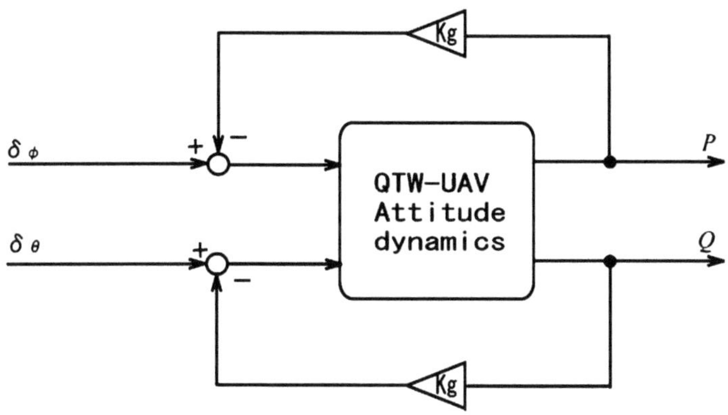

Fig. 4.9 Gyro system

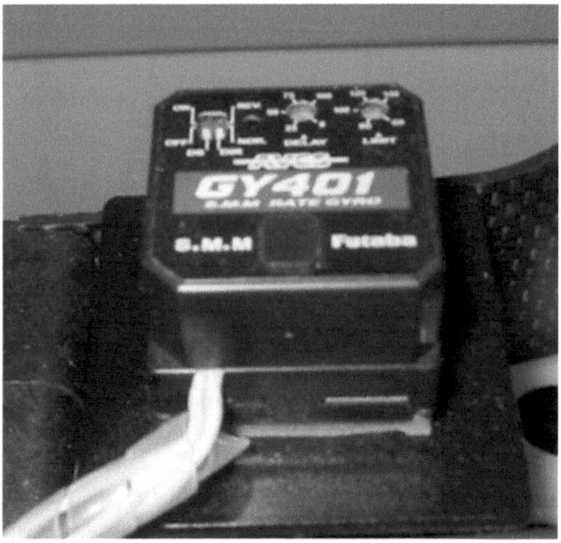

Fig. 4.10 Overview of rate gyro

Table 4.3 Parameters of roll/pitch model

Parameter	Value	Parameter	Value
a_{p1}	35.2	a_{p2}	350.23
b_{p1}	163.03	a_{q1}	10.15
a_{q2}	43.48	b_{q1}	20.35

We add an integral element, and the transfer function from the input to the attitude angle of the QTW-UAV is obtained as follows:

$$\phi = \frac{b_{p1}}{s(s^2 + a_{p1}s + a_{p2})}\delta_\phi \tag{4.23}$$

$$\theta = \frac{b_{q1}}{s(s^2 + a_{q1}s + a_{q2})}\delta_\theta \tag{4.24}$$

Here, each model parameter is obtained by a comparison between the experimental data from a manually operated flight, and the output of the model is similar to the case of a yaw model. The model parameters are shown in Table 4.3, and the comparison between the response of the model and experimental data is shown in Fig. 4.11. From Fig. 4.11, it is clear that the model output matches the experimental data.

4.4 Attitude Control System Design

In this section, the control system design for roll, pitch, and yaw attitude is introduced. A model-based control system design was performed for developing the techniques of the autonomous control of a small-scale unmanned helicopter performed in a previous study by our research group [23, 24].

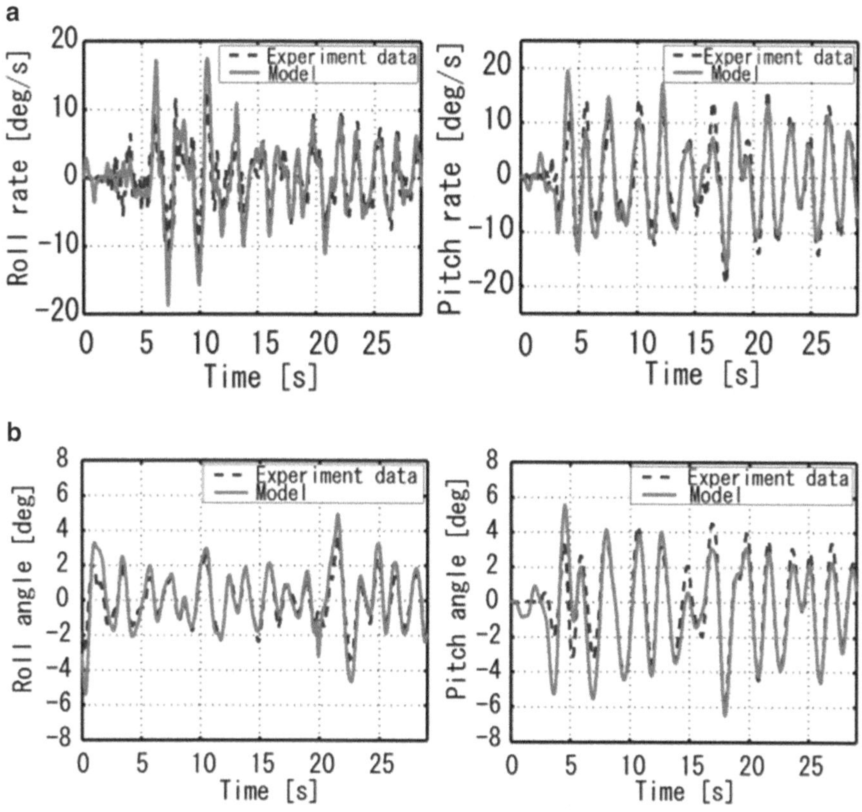

Fig. 4.11 Cross validation. (**a**) Angular rate, (**b**) Angle

4.4.1 Control System Design for Yaw Dynamics

In this section, the design of a yaw controller is introduced. Before the design, an angular rate feedback is constructed for the yaw attitude by using a rate gyro in a similar manner to the case of the roll and pitch in the previous section. The stability of the system improved significantly, and therefore, it became possible to sufficiently control the yaw angle by employing a simple control method. Hence, we designed a P control system outside the angular velocity feedback system, as shown in Fig. 4.12. In this figure, K_p is the P gain and ψ_{ref} is the reference of the yaw angle. The gyro gain is given, and the P gain was decided as $K_p = 1$ from the simulations using a closed-loop system that is constructed by combining the yaw attitude model and feedback loop by using a rate gyro system.

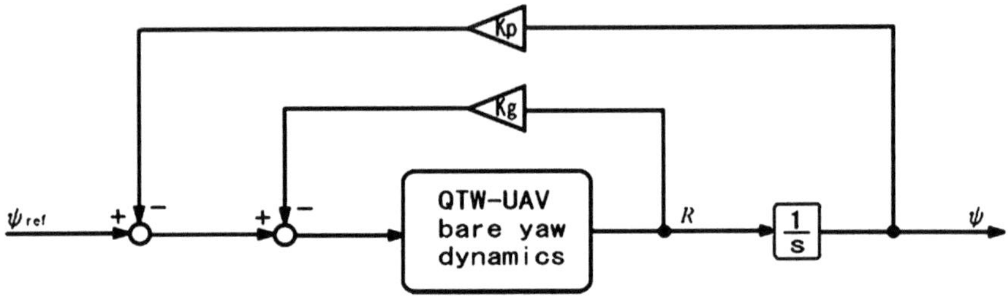

Fig. 4.12 Control system of yaw axis

4.4.2 Control System Design for Roll and Pitch Dynamics

In this section, an attitude controller for the roll and pitch is designed by using the LQI control method. Only the roll controller design is introduced here; however, the pitch controller can be designed similarly. First, the transfer function shown in (4.23) is realized as follows:

$$\begin{aligned} \dot{x}_r &= A_r x_r + B_r \delta_\phi \\ y_r &= C_r x_r \\ y_r &= \phi \end{aligned} \tag{4.25}$$

Now, the servo augmented system is constructed by using matrices in (4.25) as follows:

$$\begin{bmatrix} \dot{x}_r \\ \dot{e} \end{bmatrix} = \begin{bmatrix} A_r & O_{3\times1} \\ -C_r & 0 \end{bmatrix} \begin{bmatrix} x_r \\ e \end{bmatrix} + \begin{bmatrix} B_r \\ 0 \end{bmatrix} \delta_\phi + \begin{bmatrix} O_{3\times1} \\ 1 \end{bmatrix} \phi_{ref} \tag{4.26}$$

$$e(t) = \int_0^t \phi_{ref} - y_r(t)dt \tag{4.27}$$

Here ϕ_{ref} implies the reference of the roll attitude angle and $\bar{x}_r$ represents the state vector of this servo augmented system. The feedback gain F_r that minimizes the criterion J can be resolved by using optimal control theory. We assume the roll attitude angle to follow the reference angle without any offset by calculating the control input shown in (4.29) by using the resolved gain of F_r.

$$J = \int_0^\infty \left[\overline{x}_r^T(t) Q \overline{x}_r(t) + \delta_\phi(t) R \delta_\phi(t) \right] dt \tag{4.28}$$

$$\delta_\phi = -F_r \overline{x}_r(t) \tag{4.29}$$

$$Q = \begin{bmatrix} 1 & 0 & 0 & 0 \\ 0 & 1 & 0 & 0 \\ 0 & 0 & 1 & 0 \\ 0 & 0 & 0 & 500 \end{bmatrix} \qquad R = 2 \tag{4.30}$$

Q and R in (4.28) are weighting matrices whose values are shown in (4.30).

Next, the identity observer is designed to estimate unknown state values. We apply the Kalman filter theory to design the observer. Let the resolved observer gain be K; the controller that includes the observer is obtained as (4.31) and (4.32).

$$\begin{bmatrix} \dot{\hat{x}}_r \\ \dot{e} \end{bmatrix} = \begin{bmatrix} A_r - B_r F_{r1} - KC_r & -B_r F_{r2} \\ O_{1\times 3} & 0 \end{bmatrix} \begin{bmatrix} \hat{x}_r \\ e \end{bmatrix} + \begin{bmatrix} K & O_{3\times 1} \\ -1 & 1 \end{bmatrix} \begin{bmatrix} y_r \\ e \end{bmatrix} \tag{4.31}$$

$$\begin{bmatrix} \delta_\phi \\ e \end{bmatrix} = \begin{bmatrix} -F_{r1} & -F_{r2} \\ O_{1\times 3} & 1 \end{bmatrix} \begin{bmatrix} \hat{x}_r \\ e \end{bmatrix} \tag{4.32}$$

Here $\hat{x}_r$ is the state vector estimated by the observer and F_{r1} and F_{r2}, the feedback gain related to the state vector x_r and integral of error e, respectively.

4.5 Experiment

4.5.1 Heading Control Experiment

First, the result of the heading control experiment is shown in Fig. 4.13. The control began at 0 s in this figure, and the heading angle reference was impressed in steps at 8 s and 38 s. Since we use a simple P controller, some offsets are present; however, it is understood that the heading angle follows the reference angle.

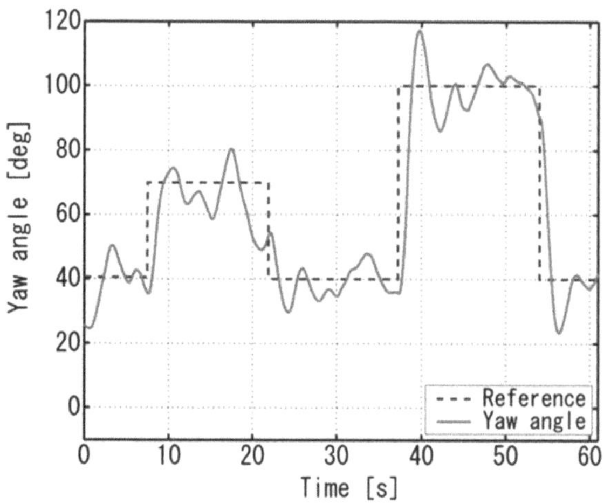

Fig. 4.13 Yaw response

4.5.2 Roll and Pitch Attitude Control Experiments

The results of the roll and pitch attitude control experiments are shown in Figs. 4.14 and 4.15. During this interval, an experiment in which the roll and pitch angle follows the 0° reference attitude was performed. By around 42 s (Fig. 4.14) and by around 7 s (Fig. 4.15), the attitude angle significantly fluctuates as a result of wind disturbances; however, it is clear that the attitude can be stabilized by control and subsequently follows a constant 0° reference attitude. From the above results, it may be said that the effectiveness of the attitude model and attitude controller for the QTW-UAV has been demonstrated.

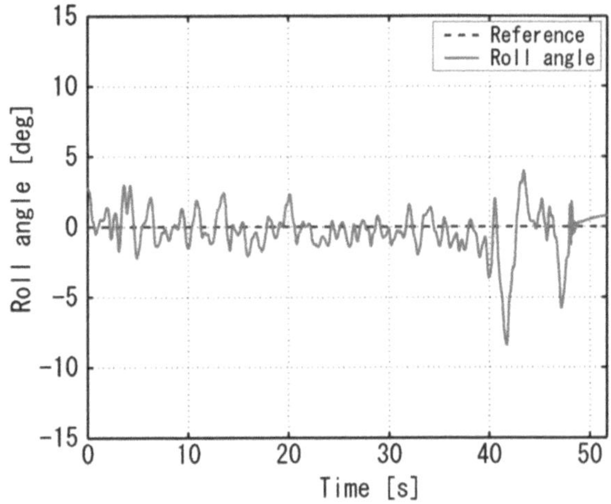

Fig. 4.14 Roll response

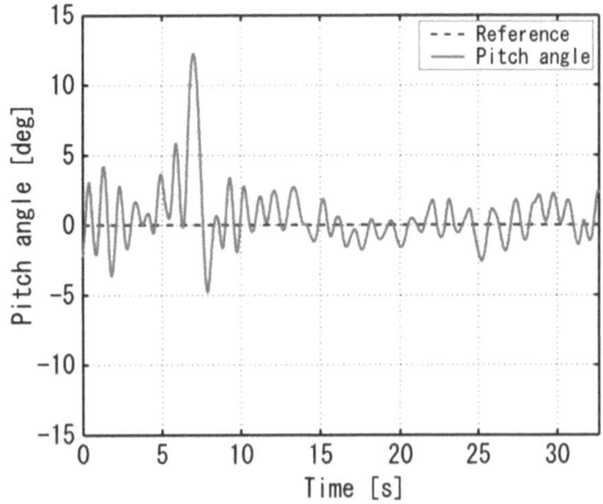

Fig. 4.15 Pitch response

4.6 Control Performance Validation at Transient State

The results shown in the previous section demonstrate that attitude control of the QTW-UAV in the helicopter mode was achieved. However, a QTW-UAV can fly like an airplane by tilting its rotors and wings. Of course, it is impossible to merely apply the designed attitude controller to the airplane mode because the model is very different, but the attitude controller can be applied to the helicopter mode in the transient state, i.e., a state between the helicopter and airplane modes. Hence, it is important for us to validate the deterioration in the control performance in the transient state. Therefore, in this chapter, we will bank the tilt angle of a rotor and wing from the helicopter mode in small steps and validate the attitude control performance at each state.

First, the response of the pitch angle when the tilt angle was 10° from the vertical is shown in Fig. 4.16. It appears that the performance deteriorates slightly as compared with the results of Fig. 4.15; however, the control almost succeeded. Next, the response of the pitch angle when the tilt angle was 20° from the vertical is shown in Fig. 4.17. In this case, the attitude angle cannot follow the reference angle, and it is clear that control was never achieved.

From the abovementioned results, we can control the QTW-UAV purely in the helicopter mode for a tilt angle from the vertical of around 10°. Furthermore, when we tilt the rotors and wings of the QTW-UAV, it is necessary to design a controller for the transient state or mix the control inputs of the helicopter-mode and airplane-mode controllers.

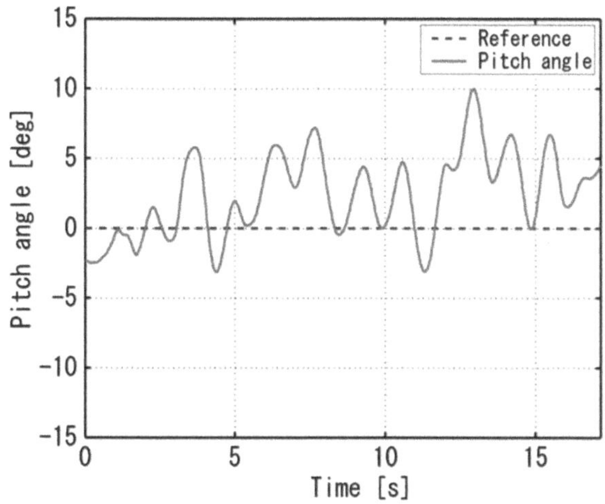

Fig. 4.16 Pitch response (tilt angle = 10°)

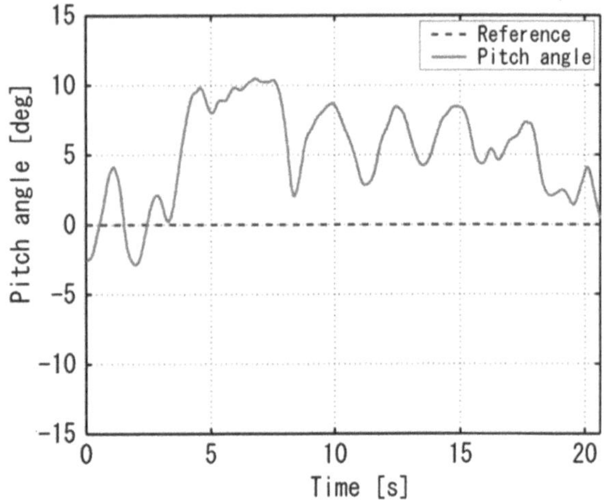

Fig. 4.17 Pitch response (tilt angle = 20°)

4.7 Summary

In this chapter, as a basic research of the autonomous control of QTW-UAVs, which have a tilt mechanism, a model-based attitude controller was designed and its effectiveness was validated by flight experiments. Moreover, the control performance in the transient state, which lies between the helicopter mode and the airplane mode, was validated; it was shown that the controller for the helicopter mode has certain limits. In a future study, we will combine a position and velocity control for the attitude control designed in this chapter, thereby completing our research on autonomous control in the helicopter mode. Moreover, we will consider the autonomous control in the airplane mode.

References

1. Watanabe K, Iwatani Y, Nonaka K, Hashimoto K (2008) A visual-servo-based assistant system for unmanned helicopter control. In: Proceedings of the IEEE/RSJ international conference on intelligent robots and systems (IROS) 2008, pp 822–827
2. Ping S, Guangping Q, Kejie L (2009) The flight control system based on multivariable pid neural network for small-scale unmanned helicopter. In: Proceedings of the international conference on information technology and computer science (ITCS) 2009, vol 1, pp 538–541
3. Jiade W, Guoyong H, Yugang F (2008) An attitude control method of unmanned helicopter based on adaptive output feedback. In: Proceedings of the 3rd international conference on intelligent system and knowledge engineering (ISKE) 2008, vol 1, pp 748–753
4. Zhou F, Jiande W, Li Ping (2008) Control system design and flight testing for a miniature unmanned helicopter. In: Proceedings of the 7th World Congress on intelligent control and automation (WCICA) 2008, pp 2315–2319
5. Mettler B, Tischler MB, Kanade K (2002) System identification modeling of a small-scale unmanned rotorcraft for flight control design. J Am Helicopter Soc 47(1):50–63

6. Esmailifar SM, Saghafi F (2009) Autonomous unmanned helicopter landing system design for safe touchdown on 6-DOF moving platform. In: Proceedings of the 5th international conference on autonomic and autonomous systems (ICAS) 2009, pp 245–250
7. Oh SR, Pathak K, Agrawal SK, Roy PH, Garrett M (2006) Approaches for a tether-guided landing of an autonomous helicopter. IEEE Trans Robot 22(3):536–544
8. Oh CK, Barlow GJ (2004) Autonomous controller design for unmanned aerial vehicles using multi-objective generic programming. In: Proceedings of congress on evolutional computation 2004, vol 2, pp 1538–1545
9. Shim DH, Kim HJ, Sastry S (2002) A flight control system for aerial robots: algorithm and experiments. 15th IFAC World Congress on automatic control
10. Ishidori A, Marconi L, Serrani A (2003) Robust nonlinear motion control of a helicopter. IEEE Trans Automat Contr 48(3):413–426
11. Sakamoto T, Katayama H, Ichikawa I (2006) Attitude control of a helicopter model by robust PID controllers. In: Proceedings of the 2006 IEEE international conference on control applications, pp 1971–1976
12. Daigo F, Jinok S, Kensaku H, Kenzo N (2004) H ∞ hovering and guidance control for autonomous small-scale unmanned helicopter. Trans Jpn Soc Mech Eng Ser C 70(694): 1708–1714
13. Kensaku H, Jinok S, Daigo F, Kazuhiro I, Delshan F, Kenzo N (2004) Autonomous flight control of hobby-class small unmanned helicopter (modeling based on experimental identification and autonomous flight control experiment). Trans Jpn Soc Mech Eng Ser C 70(691):720–727
14. Bauchau OA, Wang J (2007) Efficient and robust approaches for rotorcraft stability analysis. In: American Helicopter Society 63rd annual forum, Virginia Beach, USA, 2007
15. Benedict M, Jarugumilli T, Chopra I (2009) Design and development of a hover-capable cyclocopter micro air vehicle. In: American Helicopter Society 65th annual forum, Grapevine, USA, 2009
16. Bermes C, Schafroth D, Bouabdallah S, Siegwart R (2009) Design parameters for coaxial rotary wing mavs with passive roll and pitch stability. In: American Helicopter Society 65th annual forum, Grapevine, USA, 2009
17. Bhandari S, Colgren R (2006) 6-dof dynamic model for a raptor 50 UAV helicopter including stabilizer bar dynamics. In: AIAA modeling and simulation technologies conference and exhibit, Keystone, USA, 2006
18. Bohorquez F, Rankinsy F, Baederz J, Pines D (2003) Hover performance of rotor blades at low reynolds numbers for rotary wing micro air vehicles. An experimental and CFD study. In: AIAA applied aerodynamics conference, Orlando, USA, 2003
19. Cheviron T, Chriette A, Plestan F (2009) Generic nonlinear model of reduced scale UAVs. In: IEEE international conference on robotics and automation (ICRA), Kobe, Japan, 2009
20. Fusato D, Celi R (2003) Design sensitivity analysis for helicopter flight dynamic and aeromechanic stability. J Guid Control Dyn 26(6):918–927
21. Gress GR (2007) Lift fans as gyroscopes for controlling compact VTOL air vehicles: overview and development status of oblique active tilting. In: American Helicopter Society 63rd annual forum, Virginia Beach, USA, 2007
22. www.bellhelicopter.com
23. Jinok S, Daigo F, Kensaku H, Kenzo N (2002) Attitude control and hovering control of radio controlled helicopter. Trans Jpn Soc Mech Eng Ser C 68(675):3284–3291
24. Satoshi S, Jinok S, Makoto T, Yoshiro K, Daisuke N, Kenzo N (2007) Analytical modeling and autonomous hovering control of hobby class 5 kg unmanned helicopter. Trans Jpn Soc Mech Eng Ser C 73(726):562–569

Chapter 5
Linearization and Identification of Helicopter Model for Hierarchical Control Design

Abstract This chapter presents an analytical modeling and model-based controller design for a small unmanned helicopter. Generally, it can be said that helicopter dynamics are nonlinear, with coupling of each axis. However, for low speed flights, i.e., speeds less than 5 m/s, the dynamics can be expressed by a set of linear equations of motion as a SISO (Single Input Single Output) system. The dynamics of the helicopter are divided into several components. We derive a model for each component from either the geometric relation or equation of motion. By combining all components, we derive two linear state equations that describe the helicopter's lateral and longitudinal motion. The parameters of the model are decided by helicopter specs. Based on the derived models, we design A control system by using the linear quadratic integral (LQI). The validity of these approaches is then verified by flight tests.

Video Link:

Square trajectory following control of Eagle helicopter
http://mec2.tm.chiba-u.jp/monograph/Videos/Chapter5/1.wmv

Nomenclature

a, b	Flapping angle of the main rotor
c, d	Flapping angle of the stabilizer
p, q	Roll and pitch rate in helicopter coordinate frame
ϕ, θ	Euler angle for helicopter roll and pitch
u, v	Helicopter translational (longitudinal and lateral) speed
γ_b, γ_s	Blade Lock number
c_b, c_s	Chord length
R_b, r_b	Outside and inside radius of the main rotor
R_s, r_s	Outside and inside radius of the stabilizer
$I_{\beta b}$	Moment of inertia of the main blade
$I_{\beta s}$	Moment of inertia of the stabilizer

K. Nonami et al., *Autonomous Flying Robots: Unmanned Aerial Vehicles and Micro Aerial Vehicles*, DOI 10.1007/978-4-431-53856-1_5, © Springer 2010

C_{ab}, C_{as}	Two-dimensional lift-curve-slope
Θ_a, Θ_b	Cyclic pitch angle of the main rotor
Θ_c, Θ_d	Cyclic pitch angle of the stabilizer
Θ_{Lat}	Longitudinal tilt angle of the swash plate
Θ_{Lon}	Lateral tilt angle of the swash plate
h	Distance between C.O.G of the fuselage and vertex of the rotorhead
T	Main rotor thrust
k_β	Main blade spring constant
Ω	Rotor speed
m	Helicopter mass
ρ	Air density
g	Acceleration of gravity

5.1 Introduction

Small helicopters are one of the popular platforms for the research and development of autonomous Unmanned Aerial Vehicles (UAVs) [1–18]. The design of an autonomous control system for helicopter-based UAVs is a challenge problem because of their nonlinearities and complex dynamics that are involved. In order to evaluate a system and design reliable and efficient controllers, a precise dynamic model is an essential.

There are two basic approaches to designing control systems; model-based control method and model-free control method. In the past, both approaches have been employed for helicopter motion control [1–18]. However, once a dynamic model is derived, a designer can choose the desired control method according to the design specifications or applications, and can also use it to create an efficient simulator. Thus, the derivation of a dynamic model of a helicopter is quite beneficial. In this chapter, we introduce a modeling method and a model-based controller design for small helicopters.

In model-based controller design, accuracy of the mathematical model has a critical effect on control performance. The process of modeling is therefore quite important. Modeling methods are generally divided into two categories: the system identification (a.k.a. black box) method and the analytical modeling (a.k.a. white box) method. Although a practical model can be derived using the black box method [4], the parameters used for modeling are a helicopter-specific. The designer must therefore perform system identification experiments for every change in fuselage specification. To facilitate the efficient and rapid development of a control system for small helicopters, we constructed mathematical models using analytically derived equations of motion that reflect helicopter dynamics [19–22]. The proposed method enables the developing of models for helicopters of several sizes without performing an excitation experiment for system identification. Controller design that is efficient and safe thus becomes possible.

It must be added, however, that an overly precise and complex model is not practical for controller design. Hence, we limit the flight speed and attitude to a near-hovering state, in order to limit the complexity of the model to a practical degree. Most applications of unmanned helicopters, such as aerial surveillance, inspection are satisfactory executed by hovering; therefore, the above-mentioned limitation can also be valid in practical use. Thus, we assume that the small unmanned helicopter is a SISO system without coupling, and develop mathematical models for three types of helicopters using equations of motion. The parameters of the models are decided based upon the physical specifications of each helicopter.

We also describe controller based on the derived model. The effectiveness of the modeling and the controller design are then demonstrated by flight tests.

5.2 Modeling

In this section, we analytically derive a model of a small unmanned helicopter analytically. Figure 5.1 shows the block diagram of the dynamic system of a helicopter, which consists of the coupling of the rotor blade, a rotor-stabilizer, links and a fuselage.

We will now introduce some notations and coordinate systems that are used in the derivation of the mathematical model. Figures 5.2 and 5.3 present the notation of the parts around the rotor and the motion of the blade, respectively. We define three coordinate systems, as shown in Fig. 5.4. O_i is the navigation frame which has its origin at the location of the navigation system and has its axes aligned vertically downward along the north and east directions. The position and velocity of the helicopter are defined on the basis of this frame. O_b is the body frame, which is the orthogonal axis set X_b, Y_b, Z_i that has its origin at the center of gravity (CG) of the helicopter and is aligned with the forward, rightward and downward verticality of the fuselage. The angular velocity of the helicopter is defined on the basis of this frame. The attitude of the helicopter is defined as the rotation of O_b with respect to O_i. O_r is the rotor frame, which is equivalent to one whose origin moves from O_b to the center of the rotor head. The physical quantities of the rotor dynamics are defined based upon this frame. In this chapter, we derive a mathematical model of the

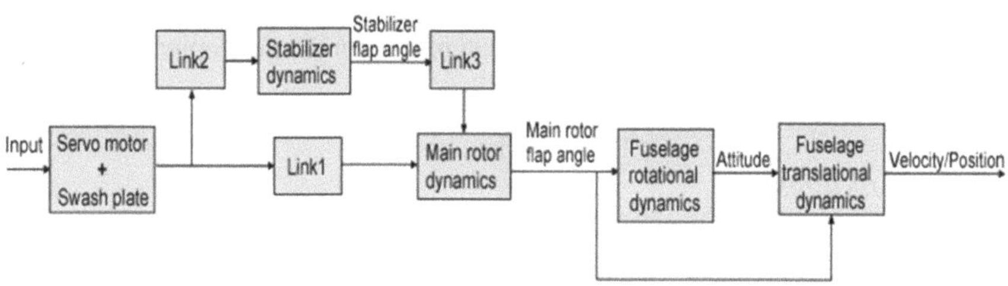

Fig. 5.1 Block diagram of helicopter system

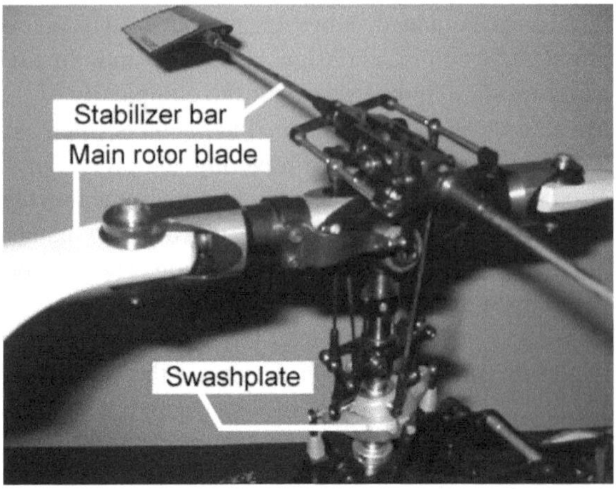

Fig. 5.2 Rotor head

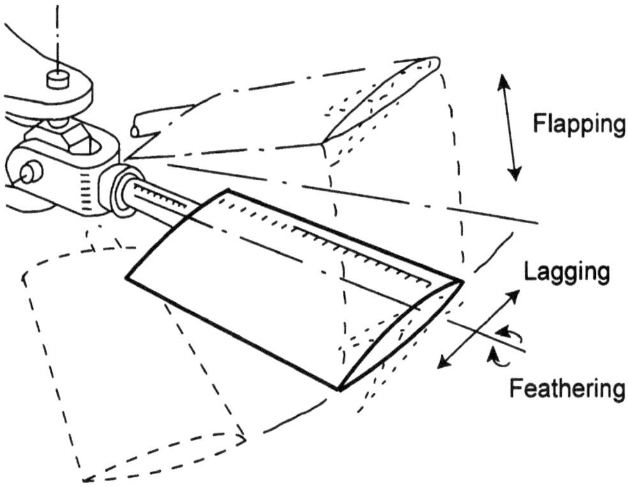

Fig. 5.3 Rotor motion

attitude and translational motion dynamics by individually modeling and integrating each block. In the derivation, we assume that the effect of the coupling of each axis is negligible. The linkages, mass of the stabilizer, etc., of the radio-controlled helicopter are adjusted such that the pilot can maneuver the helicopter easily.

5.2.1 Linkages

The control inputs are impressed into the servo motors which actuate the linkages, as illustrated in Fig. 5.5. In this section, we briefly explain about the linkages and derive linkage parameters K_{L1}, K_{L2}, K_{L3}. The structure of the linkage is composed of

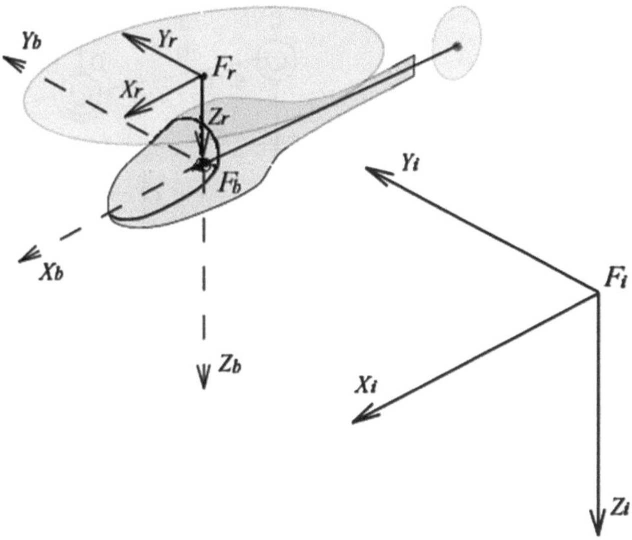

Fig. 5.4 Coordinate system

Fig. 5.5 Linkage

combinations of Link1, Link2 and Link3, as illustrated in Figs. 5.6–5.8. Link1 links the swashplate and main blade. Link2 links the swashplate to the stabilizer. Link3 links the stabilizer and main blade. In hovering flight, the tilt of the swashplate is less than 10°, so it can be assumed that $\sin\theta \approx \theta$.

On the basis of this assumption, we can approximate the ratio of the input and output angle K_{L1}, K_{L2} and K_{L3} as the following equations:

$$K_{L1} = \frac{\theta_{b-sw}}{\theta_{sw}} = \frac{l_1 l_3}{l_2 l_4} \tag{5.1}$$

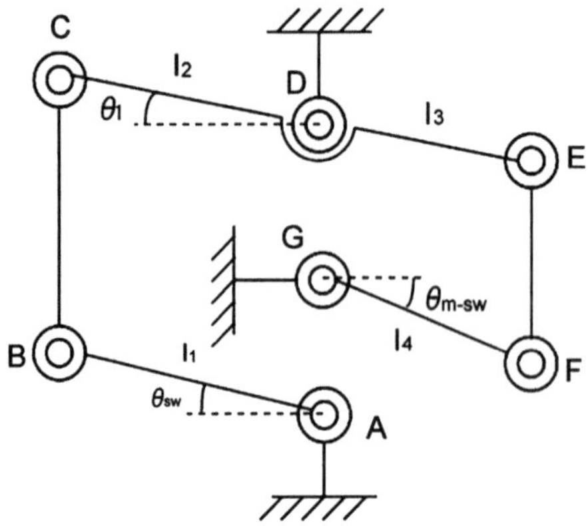

Fig. 5.6 Swashplate to main rotor link (Link1)

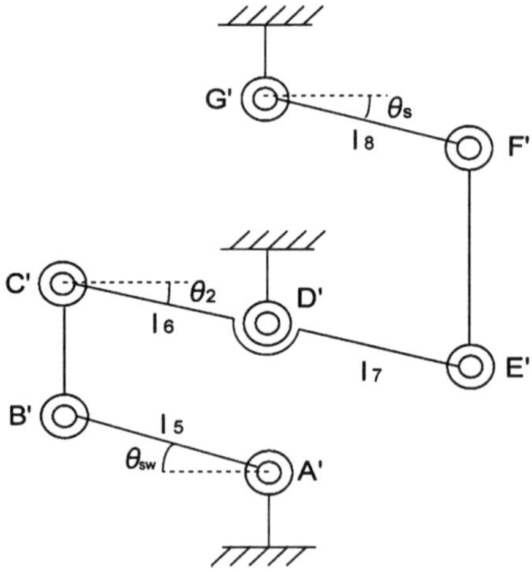

Fig. 5.7 Swashplate to stabilizer link (Link2)

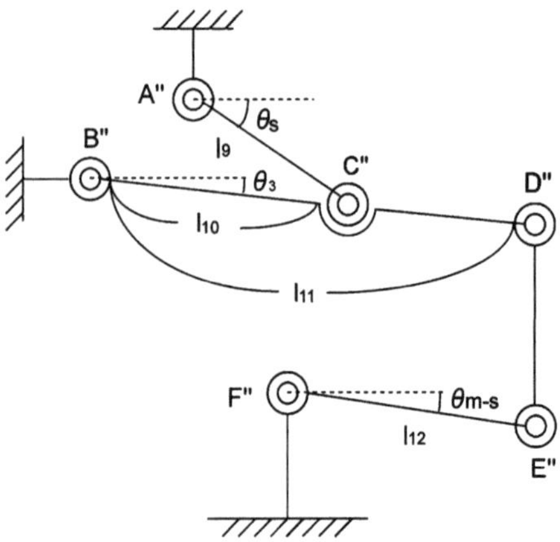

Fig. 5.8 Stabilizer to main rotor link (Link3)

$$K_{L2} = \frac{\theta_s}{\theta_{sw}} = \frac{l_5 l_7}{l_6 l_8} \tag{5.2}$$

$$K_{L3} = \frac{\theta_{b-s}}{\theta_s} = \frac{l_9 l_{11}}{l_{10} l_{12}} \tag{5.3}$$

where θ_{sw} is the tilt angle of the swashplate. θ_{b-sw} denotes the element of the pitch angle of the main blade that depends on the tilt angle of the swashplate. θ_{b-s} is the element of the pitch angle of the main blade that depends on the flap angle of the stabilizer. l denotes the length of each linkage.

5.2.2 *Dynamics of Main Rotor and Stabilizer*

There rotor is capable of three kinds of motion: feathering, flapping and lead-lag. Of these motions, the effect of the lead-lag on the total dynamic is negligible small. The effect of feathering is also small to the point of vanishing so that the motion is fixed by the swashplate. For these reasons, we consider only the flapping motion, whose dynamics we discuss below. First, we introduce a simple rigid body model. Next, flapping equations for the motion of the main rotor and the stabilizer on the rotor frame are derived by extension form of the rigid body model.

5.2.2.1 Blade Flapping Equation of Motion on Fixed Frame

In the derivation of the rotor equations of motion, we define the blade frame as that is illustrated as Fig. 5.9. First, we assume that the rigid blade rotates at rotation speed Ω and we define the blade frame which has its origin at the hinge and its axes aligned with the radial direction, blade rotative direction, and the vertical direction to XY plane. In Fig. 5.9, R is the radius of the rotor, and eR is the distance between

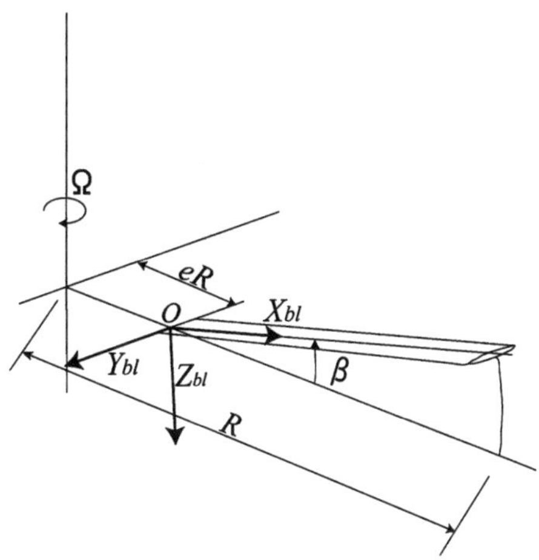

Fig. 5.9 Single flapping blade

the blade rotation axis and the origin of the blade frame. B is the flap angle on the Y_{bl} axis. Assuming that the axis of rotation is fixed in the space, the angular velocity around the axes of the blade frame is expressed as

$$(\omega_{X_{bl}}, \omega_{Y_{bl}}, \omega_{Z_{bl}}) = (-\Omega \sin \beta,\ \dot{\beta},\ \Omega \cos \beta) \tag{5.4}$$

Next, the centripetal acceleration created by the blade rotation is given by

$$(a_{oX_{bl}}, a_{oY_{bl}}, a_{oZ_{bl}}) = (-\Omega^2 eR \cos \beta,\ 0,\ -\Omega^2 \sin \beta) \tag{5.5}$$

Besides this, the position vector of the blade COG is

$$r_g = (x_g R,\ 0,\ 0) \tag{5.6}$$

The blade flapping equation of motion is given in the next equation by substituting r_g and a_o into the second equation of the extended Euler equation, which expresses the rotary motion of a rigid body as

$$\begin{aligned} &B\dot{\omega}_{Y_{bl}} - (C - A)\omega_{X_{bl}}\omega_{Z_{bl}} + M_b(r_g \times a_o)_{Y_{bl}} \\ &\quad = B\ddot{\beta} + (B \cos \beta + M_b x_g eR^2)\Omega^2 \sin \beta \\ &\quad = M_a \end{aligned} \tag{5.7}$$

where M_b is the total mass of the blade, M_a is the external moment about Y_{bl}, and A, B and C are the moment of inertia about X_{bl}, Y_{bl} and $Z_{bl}$$\cdot(r_g \times a_o)_{Y_{bl}}$ represents the component of $r_g \times a_o$ about Y_{bl}. In the transformation of the first equation of (5.7) to the second one, we simplify the equation by using the property that A + B = C. Because beta is very small, (5.7) is linearized as follows:

$$B\ddot{\beta} + (B + M_b x_g eR^2)\Omega^2 \beta = M_a \tag{5.8}$$

5.2.2.2 Blade Flapping Equation of Motion with the Rotation of Fuselage

In the preceding section, the blade flapping equation of motion is derived in the case that the blade rotates along the axes which are fixed in the space. In actual flight, however the rotation axis of the rotor is rotated by the motion of fuselage. In this section, we derive the flapping equation of motion in the case that the rotation axis rotates at roll angular speed p and pitch angular speed q, as illustrated in Fig. 5.10.

In Fig. 5.10, Ψ represents the blade position whose 0 is the forward direction of the fuselage. First, we derive the equation which takes into account the pitch angular velocity q. It is given as the next equation by adding each component of pitch angular velocity.

$$\begin{aligned} (\omega_{X_{bl}}, \omega_{Y_{bl}}, \omega_{Z_{bl}}) = (&-\Omega \sin \beta + q \sin \Psi \cos \beta,\ \dot{\beta} + q \cos \Psi, \\ &\Omega \cos \beta + q \sin \Psi \sin \beta) \end{aligned} \tag{5.9}$$

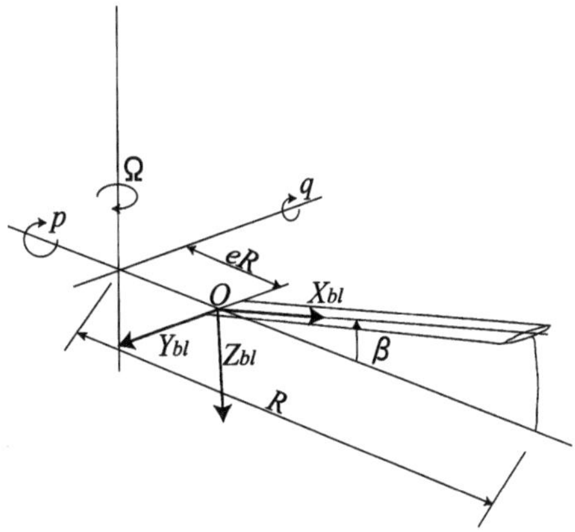

Fig. 5.10 Single flapping blade with body rotation

By assuming this, $\dot{q} = 0$ centripetal acceleration and the Coriolis acceleration by q are added to the acceleration of blade. Each acceleration is given by

$$(a_{cenX_{bl}}, a_{cenY_{bl}}, a_{cenZ_{bl}}) = (-q^2 eR\cos^2\Psi\cos\beta, q^2 eR\cos\Psi\sin\Psi, \\ -q^2 eR\cos^2\Psi\sin\beta) \quad (5.10)$$

$$(a_{colX_{bl}}, a_{colY_{bl}}, a_{colZ_{bl}}) = (-2eR\Omega q\sin\Psi\sin\beta, 0, 2eR\Omega q\sin\Psi\cos\beta) \quad (5.11)$$

where a_{cen} is the centripetal acceleration and a_{col} is the Coriolis acceleration. The suffix expresses the component for a given described axis. Summarizing the above discussion, the acceleration which results from the motion of the fuselage is

$$(a_{TX_{bl}}, a_{TY_{bl}}, a_{TZ_{bl}}) \\ = (-\Omega^2 eR\cos\beta - q^2 eR\cos^2\Psi\cos\beta - 2eR\Omega q\sin\Psi\sin\beta, \\ q^2 eR\cos\Psi\sin\Psi, \\ -\Omega^2 eR\sin\beta - q^2 eR\cos^2\Psi\sin\beta + 2eR\Omega q\sin\Psi\cos\beta) \quad (5.12)$$

By using (5.12), the component of $r_g \times a_o$ along Y_{bl} is calculated as

$$(r_g \times a_o)_{Y_{bl}} = x_g R(\Omega^2 eR\sin\beta + q^2 eR\cos^2\Psi\sin\beta - 2eR\Omega q\sin\Psi\cos\beta) \quad (5.13)$$

The flapping equation of motion which takes into account the pitching motion of fuselage is derived as (5.14) by substituting (5.9) and (5.13) into (5.7).

$$B(\ddot{\beta} - q\Omega\sin\Psi) - B\{(-\Omega\sin\beta + q\sin\Psi\cos\beta)(\Omega\cos\beta + q\sin\Psi\sin\beta)\} \\ + M_b\{x_g R(\Omega^2 eR\sin\beta + q^2 eR\cos^2\Psi\sin\beta - 2eR\Omega q\sin\Psi\cos\beta)\} = M_a \quad (5.14)$$

Based on the assumption that $\cos\beta \approx 1, \sin\beta \approx \beta$ and $\beta^2 \approx 0$, the above equation is simplified as follows

$$\begin{aligned}
&B(\ddot{\beta} - q\Omega\sin\Psi) - B\{(-\Omega\beta + q\sin\Psi)(\Omega + q\sin\Psi\beta)\}\\
&\quad + M_b\{x_g R(\Omega^2 eR\beta + q^2 eR\cos^2\Psi - 2eR\Omega q\sin\Psi)\}\\
&= B\ddot{\beta} - Bq\Omega\sin\Psi + B\Omega^2\beta + Bq\Omega\beta^2\sin\Psi - Bq\Omega\sin\Psi - Bq^2\beta\sin^2\Psi\\
&\quad + M_b x_g eR^2\Omega^2\beta + M_b x_g Rq^2 eR\beta\cos^2\Psi - 2M_b x_g eR^2 q\Omega\sin\Psi\\
&\approx B\ddot{\beta} + \Omega^2\beta(B + M_b x_g eR^2) - 2q\Omega(B + M_b x_g eR^2)\sin\Psi - Bq^2\beta\sin^2\Psi\\
&\quad + M_b x_g eR^2 q^2\cos^2\beta\\
&= M_a
\end{aligned} \tag{5.15}$$

Next, we discuss the rolling motion. The angular velocity along each axis and the acceleration which acts on the origin that includes roll angular velocity p is given as

$$\begin{aligned}
(\omega_{X_{bl}}, \omega_{Y_{bl}}, \omega_{Z_{bl}}) &= (-\Omega\sin\beta + p\cos\Psi\cos\beta,\\
&\qquad \dot{\beta} - p\sin\Psi, \Omega\cos\beta + p\cos\Psi\sin\beta)
\end{aligned} \tag{5.16}$$

$$\begin{aligned}
(a_{TX_{bl}}, a_{TY_{bl}}, a_{TZ_{bl}}) &= (-\Omega^2 eR\cos\beta - p^2 eR\sin^2\Psi\cos\beta\\
&\qquad -2eR\Omega p\cos\Psi\sin\beta, p^2 eR\cos\Psi\sin\Psi,\\
&\qquad -\Omega^2 eR\sin\beta - p^2 eR\sin^2\Psi\sin\beta + 2eR\Omega p\cos\Psi\cos\beta)
\end{aligned} \tag{5.17}$$

where we assume that $\dot{p} = 0$. From this assumption, we obtain the Y_{bl} component of $r_g \times a_o$ as

$$(r_g \times a_o)_{Y_{bl}} = x_g R(\Omega^2 eR\sin\beta + p^2 eR\sin^2\Psi\sin\beta - 2eR\Omega p\cos\Psi\cos\beta) \tag{5.18}$$

By substituting (5.17) and (5.18) into (5.7), we obtain the next flapping equation of motion.

$$\begin{aligned}
&B(\ddot{\beta} - p\Omega\sin\Psi) - B\{(-\Omega\sin\beta + p\cos\Psi\cos\beta)(\Omega\cos\beta + p\cos\Psi\sin\beta)\}\\
&\quad + M_b\{x_g R(\Omega^2 eR\sin\beta + p^2 eR\sin^2\Psi\sin\beta - 2eR\Omega p\cos\Psi\cos\beta)\}\\
&\approx B(\ddot{\beta} - p\Omega\cos\Psi) - B\{(-\Omega\beta + p\cos\Psi)(\Omega + p\cos\Psi\beta)\}\\
&\quad + M_b\{x_g R(\Omega^2 eR\beta + p^2 eR\beta\sin^2\Psi - 2eR\Omega p\cos\Psi)\}\\
&= B\ddot{\beta} - Bp\Omega\cos\Psi + B\Omega^2\beta + Bp\Omega\beta^2\cos\Psi - Bp\Omega\cos\Psi - Bp^2\beta\cos^2\Psi\\
&\quad + M_b x_g eR^2\Omega^2\beta + M_b x_g eR^2 p^2\beta\sin^2\Psi - 2M_b x_g eR^2 p\Omega\cos\Psi\\
&\approx B\ddot{\beta} + \Omega^2\beta(B + M_b x_g eR^2) - 2p\Omega(B + M_b x_g eR^2)\cos\Psi - Bp^2\beta\cos^2\Psi\\
&\quad + M_b x_g eR^2 p^2\sin^2\beta\\
&= M_a
\end{aligned} \tag{5.19}$$

where we assume that $\cos\beta \approx 1, \sin\beta \approx \beta$ and $\beta^2 \approx 0$.

By summarizing (5.15) and (5.19), the flapping equation of motion which includes the rolling and pitching rotation is derived as follows:

$$\begin{aligned} &B\ddot{\beta} + \Omega^2\beta(B + M_b x_g eR^2) - 2q\Omega(B + M_b x_g eR^2)\sin\Psi \\ &\quad -2p\Omega(B + M_b x_g eR^2)\cos\Psi - B\beta(q^2\sin^2\Psi + p^2\cos^2\Psi) \\ &\quad +M_b x_g eR^2(q^2\cos^2\Psi + p^2\sin^2\Psi) \\ &= M_a \end{aligned} \tag{5.20}$$

5.2.2.3 Derivation of External Force Term

The moment by external force M_a is consists mainly of following two components: one is the moment that comes from aerodynamic force, the other is the moment that results from the flexural rigidity of the blade.

First, we derive the first component. A blade element dr at a distance r from the origin of the blade frame is defined as shown in Fig. 5.11. According to the blade element theory, the minute lift force dL which works on dr can be represented as

$$dL = \frac{1}{2}\rho ac\alpha V_r^2 dr \tag{5.21}$$

where α is the blade attack angle (rad), which is defined as shown in Fig. 5.12, V_r is the air speed (m/s), c is the length of chord (m), and a is the two-dimensional lift-curve-slope (1/rad). In the case of hovering, V_r can be expressed as follows:

$$V_r = \sqrt{\{\Omega(r + eR)\}^2 + \{r\dot{\beta} + (r + eR)(q\cos\Psi - p\sin\Psi)\}^2} \tag{5.22}$$

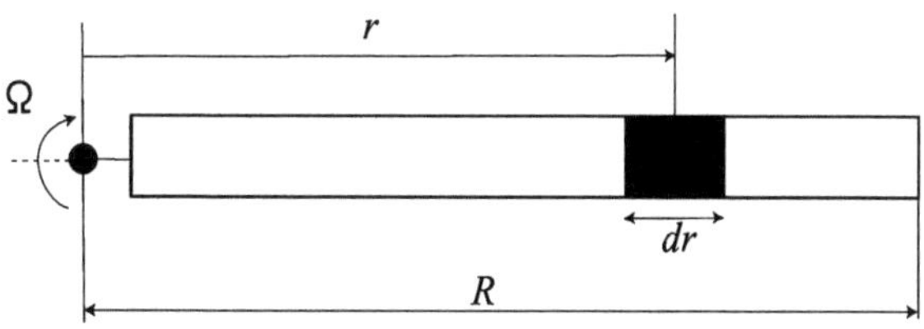

Fig. 5.11 Blade element

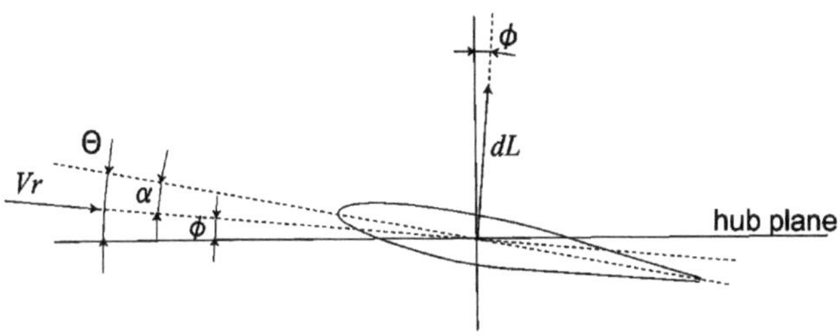

Fig. 5.12 Angle of attack

where the induced velocity or velocity imparted to the mass of air contained in the control volume at the rotor disk [21] is ignored because the induced velocity affects to the constant term and has little effect on the periodic flapping motion. The blade attack angle α can be described by the blade pitch angle Θ (rad) and the rolling and pitching motion of the fuselage, and can be written as

$$\begin{aligned}\alpha &= \Theta - \frac{1}{(r+eR)\Omega}\{r\dot{\beta} + (r+eR)(q\cos\Psi - p\sin\Psi)\} \\ &= \Theta - \frac{r}{r+eR}\frac{\dot{\beta}}{\Omega} - \frac{1}{\Omega}(q\cos\Psi - p\sin\Psi) \end{aligned} \tag{5.23}$$

By summarizing the above equations, the minute lift force dL in the case of hovering is written as

$$\begin{aligned} dL = \frac{1}{2}\rho ac &\left\{\Theta - \frac{r}{r+eR}\frac{\dot{\beta}}{\Omega} - \frac{1}{\Omega}(q\cos\Psi - p\sin\Psi)\right\} \\ &[\{\Omega(r+eR)\}^2 + \{r\dot{\beta} + (r+eR)(q\cos\Psi - p\sin\Psi)\}^2]dr \end{aligned} \tag{5.24}$$

We assume that $e \approx 0, \Omega^2 >> (\beta + q\cos\Psi - p\sin\Psi)^2$. With these assumptions, (5.24) can be simplified as follows:

$$dL = \frac{1}{2}\rho ac\left\{\Theta - \frac{\dot{\beta}}{\Omega} - \frac{1}{\Omega}(q\cos\Psi - p\sin\Psi)\right\} r^2\Omega^2 dr \tag{5.25}$$

The minute moment dM generated by dL can be written as

$$\begin{aligned} dM &= r dL\cos\alpha \\ &\approx \frac{1}{2}\rho ac\left\{\Theta - \frac{\dot{\beta}}{\Omega} - \frac{1}{\Omega}(q\cos\Psi - p\sin\Psi)\right\} r^3\Omega^2 dr \end{aligned} \tag{5.26}$$

where we assume that $\cos\alpha \approx 1$. By integrating dM radially, the aerodynamic force component of the external force moment $(M_a)_{air}$ can be expressed as (5.28), where we put the assumption on a two-dimensional lift-curve-slope as follows:

$$\begin{cases} a = 0 & (r < eR) \\ a = a_r & (r \geq eR) \end{cases} \tag{5.27}$$

$$\begin{aligned} (M_a)_{air} &= \frac{1}{2}\rho a_r c\left\{\Theta - \frac{\dot{\beta}}{\Omega} - \frac{1}{\Omega}(q\cos\Psi - p\sin\Psi)\right\}\Omega^2\int_{eR}^{R} r^3 dr \\ &= \frac{1}{8}\rho a_r c\left\{\Theta - \frac{\dot{\beta}}{\Omega} - \frac{1}{\Omega}(q\cos\Psi - p\sin\Psi)\right\}\Omega^2\{R^4 - (eR)^4\} \end{aligned} \tag{5.28}$$

Secondly, the moment generated by the flexural rigidity of the blade $(M_a)_k$ can be written as

$$(M_a)_k = -k_\beta \beta \tag{5.29}$$

By adding (5.28) and (5.29), the external force moment M_a is written as follows:

$$\begin{aligned} M_a &= (M_a)_{air} + (M_a)_k \\ &= \frac{1}{8}\rho a_r c \left\{ \Theta - \frac{\dot{\beta}}{\Omega} - \frac{1}{\Omega}(q \cos \Psi - p \sin \Psi) \right\} \Omega^2 \{R^4 - (eR)^4\} - k_\beta \beta \end{aligned} \tag{5.30}$$

5.2.2.4 Motion of Blade Pitch and Flapping Angle

Flapping motion can be divided into a constant component and a periodic compo-nent that is coincides with the rotor frequency. We described how the feathering motion is constrained by the swash plate. In fact, this feathering motion is also consists of a constant and a periodic component. We will explain the feathering and flapping motions as preparation for the derivation of the flapping equation of motion on the rotor frame in the next section.

The blade pitch angle Θ responds synchronously with the rotor revolution as follows:

$$\Theta = \Theta_0 + \Theta_1 \cos \Psi + \Theta_2 \sin \Psi \tag{5.31}$$

In (5.31), Θ_0 is the constant component of the blade pitch angle, which is called "collective pitch". We should note that the periodic component, excluding the collective pitch, is called the cyclic pitch. Θ_1 and Θ_2 represent the cyclic pitch component at $\Psi = 0$ and $\Psi = \pi/2$, respectively. Figure 5.13 illustrates these angles. Thus, the flapping angle can also be described by the constant and periodic components that are based on the periodic motion of the blade pitch.

$$\beta = \beta_0 + a_\beta \cos \Psi - b_\beta \sin \Psi \tag{5.32}$$

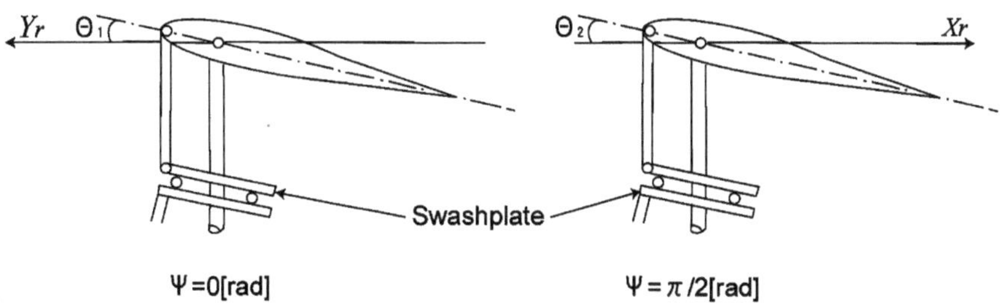

Fig. 5.13 Cyclic pitch angle

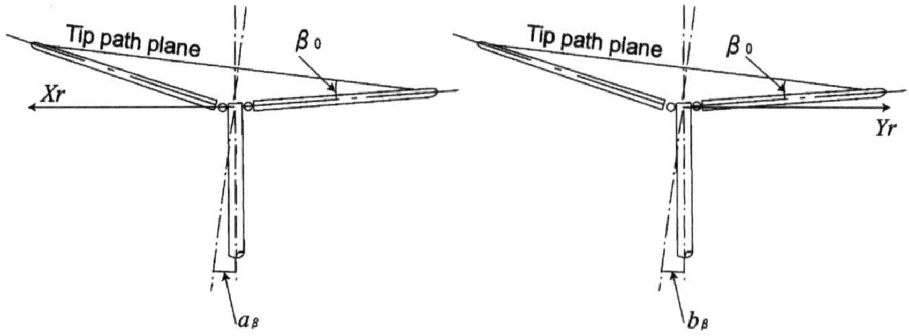

Fig. 5.14 Blade coning

where β_0 is the constant component of the flapping angle which is called the coning angle. a_β and b_β are the tilt angle of the rotor disk and are measured from the left and backside direction of fuselage. These angles are illustrated in Fig. 5.14. The constant components such as the collective pitch and the coning angle are important in modeling of the heave motion. However, these components have little effect on the dynamics of the attitude and translational motion, and can therefore be disregarded here. The blade pitch and flapping angles can then be written as the cyclic motion as follows:

$$\Theta \approx \Theta_1 \cos\Psi + \Theta_2 \sin\Psi \tag{5.33}$$

$$\beta \approx a_\beta \cos\Psi - b_\beta \sin\Psi \tag{5.34}$$

5.2.2.5 Flapping Equation of Motion of the Rotor Frame

We derive the flapping equation of motion of the rotor frame by using the results of the above discussion. By the differentiation of (5.34) in the time domain, we obtain

$$\dot{\beta} = \dot{a}_\beta \cos\Psi - a_\beta\Omega \sin\Psi - \dot{b}_\beta \sin\Psi - b_\beta\Omega \cos\Psi \tag{5.35}$$

By again differentiating (5.35), we obtain

$$\begin{aligned}\ddot{\beta} &= \ddot{a}_\beta \cos\Psi - 2\dot{a}_\beta\Omega \sin\Psi - a_\beta\Omega^2 \cos\Psi - \ddot{b}_\beta \sin\Psi \\ &\quad -2\dot{b}_\beta\Omega \cos\Psi + b_\beta\Omega^2 \sin\Psi\end{aligned} \tag{5.36}$$

Substituting (5.30), (5.33), (5.34), (5.35) and (5.36) into (5.20), we obtain the next equation, based on the assumption $e \approx 0$, $\dot{q} = 0$ and $\dot{p} = 0$.

$$\begin{aligned}&B(\ddot{a}_\beta \cos\Psi - 2\dot{a}_\beta\Omega \sin\Psi - a_\beta\Omega^2 \cos\Psi - \ddot{b}_\beta \sin\Psi - 2\dot{b}_\beta\Omega \cos\Omega \\ &\quad +b_\beta\Omega^2 \sin\Psi) + \Omega^2 B(a_\beta \cos\Psi - b_\beta \sin\Psi) - 2Bq\Omega \sin\Psi - 2Bp\Omega \cos\Psi \\ &\quad -B(q^2 \sin^2\Psi + p^2 \cos^2\Psi)(a_\beta \cos\Psi - b_\beta \sin\Psi)\end{aligned}$$

$$
\begin{aligned}
&= \frac{1}{8}\rho a_r c \left\{ \Theta_1 \cos\Psi + \Theta_2 \sin\Psi - \frac{1}{\Omega}(\dot{a}_\beta \cos\Psi - a_\beta \Omega \sin\Psi - \dot{b}_\beta \sin\Psi \right. \\
&\quad \left. - b_\beta \Omega \cos\Psi) - \frac{1}{\Omega}(q\cos\Psi - p\sin\Psi) \right\} \Omega^2 \{R^4 - (eR)^4\} \\
&\quad - k_\beta (a_\beta \cos\Psi - b_\beta \sin\Psi)
\end{aligned} \tag{5.37}
$$

Dividing (5.37) by Ω^2 and assuming $(q/\Omega)^2 \approx 0$ and $(p/\Omega)^2 \approx 0$, we obtain

$$
\begin{aligned}
&\frac{B}{\Omega^2}(\ddot{a}_\beta \cos\Psi - 2\dot{a}_\beta \Omega \sin\Psi - a_\beta \Omega^2 \cos\Psi - \ddot{b}_\beta \sin\Psi - 2\dot{b}_\beta \Omega \cos\Psi \\
&\quad + b_\beta \Omega^2 \sin\Psi) + B(a_\beta \cos\Psi - b_\beta \sin\Psi) - \frac{2Bq}{\Omega}\sin\Psi - \frac{2Bp}{\Omega}\cos\Psi \\
&\quad = \frac{1}{8}\rho a_r c \left\{ \Theta_1 \cos\Psi + \Theta_2 \sin\Psi - \frac{1}{\Omega}(\dot{a}_\beta \cos\Psi - a_\beta \Omega \sin\Psi - \dot{b}_\beta \sin\Psi \right. \\
&\qquad \left. - b_\beta \Omega \cos\Psi) - \frac{1}{\Omega}(q\cos\Psi - p\sin\Psi) \right\} \{R^4 - (eR)^4\} \\
&\qquad - \frac{k_\beta}{\Omega^2}(a_\beta \cos\Psi - b_\beta \sin\Psi)
\end{aligned} \tag{5.38}
$$

Summarizing (5.38) by $\sin\Psi$ and $\cos\Psi$, we obtain

$$
\begin{aligned}
&\left(\frac{B}{\Omega^2}\ddot{a}_\beta - Ba_\beta - \frac{2B\dot{b}_\beta}{\Omega} + Ba_\beta - \frac{2Bp}{\Omega} \right) \cos\Psi \\
&\quad + \left(-\frac{B}{\Omega^2}\ddot{b}_\beta + Bb_\beta - \frac{2B\dot{a}_\beta}{\Omega} - Bb_\beta - \frac{2Bq}{\Omega} \right) \sin\Psi \\
&\quad = \left[\frac{1}{8}\rho a_r c\{R^4 - (eR)^4\} \left\{ \Theta_1 - \frac{\dot{a}_\beta}{\Omega} + b_\beta - \frac{q}{\Omega} \right\} - \frac{k_\beta}{\Omega^2} a_\beta \right] \cos\Psi \\
&\qquad + \left[\frac{1}{8}\rho a_r c\{R^4 - (eR)^4\} \left\{ \Theta_2 + \frac{\dot{b}_\beta}{\Omega} + a_\beta + \frac{p}{\Omega} \right\} + \frac{k_\beta}{\Omega^2} b_\beta \right] \sin\Psi
\end{aligned} \tag{5.39}
$$

In the above equation, $\Psi = 0$ is substituted as the flapping equation of motion along the Y_r axis on the rotor frame, and that $\Psi = \pi/2$ is substituted as the equation along the X_r axis. Each equation then becomes as follows:

$$
\begin{aligned}
&\frac{B}{\Omega^2}\ddot{a}_\beta - \frac{2B\dot{b}_\beta}{\Omega} - \frac{2Bp}{\Omega} \\
&\quad = \frac{1}{8}\rho a_r c\{R^4 - (eR)^4\} \left\{ \Theta_1 - \frac{\dot{a}_\beta}{\Omega} + b_\beta - \frac{q}{\Omega} \right\} - \frac{k_\beta}{\Omega^2} a_\beta
\end{aligned} \tag{5.40}
$$

$$
\begin{aligned}
&-\frac{B}{\Omega^2}\ddot{b}_\beta - \frac{2B\dot{a}_\beta}{\Omega} - \frac{2Bq}{\Omega} \\
&\quad = \frac{1}{8}\rho a_r c\{R^4 - (eR)^4\} \left\{ \Theta_2 + \frac{\dot{b}_\beta}{\Omega} + a_\beta + \frac{p}{\Omega} \right\} + \frac{k_\beta}{\Omega^2} b_\beta
\end{aligned} \tag{5.41}
$$

For the above equations, we define the blade Lock number as $\gamma = \rho ac\{R^4 - (eR)^4\}/B$ which represents the ratio between the aerodynamic and inertial forces acting on the blade, and obtain

$$\frac{8}{\gamma\Omega^2}\ddot{a}_\beta - \frac{16}{\gamma\Omega}\dot{b}_\beta - \frac{16}{\gamma\Omega}p + \frac{1}{\Omega}\dot{a}_\beta - b_\beta + \frac{1}{\Omega}q = \Theta_1 - \frac{8k_\beta}{\gamma B\Omega^2}a_\beta \quad (5.42)$$

$$-\frac{8}{\gamma\Omega^2}\ddot{b}_\beta - \frac{16}{\gamma\Omega}\dot{a}_\beta - \frac{16}{\gamma\Omega}q - \frac{1}{\Omega}\dot{b}_\beta - a_\beta - \frac{1}{\Omega}p = \Theta_2 + \frac{8k_\beta}{\gamma B\Omega^2}b_\beta \quad (5.43)$$

In (5.42) and (5.43), $\ddot{a}_\beta/\Omega^2$, $\ddot{b}_\beta/\Omega^2$, a_β/Ω and b_β/Ω are minute and negligible. Disregarding these terms, then, the terms depend on the gyro effect that remains, so that the following coupled equations of motion are obtained:

$$\frac{16}{\gamma\Omega}\dot{b}_\beta = -b_\beta - \frac{16}{\gamma\Omega}p + \frac{1}{\Omega}q - \frac{8k_\beta}{\gamma B\Omega^2}a_\beta - \Theta_1 \quad (5.44)$$

$$\frac{16}{\gamma\Omega}\dot{a}_\beta = -a_\beta - \frac{16}{\gamma\Omega}q - \frac{1}{\Omega}p + \frac{8k_\beta}{\gamma B\Omega^2}b_\beta - \Theta_2 \quad (5.45)$$

5.2.2.6 Main Rotor and Stabilizer Equation of Motion

In (5.44) and (5.45), by replacing the parameters with those for the main rotor, the main rotor equations of motion are obtained as follows:

$$\frac{16}{\gamma_b\Omega}\dot{a} = -a - \frac{16}{\gamma_b\Omega}q - \frac{1}{\Omega}p - \frac{8k_\beta}{\gamma_b\Omega^2 I_{\beta b}}b - \Theta_a \quad (5.46)$$

$$\frac{16}{\gamma_b\Omega}\dot{b} = -b - \frac{16}{\gamma_b\Omega}p + \frac{1}{\Omega}q - \frac{8k_\beta}{\gamma_b\Omega^2 I_{\beta b}}a - \Theta_b \quad (5.47)$$

where γ_b is as follows:

$$\gamma_b = \frac{\rho c_b C_{ab}\left(R_b^4 - r_b^4\right)}{I_{\beta b}} \quad (5.48)$$

We assume that the coupling on each axis is minute. Neglecting the coupling terms in (5.46) and (5.47), we obtain

$$\frac{16}{\gamma_b\Omega}\dot{a} = -a - \frac{16}{\gamma_b\Omega}q - \Theta_a \quad (5.49)$$

$$\frac{16}{\gamma_b\Omega}\dot{b} = -b - \frac{16}{\gamma_b\Omega}p - \Theta_b \quad (5.50)$$

At the same time, in the case of the Bell-Hiller mixer, the relation of the blade cyclic pitch angle of the main rotor Θ_a and Θ_b to the tilt angle of the swashplate and the flap angle of the stabilizer is as follows:

$$\Theta_a = -K_{L1}\Theta_{Lat} - K_{L3}c \quad (5.51)$$

$$\Theta_b = -K_{L1}\Theta_{Lon} - K_{L3}d \quad (5.52)$$

where K_{L1} and K_{L2} are the link gain described in (5.1) and (5.3), respectively. On the other hand, the flapping equations of motion of the stabilizer are also obtained from (5.44) and (5.45) in same manner, as follows:

$$\frac{16}{\gamma_s \Omega}\dot{c} = -c - \frac{16}{\gamma_s \Omega}q - \Theta_c \tag{5.53}$$

$$\frac{16}{\gamma_s \Omega}\dot{d} = -d - \frac{16}{\gamma_s \Omega}p - \Theta_d \tag{5.54}$$

where γ_s is as follows:

$$\gamma_s = \frac{\rho c_s C_{as}\left(R_s^4 - r_s^4\right)}{I_{\beta s}} \tag{5.55}$$

and the relation between the cyclic pitch of the stabilizer and the tilt angle of the swashplate is as follows:

$$\Theta_c = -K_{L2}\Theta_{Lat} \tag{5.56}$$

$$\Theta_d = -K_{L2}\Theta_{Lon} \tag{5.57}$$

By substituting (5.51), (5.52), (5.56) and (5.57) into (5.49), (5.50) and (5.53), these simplifications produce the following flapping equations of motion for the main blade and the stabilizer:

$$\frac{16}{\gamma_b \Omega}\dot{a} = -a - \frac{16}{\gamma_b \Omega}q + K_{L3}c + K_{L1}\Theta_{Lat} \tag{5.58}$$

$$\frac{16}{\gamma_b \Omega}\dot{b} = -b - \frac{16}{\gamma_b \Omega}p + K_{L3}d + K_{L1}\Theta_{Lon} \tag{5.59}$$

$$\frac{16}{\gamma_s \Omega}\dot{c} = -c - \frac{16}{\gamma_s \Omega}q + K_{L2}\Theta_{Lat} \tag{5.60}$$

$$\frac{16}{\gamma_s \Omega}\dot{d} = -d - \frac{16}{\gamma_s \Omega}p + K_{L2}\Theta_{Lon} \tag{5.61}$$

5.2.3 Dynamics of Fuselage Motion

In this section, we discuss the dynamics of the attitude and the translational motion. The forces and moments that act upon the fuselage are illustrated in Fig. 5.15. In this figure, we assume that the azimuth of the nose faces north. The roll moment which makes the fuselage rotate is written as the sum of the moment by the thrust of the main rotor and the reaction moment due to the flexural rigidity of the rotor blade.

$$\begin{aligned} L &= hT\sin b + k_\beta b \\ &\approx hTb + k_\beta b \end{aligned} \tag{5.62}$$

where we assume that b is small to the point of vanishing, and $\sin b = b$.

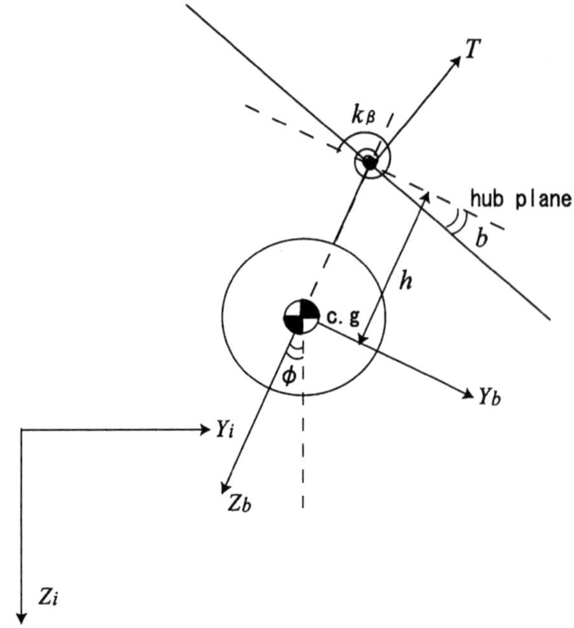

Fig. 5.15 Forces and moments

The pitch moment is also expressed as follows:

$$M = hTa + k_\beta a \tag{5.63}$$

Using the flapping angles, the Y_i axis component of the translational force is written as the sum of the horizontal component of the thrust of the rotor and the air drag.

$$\begin{aligned} f_Y &= T\sin(b+\phi) - D_v v \\ &\approx T(b+\phi) - D_v v \end{aligned} \tag{5.64}$$

The X_i axis component is also written as follows:

$$\begin{aligned} f_X &= -T\sin(a+\theta) - D_u u \\ &\approx -T(a+\theta) - D_u u \end{aligned} \tag{5.65}$$

where D_v and D_u are the coefficients of the air drag.

By using above derived moments and forces and the assumption that the thrust T is equal to the gravity mg which acts on the CG of the fuselage during hovering, the dynamics of the attitude angles and the translational velocity are written as follows:

$$\dot{p} = \frac{L}{I_{xx}} = \frac{hT + k_\beta}{I_{xx}} b \tag{5.66}$$

$$\dot{q} = \frac{L}{I_{yy}} = \frac{hT + k_\beta}{I_{yy}} a \tag{5.67}$$

$$\dot{v} = \frac{f_Y}{m} = g(b+\phi) - \frac{D_v}{m}v \tag{5.68}$$

$$\dot{u} = \frac{f_X}{m} = -g(a+\theta) - \frac{D_u}{m}u \tag{5.69}$$

5.2.4 Small Helicopter Model

We assume that $\dot{\phi} = p$ and $\dot{\theta} = q$ during hovering. By summarizing the rotor equations ((5.58) through (5.61)), the fuselage equations ((5.66) through (5.69)), and the above assumptions, we obtain following state equations:

$$\begin{bmatrix} \dot{v} \\ \dot{p} \\ \dot{\phi} \\ \dot{b} \\ \dot{d} \end{bmatrix} = \begin{bmatrix} -\frac{D_v}{m} & 0 & g & g & 0 \\ 0 & 0 & 0 & \frac{hT+k_\beta}{I_{xx}} & 0 \\ 0 & 1 & 0 & 0 & 0 \\ 0 & -1 & 0 & \frac{-\gamma_b\Omega}{16} & \frac{\gamma_b\Omega}{16}K_{L3} \\ 0 & -1 & 0 & 0 & \frac{-\gamma_b\Omega}{16} \end{bmatrix} \begin{bmatrix} v \\ p \\ \phi \\ b \\ d \end{bmatrix} + \begin{bmatrix} 0 \\ 0 \\ 0 \\ \frac{\gamma_b\Omega}{16}K_{L1} \\ \frac{\gamma_b\Omega}{16}K_{L2} \end{bmatrix} \Theta_{Lon} \tag{5.70}$$

$$\begin{bmatrix} \dot{u} \\ \dot{q} \\ \dot{\theta} \\ \dot{a} \\ \dot{c} \end{bmatrix} = \begin{bmatrix} -\frac{D_u}{m} & 0 & -g & -g & 0 \\ 0 & 0 & 0 & \frac{hT+k_\beta}{I_{yy}} & 0 \\ 0 & 1 & 0 & 0 & 0 \\ 0 & -1 & 0 & \frac{-\gamma_b\Omega}{16} & \frac{\gamma_b\Omega}{16}K_{L3} \\ 0 & -1 & 0 & 0 & \frac{-\gamma_b\Omega}{16} \end{bmatrix} \begin{bmatrix} u \\ q \\ \theta \\ a \\ c \end{bmatrix} + \begin{bmatrix} 0 \\ 0 \\ 0 \\ \frac{\gamma_b\Omega}{16}K_{L1} \\ \frac{\gamma_b\Omega}{16}K_{L2} \end{bmatrix} \Theta_{Lat} \tag{5.71}$$

These two equations are the mathematical models that include the dynamics of the attitude and translational motion.

The control inputs of (5.70) and (5.71) are the tilt angle of the swashplate. The actual control inputs, however, are the aileron input δ_{ail} and the elevator input δ_{ele}, which constitute the pulse signal to the servo motors. The tilt angle of the swashplate is proportional to the pulse inputs. Therefore, by using the gains K_{ail} and K_{ele}, the relationship can be written as follows:

$$\Theta_{Lon} = K_{ail}\delta_{ail} \tag{5.72}$$

$$\Theta_{Lat} = K_{ele}\delta_{ele} \tag{5.73}$$

5.2.5 Parameter Identification and Validation

In this section, we substitute the parameters of helicopters such as the SST-Eagle-FREYA-Ex2, SF40, and SKY SURVEYOR SF125 (see Figs. 5.16 through 5.18) into (5.70) and (5.71), and validate the derived model by a comparison of the experimental data with the simulation data. Each helicopter was constructed by Hirobo Limited. The SST-Eagle is a general radio-controlled helicopter for hobby use.

Fig. 5.16 SST-EAGLE-FREYA-Ex2

Fig. 5.17 SF40

Fig. 5.18 SKY SURVEYOR

The SF40 is a test platform for autonomous control and aerial monitoring. SKY SURVEYOR is a helicopter for practical applications such as power line monitoring or aerial surveying.

The parameters of each helicopter are shown in Table 5.1. We used the actually measured values for parameters that are easy to measure, such as the weight and length. For those parameters that are difficult to determine, however, such as moment of inertia, we measured the values roughly and tuned each of them manually for our comparison of the experimental and simulation results.

The results of the cross validation are shown in Fig. 5.19 through Fig. 5.21. These figures show the responses of attitude angle and angular velocity in rolling and pitching motion, and in term, the model outputs are consistent with those of actual motion.

Additionally, from these results, it is shown that the mathematical models (5.70) and (5.71) can be adapted for application to helicopters of various scales.

5.3 Controller Design

We next describe a model-based controller design. In this section, only the roll angle and the longitudinal controller design methods are shown. However, the pitch angle and latitudinal controllers can also be designed in same approach.

5.3.1 Configuration of Control System

First, we will explain the configuration of the control system, which is shown in a schematic diagram in Fig. 5.22. The system is constructed as a serialized control system that sequentially addresses the attitude, the velocity and the position controller.

Table 5.1 Model parameters

Parameter	Unit	SST-Eagle	SF40	SF125
$I_{\beta b}$	kgm^2	0.2818	0.414	0.0442
$I_{\beta s}$	kgm^2	0.007424	0.009424	0.00107
R_b	m	0.78	0.93	1.325
R_s	m	0.315	0.3425	0.45
r_b	m	0.15	0.165	0.295
r_s	m	0.19	0.23	0.3
c_b	m	0.06	0.06	0.1
c_s	m	0.06	0.06	0.09
C_{ab}	1/rad	3.12	3.9	0.22
C_{as}	1/rad	7.2	7.162	0.12
k_β	Nm/rad	23.4	38.4	230.4
m	kg	6.89	12.0	50.0
h	m	0.2481	0.302	0.45
I_{xx}	kgm^2	0.2882	0.405	2.029
I_{yy}	kgm^2	0.6883	0.940	5.216
ρ	kg/m^3	1.125	1.125	1.125
g	m/s^2	9.8	9.8	9.8
Ω	rad/s	151.8	152.89	130.89
D_v	Ns/m	1.723	1.4	2.32
D_u	Ns/m	1.516	1.2	2.08
K_{L1}	None	0.4684	0.2578	0.3822
K_{L2}	None	2.2727	1.3293	1.5732
K_{L3}	None	1.0848	1.038	0.6689
K_{ail}	$rad/s \times 10^6$	0.013	0.020	0.025
K_{ele}	$rad/s \times 10^6$	0.012	0.023	0.020

The control inputs calculated by the outer controller serve as the commands to the inner controller, and the attitude controller's outputs are the actual pulse inputs to the servo motors. With this configuration, it is easy to limit each command to prevent an excessive velocity or attitude angle. In the following, we will explain each controller design.

5.3.2 *Attitude Controller Design*

Based on the derived model, we designed the roll angle controller by using the LQI control method.

First, we extract the dynamics of the roll angle from (5.70) and define the roll angle model as follows:

$$\dot{x}_r = A_r x_r + B_r \Theta_{Lon} \tag{5.74}$$

$$y_r = C_r x_r \tag{5.75}$$

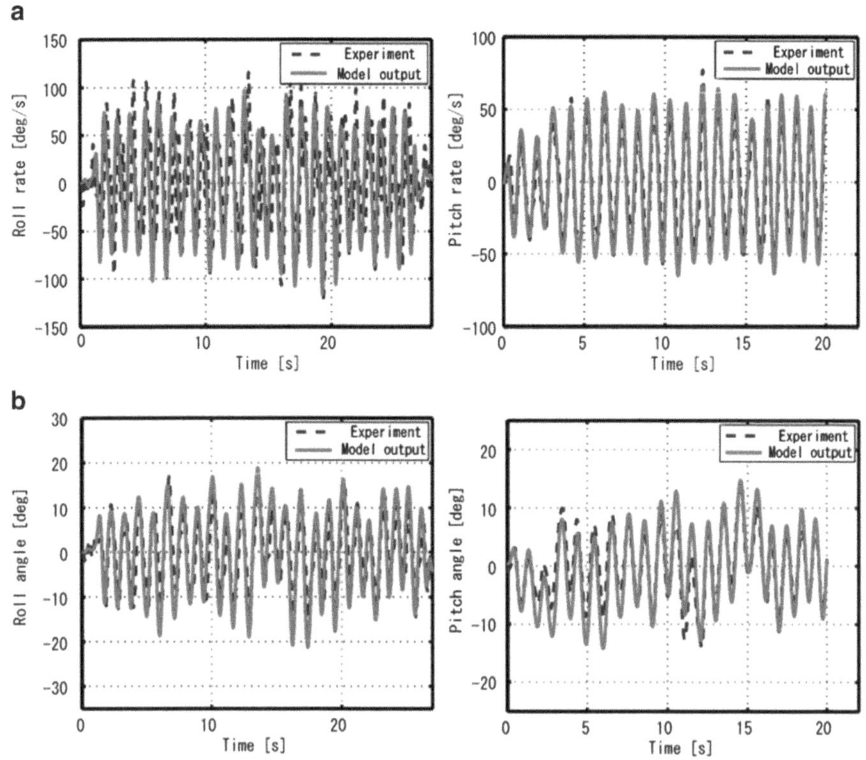

Fig. 5.19 Cross validation result of SST-Eagle-FREYA-Ex2. (**a**) Angular rate, (**b**) Angle

$$x_r = \begin{bmatrix} p & \phi & b & d \end{bmatrix}^T$$
$$y_r = \phi$$

From (5.74) and (5.75), we obtain the following augmented system:

$$\begin{bmatrix} \dot{x}_r \\ \dot{e} \end{bmatrix} = \begin{bmatrix} A_r & 0_{4\times1} \\ -C_r & 0 \end{bmatrix} \begin{bmatrix} x_r \\ e \end{bmatrix} + \begin{bmatrix} B_r \\ 0 \end{bmatrix} \Theta_{Lon} + \begin{bmatrix} 0_{4\times1} \\ 1 \end{bmatrix} \phi_r \tag{5.76}$$

$$e(t) = \int_0^t (\phi_r - y_r(\tau))d\tau \tag{5.77}$$

where ϕ_r is the roll angle command.

We define the state-variable vector of (5.76) as $\bar{x}_r$, and calculate the feedback gain F_r, which minimizes the performance index J based on the optimal control theory. By using F_r, the control input can be calculated as (5.79). Q and R are weighting matrices whose values are tuned as (5.80).

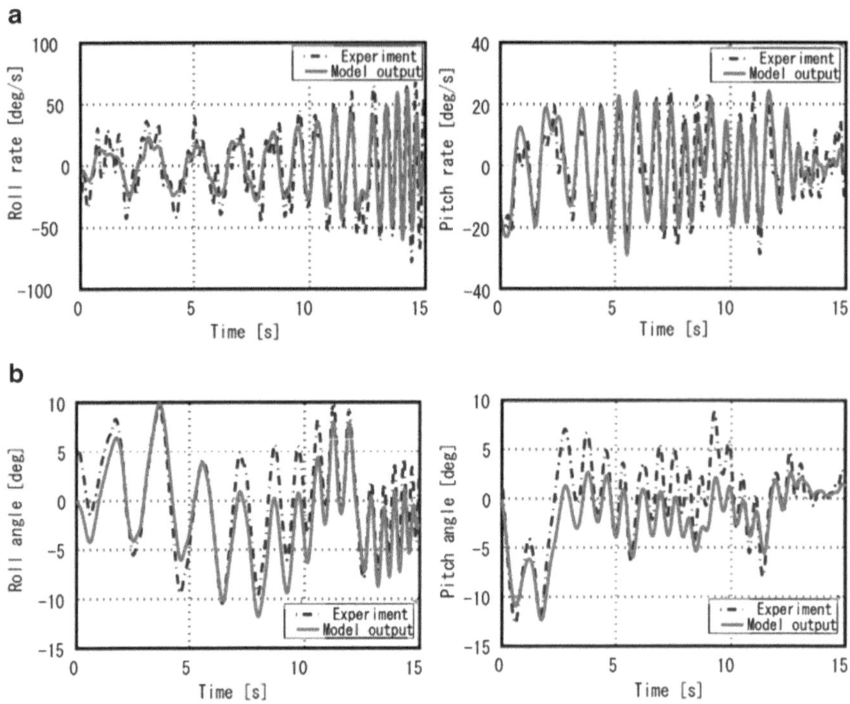

Fig. 5.20 Cross validation result of SF40. (**a**) Angular rate, (**b**) Angle

$$J = \int_0^\infty \left[\bar{x}_r^T(t) Q \bar{x}_r(t) + R\Theta_{Lon}^2(t) \right] dt \tag{5.78}$$

$$\Theta_{Lon} = -F_r \bar{x}_r(t) \tag{5.79}$$

$$Q = \mathrm{diag}[\,1.5 \;\; 1 \;\; 1 \;\; 1 \;\; 9.0 \times 10^3\,] \qquad \mathrm{R} = 0.50 \tag{5.80}$$

Although in state feedback control it is necessary to measure all state variables, there are unmeasurable components in the state variable vector, such as the flap angle of the main rotor and the stabilizer. To estimate such unmeasurable state variables, we designed an observer. In order to reduce the effect of the sensor noise, we designed an identity observer by using a Kalman filter, and estimated all state variables. The controller with the observer is written as follows:

$$\begin{bmatrix} \dot{\hat{x}}_r \\ \dot{e} \end{bmatrix} = \begin{bmatrix} A_r - B_r F_{r1} - K C_r & -B_r F_{r2} \\ 0_{1\times 4} & 0 \end{bmatrix} \begin{bmatrix} \hat{x}_r \\ e \end{bmatrix} + \begin{bmatrix} K & 0_{4\times 1} \\ -1 & 1 \end{bmatrix} \begin{bmatrix} y_r \\ \phi_r \end{bmatrix} \tag{5.81}$$

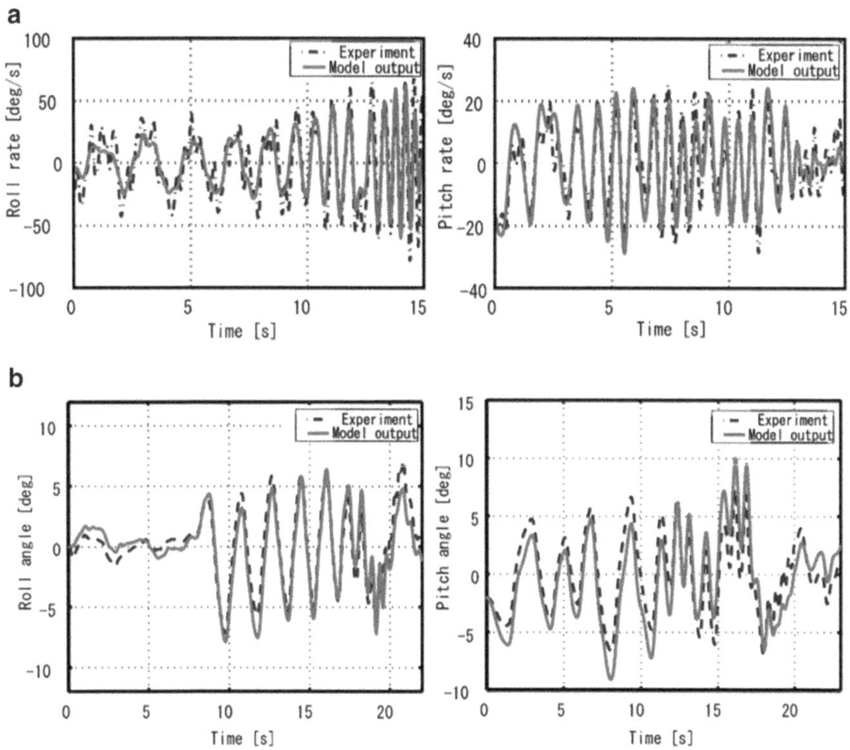

Fig. 5.21 Cross validation result of SF125. (**a**) Angular rate, (**b**) Angle

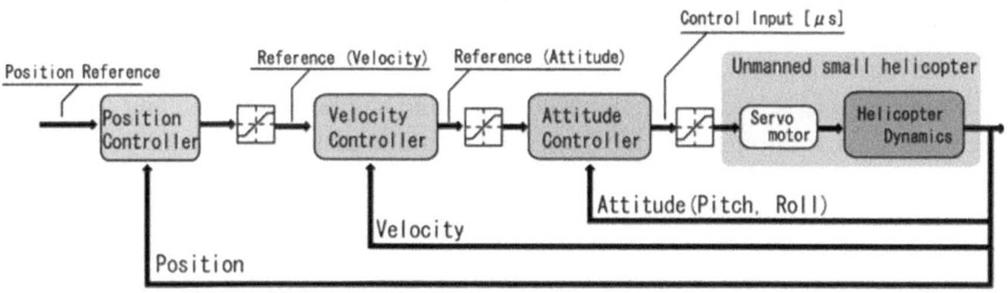

Fig. 5.22 Configuration of control system

$$\begin{bmatrix} \phi_r \\ e \end{bmatrix} = \begin{bmatrix} -F_{r1} & -F_{r2} \\ 0_{1\times 4} & 1 \end{bmatrix} \begin{bmatrix} \hat{x}_r \\ e \end{bmatrix} \tag{5.82}$$

where K, F_{r1}, F_{r2} are the observer gain, the feedback gain on the state variables and the gain on the integral of the tracking error, respectively.

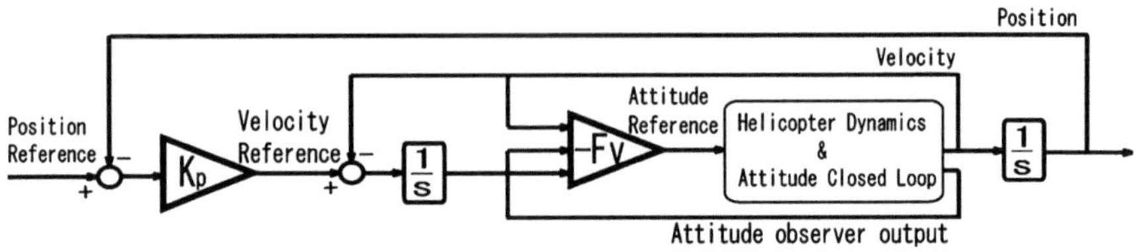

Fig. 5.23 Control System

5.3.3 *Translational Motion Control System*

We designed a translational velocity controller which includes an attitude control system as the inner loop. The velocity model that includes the dynamics of the attitude control system is defined as (5.83) through (5.85).

$$\dot{x}_v = A_v x_x + B_v \phi_r \tag{5.83}$$

$$y_v = C_v x_v \tag{5.84}$$

$$\begin{aligned} A_v &= \begin{bmatrix} Y_v & 0 \quad g \qquad g & 0_{1\times 2} \\ 0_{5\times 1} & A_s - B_s F_r & \end{bmatrix} \\ B_v &= \begin{bmatrix} 0_{5\times 1} \\ 1 \end{bmatrix} \\ C_v &= \begin{bmatrix} 1 & 0_{1\times 5} \end{bmatrix} \\ x_v &= \begin{bmatrix} v & x_r^T & e \end{bmatrix}^T \end{aligned} \tag{5.85}$$

where A_s and B_s are the system matrix and the input matrix of (5.76), and we assume that $x_r = \hat{x}_r$ because the dynamics of the observer in the attitude controller is more rapid than the dynamics of the velocity.

Based on this velocity model, we construct the augmented model and design the feedback gain F_v along the same lines as the attitude controller. The velocity can be measured by GPS, and other state variables are estimated by the attitude controller.

The position controller is constructed as the outer loop of the velocity controller. This controller is designed by the P control method. Figure 5.23 shows the entire control system, which is consists of a triple closed-loop system.

5.4 Experiment

In this section, some experimental results of the designed controller are presented. To show the effectiveness of the method, the test flights ware carried out for several sizes of helicopter.

5.4.1 *Avionics Architecture*

Before describing the results of the experiment, we will explain the two forms of avionics architecture in the helicopter control that we built. We refer to one of these as "Host-based Control System (HCS)" and to the other as "Embedded Control System (ECS)".

A schematic diagram of the HCS is presented in Fig. 5.24. In HCS, first, the helicopter's motion is sensed by GPS and IMU. The collected information is processed by the embedded computer according to the communication protocol and transmitted to the Ground Control Station (GCS) by a wireless modem. In the GCS, the flight data is downloaded by a laptop on which the software for the flight information display and logging, the control calculation, and the uploading of the various operating commands is implemented using C++. The control input is transmitted to a device called a "Pulse Generator" and converted to PWM signals. The Pulse Generator is connected to the RC transmitter by a trainer cable. The pilot can switch the manual operation to an autonomous mode by using the "Trainer Function".

The ECS, on the other hand, is constructed as shown in Fig. 5.25. Unlike the HCS, in the ECS the control calculation is executed on the embedded computer. The calculated control input is then transmitted to a "Servo Pulse Processing Unit" (SPPU) on which the control input is converted to PWM signals. The SPPU also takes on the switching of the manual and autonomous flight, according to the state of a designated channel (i.e., control switch) of the RC transmitter. The role of the GCS is same as it is for the HCS, except for the control calculation.

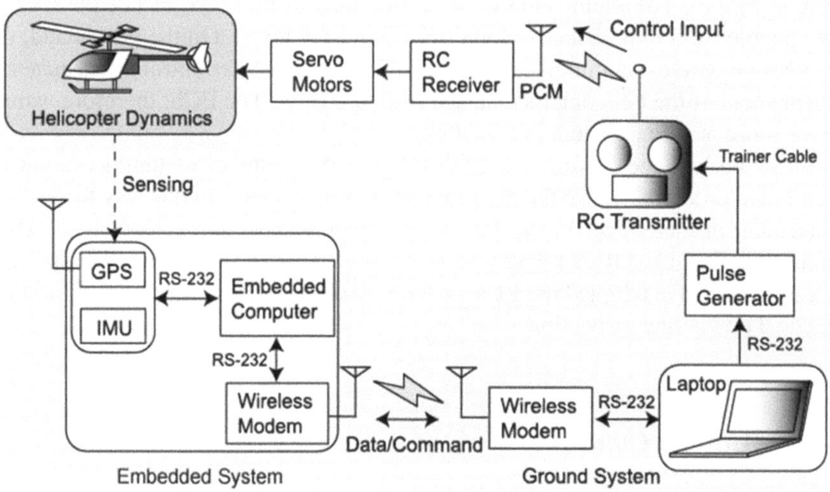

Fig. 5.24 Architecture of the Host-based Control System

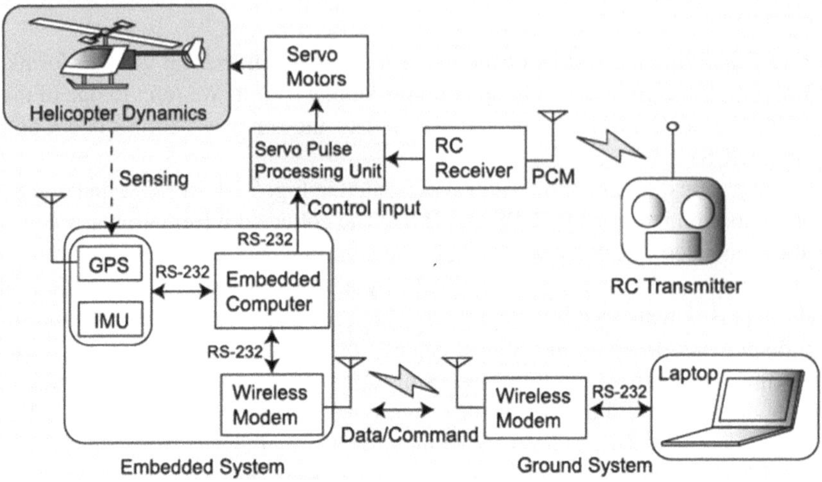

Fig. 5.25 Architecture of the Embedded Control System

The advantage of HCS is that the process of tuning the controller and debugging is simpler than it is for the ECS. In the ECS, when we would like to change the controller gain, we have to rewrite the source code of the embedded program or transmit the "new" gain by wireless communication, or some other means. In the HCS, the control calculation is implemented on the laptop, so we can easily to change the code and debug the program without wireless communication. However, there is wireless communication in the closed loop in the HCS. Hence the HCS is not suitable for a long distance flight over the visual range. On the other hand, the ECS has no wireless communication in the closed loop, so its control performance is independent of the helicopter's distance from the GCS. The ECS, therefore, would be the more practical system in actual use.

In the early stage of research or development, a number of tunings might be required in order to achieve the desired control performance. In this way the HCS is convenient in that stage. In the following experiment we have adopted the HCS on an SST-EAGLE-FREYA-Ex2 and SF40, which are the platforms for our UAV research, and have adopted the ECS on SKY SURVEYOR, which is a platform for practical power line inspection.

5.4.2 *Attitude Control*

In this section, we show the experimental results of the attitude control on an SST-Eagle-FREYA-Ex2, SF40, and SKY SURVEYOR SF125. Figures 5.26 through 5.31 show the response of the roll and pitch angles of each helicopter. The attitude angle

Fig. 5.26 Roll response of SST-Eagle

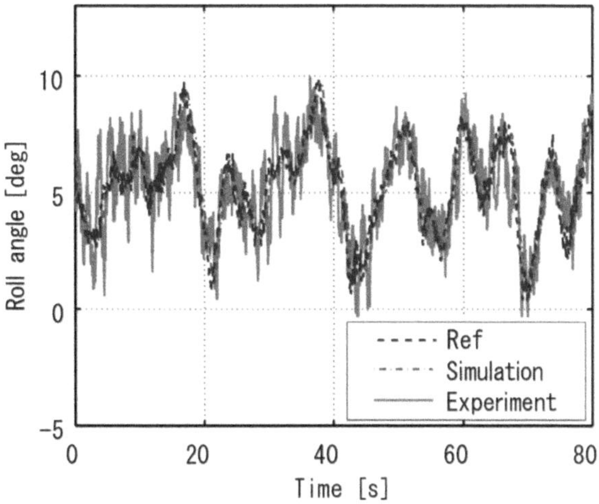

Fig. 5.27 Pitch response of SST-Eagle

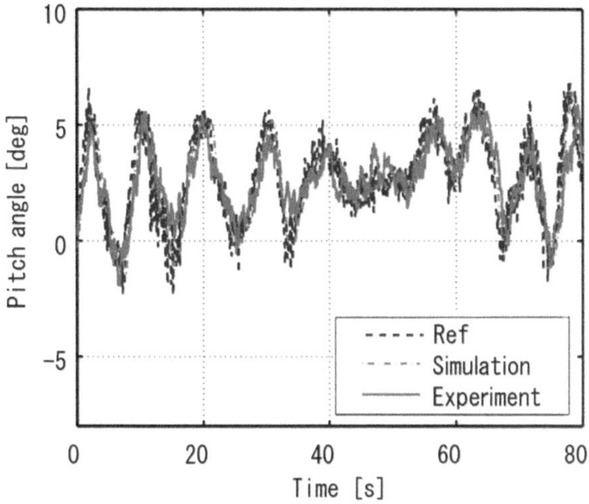

Fig. 5.28 Roll response of SF40

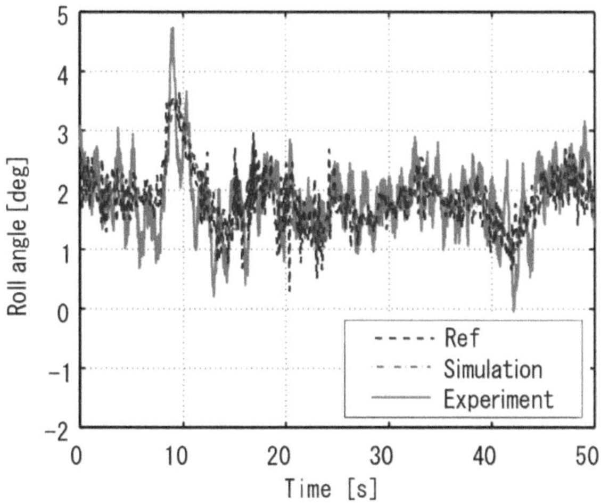

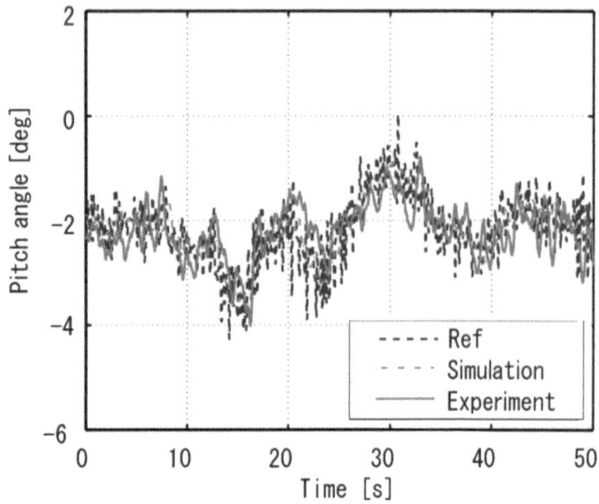

Fig. 5.29 Pitch response of SF40

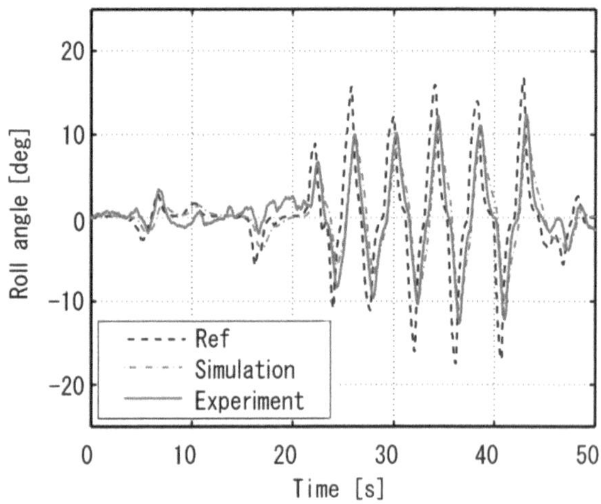

Fig. 5.30 Roll response of SF125

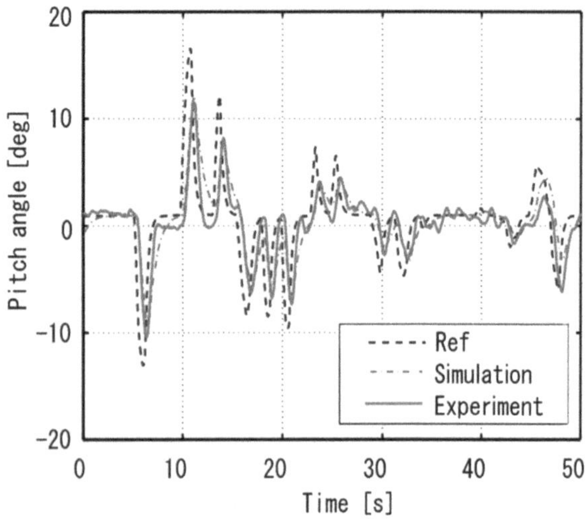

Fig. 5.31 Pitch response of SF125

references are generated by the velocity controller on the outer loop except for the results from SKY SURVEYOR SF125. For this reason, the references to SST-Eagle-FREYA-Ex2 and SF40 are somewhat noisy because of the noise made by the GPS. The reference to SF125 is given by the pilot by using the RC transmitter.

That is, the attitude references are calculated in proportion to the tilt angle of the stick of the RC transmitter.

We can see that the attitude angles track to the references precisely. In the case of SF125, there is somewhat of a time delay between the reference and attitude angle, even as the experimental result corresponds to the simulation result. Its performance, then, can be improved by tuning on the controller.

From the results of our experiment, we have been able to demonstrate the effectiveness of the derived attitude models and the controller design.

5.4.3 Hovering and Translational Flight Control

In this section we show the flight data for hovering and translational flight control. Note, however, that we have not yet carried out the hovering control test using the proposed method on SKY SURVEYOR SF125. We therefore show only the results for the SST-Eagle-FREYA-Ex2 and SF40. Also, we have controlled only the translational motion, because, in small-attitude flight such as hovering, the effect of the collective pitch input upon the translational motion is small. Thus, we can demonstrate the translational control performance all the same.

Figures 5.32 through 5.35 show the response of the forward and rightward velocity. When the velocity reference approaches zero, the flight data has not converged to the reference because of some disturbance such as wind. The velocity, however, has converged to the reference in translational motion flight. We can see this not

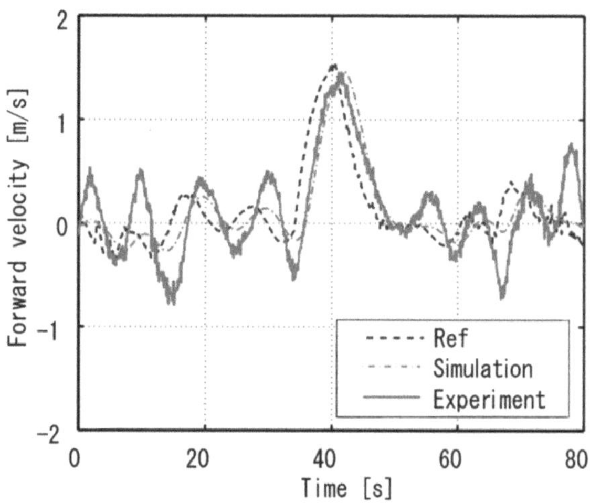

Fig. 5.32 Forward velocity of SST-Eagle

Fig. 5.33 Rightward velocity of SST-Eagle

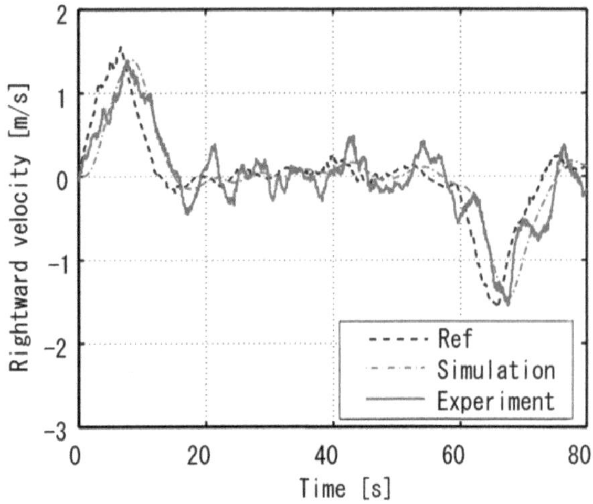

Fig. 5.34 Forward velocity of SF40

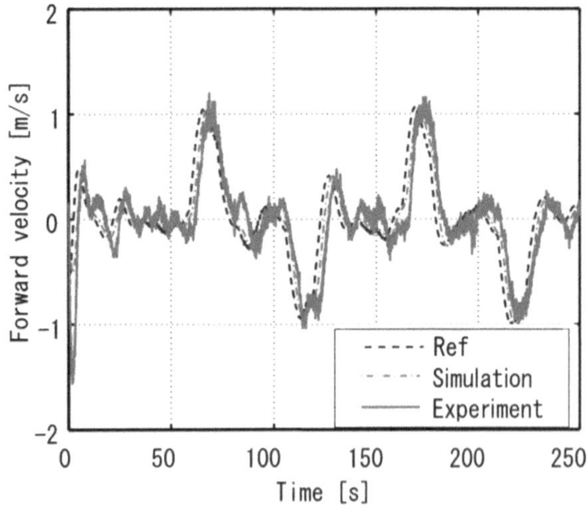

Fig. 5.35 Rightward velocity of SF40

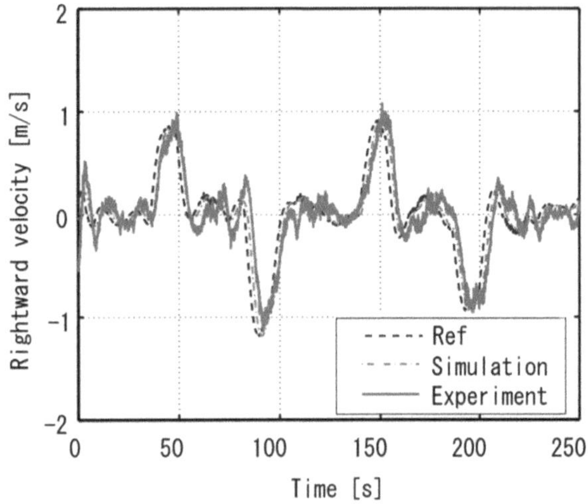

only in the experimental data but also in the fact that the simulation results tracks to the reference. From these results, we can verify the effectiveness of our modeling and controller design.

Figures 5.36 and 5.37 show the results of the hovering test. These figures illustrate the flight trajectory at a zero position reference. We can see that hovering control within about 1 m of accuracy has been achieved.

Figures 5.38 and 5.39 show the translational flight results for a square with a trajectory of about 10 m. The reference has been given in real time from the ground station in a "point-to-point" manner. In these figures we can see that the flight along the square has been achieved.

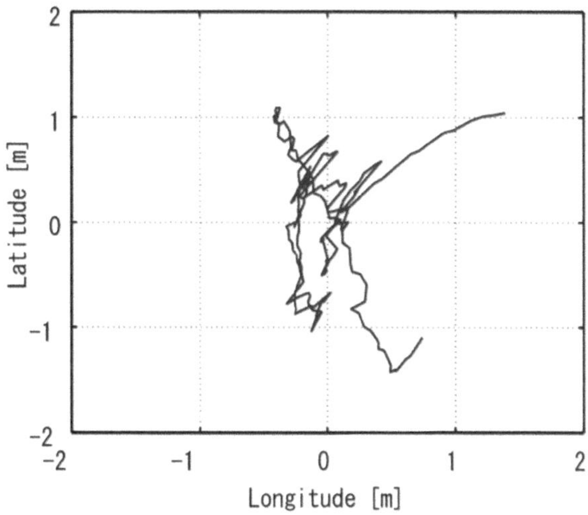

Fig. 5.36 Hovering result of SST-Eagle

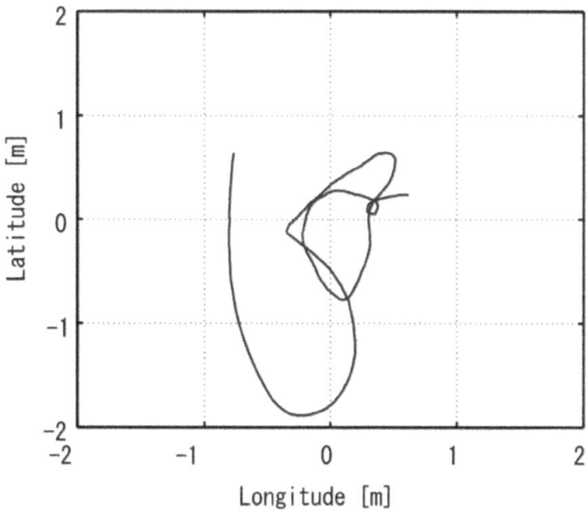

Fig. 5.37 Hovering result of SF40

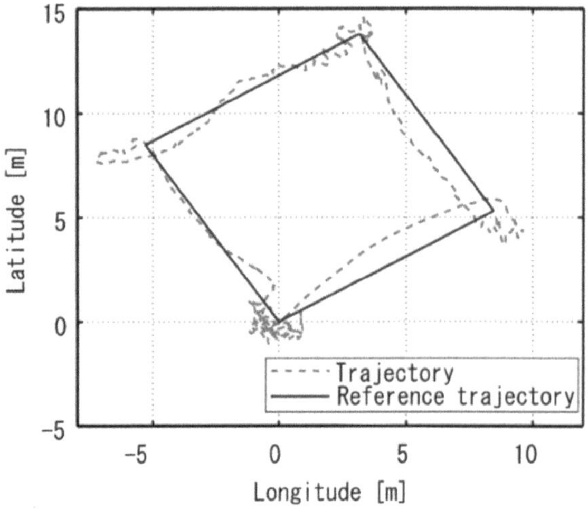

Fig. 5.38 Horizontal movement of S-ST-Eagle

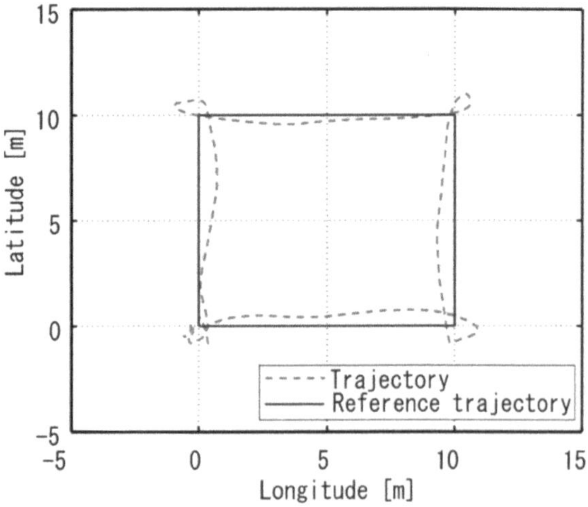

Fig. 5.39 Horizontal movement of SF40

5.5 Summary

This chapter has presented a modeling and controller design method for a small unmanned helicopter. Two linear models have been derived by an analytical method referred to as the "white box modeling" method. In the derived models, all parameters come from the spec of the helicopter or from physical information. They can therefore be easily applied to several sizes of helicopter. Unlike the system identification method, the present approach does not require an excitation experiment to be performed. Along with its safety, this will decrease the workload of the designer. The validity of the models and the controller design has been demonstrated by test flights using three kinds of helicopters.

References

1. Rahideh A, Shaheed MH (2008) Dynamic modeling of a twin rotor MIMO system using grey box approach. In: Proceedings of the 5th international symposium on mechatronics and its applications, Amman, Jordan, pp 1–6
2. Kato A, Kobayashi T, Katayama H, Ichikawa A (2003) Attitude control of a helicopter model by nonlinear sampled-data H∞ control. In: Proceedings of the SICE annual conference, Fukui, Japan, pp 930–935
3. Casilo CL, Moreno W, Valavanis KP (2007) Unmanned helicopter waypoint trajectory tracking using model predictive control. In: Proceedings of the Mediterranean conference on control and automation, Athens, Greece, pp 1–8
4. Fujiwara D, Nonami K, Shin J, Hazawa K (2004) H∞ hovering and guidance control for autonomous small-scale unmanned helicopter. In: Proceedings of the 2004 IEEE/RSJ international conference on intelligent robots and systems, Sendai, Japan, pp.2463–2468
5. Song D, Qi J, Dai L, Han J, Liu G (2008) Modeling a small-size unmanned helicopter using optimal estimation in the frequency domain. In: Proceedings of the 15th international conference on mechatronics and machine vision in practice, Auckland, New Zealand, pp 97–102
6. Yinzhe D, Liang L, Zhenbarg G (2005) Modeling analysis and designing the flight control system on a sub-mini unmanned helicopter robot. In: Proceedings of the conference on high density microsystem design and packaging and component failure analysis, Shanghai, Japan, pp 1–4
7. Yinzhe D, Liang L, Zhenbang G (2005) Modeling analysis of vibration for subminiature unmanned helicopter. In: Proceedings of the 2005 IEEE international workshop on safety and rescue robotics, Kobe, Japan, pp 131–136
8. Johnson EN, Kannan SK (2005) Adaptive trajectory control for autonomous helicopters. AIAA J Guid Cont Dynam 28(3):524–538
9. Wang H, Duan H (2007) Modeling and hover control of a novel unmanned coaxial rotor/ducted-fan helicopter. In: Proceedings of the IEEE international conference on automation and logistics, Jinan, China, pp 1768–1773
10. Carro J, Valero J, Barrientos A (2005) Identification of a small unmanned helicopter model using genetic algorithms. In: Proceedings of the 2005 IEEE/RSJ international conference on intelligent robots and systems, Alberta, Canada, pp 3360–3365
11. Civita ML, Papageorgiou G, Messner WC, Kanade T (2006) Design and flight testing of an H∞ controller for a robotic helicopter. J Guid Contr Dynam 29(2):485–494
12. Song P, Qi G, Li K (2009) The flight control system based on multivariable PID neural network for small-scale unmanned helicopter. In: Proceedings of the international conference on information technology and computer science, Kiev, Ukraine, pp 538–541
13. Rebeschiess S, Roloff M (1999) Position control by feedback linearization for a simplified helicopter model. In: Proceedings of the 1999 IEEE international conference on control applications, Kohala Coast, Hawaii, pp 143–145
14. Gen SS, Red B, Tee KP, Lee TH (2009) Approximation-based control of uncertain helicopter dynamics. IET Contr Theor Appl 3(7):941–956
15. Westerberg S, Mettin U, SHriaev AS, Freidvich LB, Orlov Y (2009) Motion planning and control of a simplified helicopter model based on virtual holonomic constraints. In: Proceedings of the international conference on advanced robotics, Munich, Germany, pp 1–6
16. Benjanarasuth T (2008) Experimental study on servo adaptive pitch control of a model helicopter. In: SICE Annual Conference, Chofu, Tokyo, Japan, pp 209–213
17. Xianpqing W, Yimin H (2008) Model system design for small-size unmanned helicopter based on moment theory, blade element theory and FlightLab software. In: Proceedings of the 2nd international symposium on systems and control in aerospace and astronautics, Shenzhen, China, pp 1–4
18. Jiang Z, Han J, Wang Y, Song Q (2006) Enhanced LQR control for unmanned helicopter in hover. In: Proceedings of the 1st International Symposium on Systems and Control in Aerospace and Astronautics, Harbin, China, pp 1438–1443

19. Mettler B (2003) Identification modeling and characteristics of miniature rotorcraft. Kluwer Academic Publishers, Norvell
20. Done G, Balmford D (2001) Bramwell's helicopter dynamics, 2nd edn. Butterworth Heinemann, Boston
21. Leishman JG (2000) Principles of helicopter aerodynamics, Cambridge Aerospace Series. Cambridge University Press, New York
22. Kato K, Imanaga I (1988) Introduction to helicopters. The University of Tokyo Press, Tokyo

Part II
Advanced Flight Control Systems for Rotorcraft UAVs and MAVs

Chapter 6
Analysis of the Autorotation Maneuver in Small-Scale Helicopters and Application for Emergency Landing

Abstract In this chapter, we derive linear and nonlinear models of a helicopter autorotation maneuver focusing on main rotor revolutions and descent rate dynamics. The nonlinear model for descent rate dynamics is derived by combining blade element theory (BET) with an induced velocity linear approximation proposed in this chapter. The linear model is derived by implementing a neural network optimization method based on the experimental data. The derived nonlinear and linear models are verified by comparing the simulated data of the model with experimental data. We also perform a preliminary experiment on autorotation landing with a simple PI controller. Rotor revolutions and descent rates are found to be efficiently controlled by the designed controller.

Video Link:

Fully autonomous autorotation flight control
http://mec2.tm.chiba-u.jp/monograph/Videos/Chapter6/1.wmv

Nomenclature

α	Angle of attack
δ	Average drag coefficient
φ	Inflow angle
θ	Collective pitch angle
θ_0	Collective pitch angle at blade tip
ρ	Air density
Ω	Angular velocity of main rotor
a	2D lift curve slope
b	Number of blades
c	Cord length
Cd0	Profile drag coefficient
dD	Drag at blade element
dL	Lift at blade element
D_i	Induced drag at blade element

K. Nonami et al., *Autonomous Flying Robots: Unmanned Aerial Vehicles and Micro Aerial Vehicles*, DOI 10.1007/978-4-431-53856-1_6, © Springer 2010

Dp	Profile drag at blade element
F	Force
Q	Toque
R	Main rotor radius
T	Thrust
tw	Twist angle
UT	Relative wind
vd	Induced velocity
vh	$\sqrt{T/2\rho\pi R^2}$
Vh	Horizontal velocity of fuselage
Vz	Vertical velocity of fuselage

6.1 Introduction

When operating small unmanned helicopters, high reliability and safety are required. High reliability is achieved by constructing a system that is similar to manned aircraft systems. However, since small unmanned helicopters have payload limitations, it is almost impossible to realize a parallel system for all system components. This is true for engines in particular, which account for the biggest percentage of the total helicopter weight. Since it is absolutely impossible to have two engines, we tried to use parachutes and airbags to avoid crashes due to engine failures. However parachutes and airbags become additional payloads and cannot be placed on the fuselage; therefore, we stopped using them.

In addition, manned helicopters can maneuver to perform autorotation landing during engine emergencies. The autorotation maneuver is defined as a self-sustained rotation of the rotor without the application of any shaft torque from the engine. Under these conditions, the power to drive the rotor is a result of the relative airstream through the rotor as the helicopter descends through the air. Autorotation is used as a means of recovering the safe flight of the helicopter in the event of catastrophic mechanical failures; such as engine, transmission, or tail rotor failures. The autorotation maneuver is performed for both engine emergencies and in competition in radio control (R/C).

Our unmanned helicopter has a mechanical setup for autorotation. If autonomous autorotation is realized, the autonomous fight control system becomes more reliable.

Autorotation landings have been performed based on operators' experience with flight techniques. There are no reports focusing on autonomous autorotation in the case of either manned or unmanned helicopters. The objective of this study is to realize autonomous autorotation landing. In this chapter, we derive a mathematical autorotation model based on aerodynamic analysis and focus on main rotor revolutions and descent rate. Moreover, we perform some preliminary experiments for autonomous autorotation landing.

6.2 Autorotation

6.2.1 Aerodynamic Force at Blade Element

The two aerodynamic forces produced by a blade element can be divided into two main forces. One force is the lift that acts perpendicular to the resultant flow velocity, and the other force is the drag that acts parallel to the resultant flow velocity. The relationship between these two forces is shown in Fig. 6.1. UT is the difference in the induced and vertical velocities of the fuselage.

$$UT = v_d + V_Z \tag{6.1}$$

6.2.2 Aerodynamics in Autorotation

The relative wind velocity Ωr changes with the distance between the rotor hub and blade element. In an autorotation maneuver, the signs of the resultant force, which consists of lift and drag, are in opposition to the positions of r1 and r2 in Fig. 6.2, because lift, drag, and the angle of attack change with changes in the relative wind velocity.

Therefore, the autorotation situation can be described by the equivalence of two kinds of forces: the force rotating the main rotor and the opposing force.

In general, the blade region where the force is produced for driving the rotor is called the 'Driving Region,' while the region of the blade where power is absorbed is called the 'Driven Region.' Since rΩ becomes 0, the angle of attack increases around the rotor hub. Since a stall can occur, this region is called the 'Stall Region.'

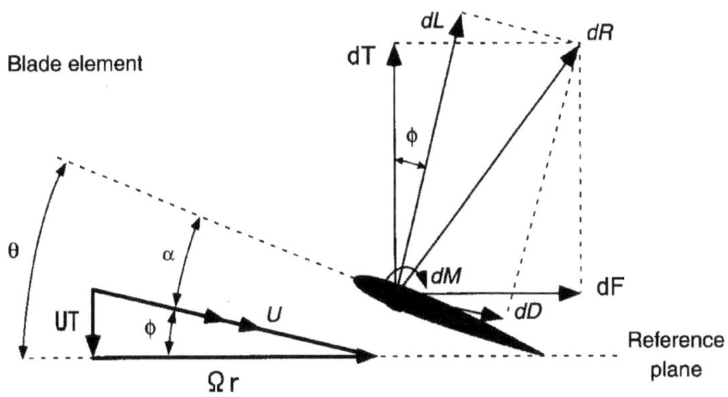

Fig. 6.1 Incident velocity and aerodynamic environment at a typical blade element

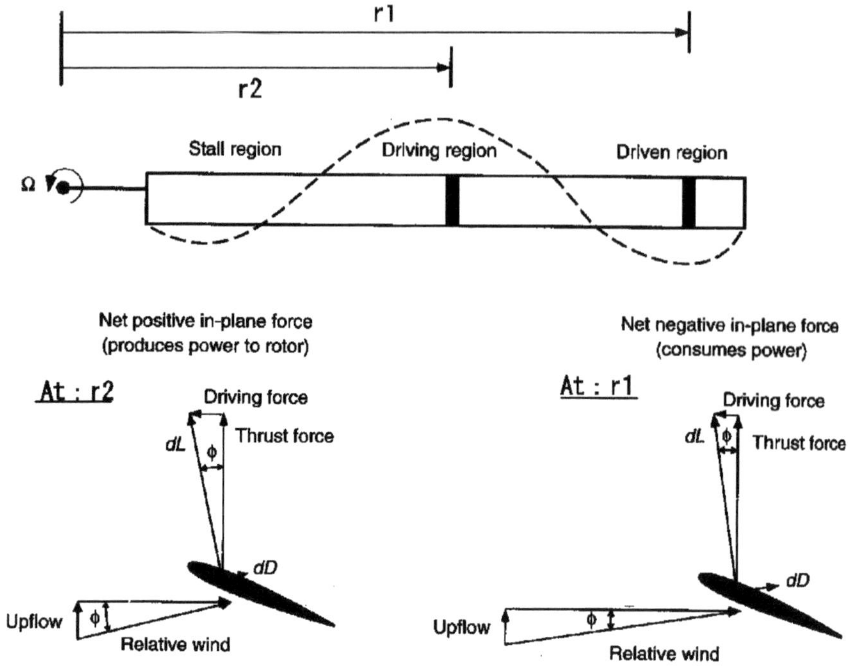

Fig. 6.2 Aerodynamic forces of two different blade elements in an autorotation flight

6.3 Nonlinear Autorotation Model Based on Blade Element Theory

In particular, two theories have been applied to helicopter aerodynamic analysis, momentum theory and blade element theory (BET). Momentum theory macroscopically deals with the airstream flow through the main rotor, on the basis of changes in the momentum of air. In contrast, BET deals with the micro-lift and drag at a blade element, as shown in Fig. 6.1, and the total thrust and power are analyzed by integrating them. Since it is difficult to analyze profile drag and dynamics due to change in the angle of attack, in this chapter, we apply BET to the autorotation analysis.

As shown in Fig. 6.1, the normal component to the rotor disc (lift) and horizontal component (drag) are as follows:

$$\begin{aligned} dT &= b\,(dL\cos\phi - dD\sin\phi) \\ &= b\,dL \end{aligned} \tag{6.2}$$

$$\begin{aligned} dF &= b\,(dL\sin\phi - dD\cos\phi) \\ &= b\,(dL\phi + dD) \\ &= dD_i + dD_P \end{aligned} \tag{6.3}$$

Since φ is very small, we assume

$$\cos\phi = 1,\ \sin\phi = \phi \tag{6.4}$$

Since, in general, dD is relatively smaller than dL to the 10^{-1} order, we neglect the dD component in (6.2).

6.3.1 Thrust

According to BET

$$dL = \frac{1}{2}\rho(r\Omega)^2 bca(\theta(r) - \phi) \tag{6.5}$$

Here θ is the collective pitch angle. The twist of our helicopter is described by the following equation.

$$\theta(r) = \theta_0 - \theta_{tw} r,\ \theta_{tw} = \frac{tw}{R} \tag{6.6}$$

Here tw is the twist angle and it is equal to 4°. φ is the inflow angle,

$$\phi = \tan^{-1}\left(\frac{vd + V_z}{r\Omega}\right) \cong \frac{vd + V_z}{r\Omega} \tag{6.7}$$

Substituting φ and θ in (6.5) and using (6.6) and (6.7),

$$dT = \frac{1}{2}\rho cab\Omega^2\left(\theta_0 r^2 - \theta_{tw} r^3 - \frac{v_d - V_z}{\Omega} r\right) \tag{6.8}$$

Since the induced velocity changes along the blade radius, it has to be expressed as a function of r. However, to simplify issues, we assume that it is constant. By integrating (6.8),

$$T = \int_0^R dT = \frac{1}{2}\rho cba\Omega^2 R^3\left(\frac{1}{3}\theta_0 - \frac{1}{4}tw - \frac{vd + V_z}{2R\Omega}\right) \tag{6.9}$$

6.3.2 Torque

First, we calculate the first term of (6.3) (induced drag).

$$\begin{aligned} dD_i &= bdL\phi \\ &= \frac{1}{2}\rho cab\Omega^2\left(\theta_0 r^2 - \theta_{tw} r^3 - \frac{v_d + V_Z}{\Omega} r\right)\frac{v_d + V_Z}{r\Omega} \end{aligned} \tag{6.10}$$

Then, the torque due to the induced drag becomes

$$Q_i = \int_0^R r dD_i$$
$$= \frac{1}{2}\rho cba\Omega(vd + V_Z)R^3\left(\frac{1}{3}\theta_0 - \frac{1}{4}tw - \frac{vd + V_Z}{2R\Omega}\right) \qquad (6.11)$$

The second term of (6.3) (profile drag) is

$$dD_p = \frac{1}{2}\rho r^2\Omega^2 cbC_{d0} \qquad (6.12)$$

Here Cd0 is the profile drag coefficient, which is not constant along the blade. However, in general, it can be assumed to be constant by introducing δ, which is called the 'average profile drag coefficient [1].' Hence the torque due to profile drag becomes

$$Q_p = \frac{1}{8}\rho cbR^4\delta\Omega^2 \qquad (6.13)$$

6.3.3 Induced Velocity

Using (6.2)–(6.13), the total thrust and torque are derived based on BET and it includes an induced velocity. The induced velocity must be approximated by some theoretical or experimental equations because we cannot directly measure it.

From momentum theory, the induced velocity is expressed by the following equations [2].

$$\frac{v_d}{v_h} = -\left(\frac{V_z}{2v_h}\right) + \sqrt{\left(\frac{V_z}{2v_h}\right) + 1} \quad \text{for } V_z > 0 \qquad (6.14)$$

$$\frac{v_d}{v_h} = -\left(\frac{V_z}{2v_h}\right) - \sqrt{\left(\frac{V_z}{2v_h}\right) - 1} \quad \text{for } V_z < -2v_h \qquad (6.15)$$

However, when −2vh < Vz < 0, the induced velocity cannot be calculated by momentum theory because the control volume required for momentum analysis cannot be defined around the rotor disk where the vortex exists. This situation is called the 'vortex ring state' (Fig. 6.3).

The induced velocity at the vortex ring state cannot be calculated theoretically. However, it is possible to measure it with a wind tunnel experiment. Figure 6.4 shows the experimental results. The experimentally measured induced velocity is approximated by several equations. For example [2],

Fig. 6.3 Vortex ring state

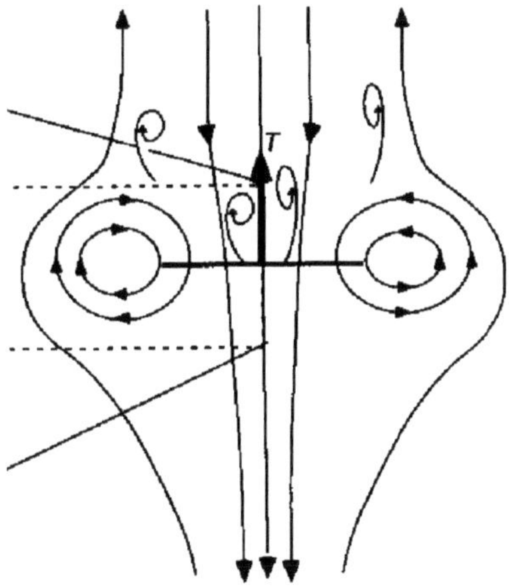

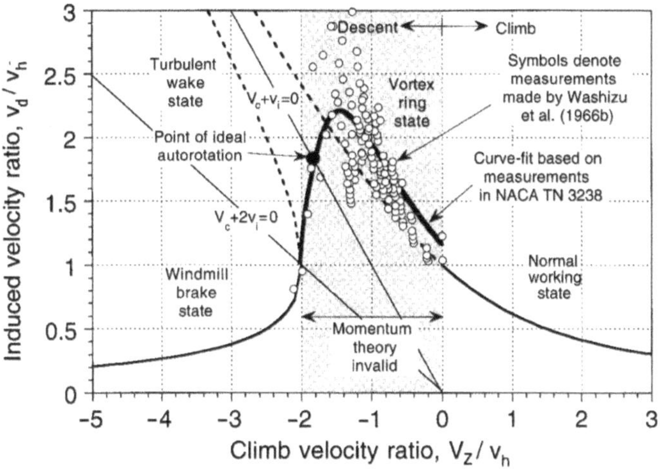

Fig. 6.4 Induced velocity variation as a function of the ascent and descent velocities

$$\frac{v_d}{v_h} = \kappa - \frac{V_z}{v_h} \text{ for } -1.5 < \frac{V_z}{v_h} < 0 \tag{6.16}$$

$$\frac{v_d}{v_h} = \kappa\left[7 + 3\frac{V_z}{v_h}\right] \text{ for } -2.0 < \frac{V_z}{v_h} < 1.5 \tag{6.17}$$

Since the autorotation maneuver occurs primarily in the region of −2vh < Vz < 0, approximations using the above equations are very important.

6.4 Validity of Autorotation Model

6.4.1 Experimental Data

To verify the autorotation model, we performed some experiments. Since our ultimate purpose is autorotation landing, experimental data near the ground is required.

However, it is extremely dangerous to perform experiments near the ground, even when the autorotation maneuver is performed manually. Hence, the experimental data shown in Fig. 6.5 is the data for a helicopter landing at 10 m above the ground, i.e., the vertical velocity of the fuselage is controlled by an operator to be 0 m/s.

The following analysis is performed based on the experimental data shown in Fig. 6.5.

6.4.2 Verification of Autorotation Model

For the calculation of thrust, we use (6.9), while, for torque, we use (6.11) and (6.13). The induced velocity is calculated by (6.14)–(6.17). To compare the theoretical and experimental data, we first calculate the derivative of the measured velocity and angular velocity. The thrust is derived by the velocity derivative, because we know the weight of the fuselage. Since we also know the moment of inertia of each blade, we can calculate the theoretical angular acceleration using (6.11) and (6.13).

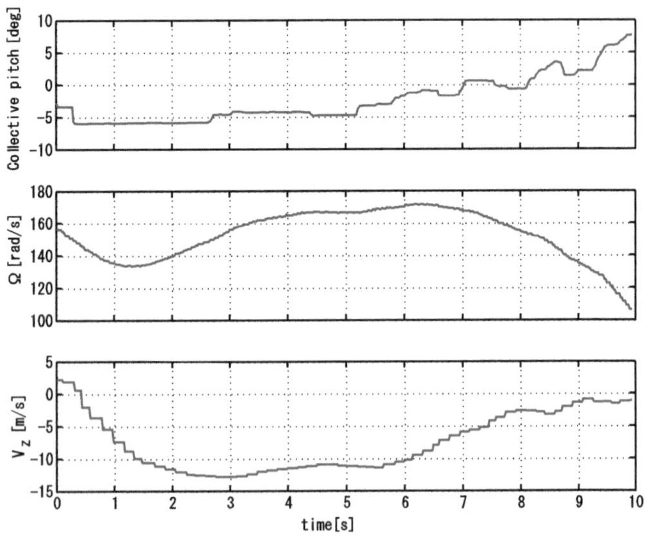

Fig. 6.5 Experimental data of autorotation

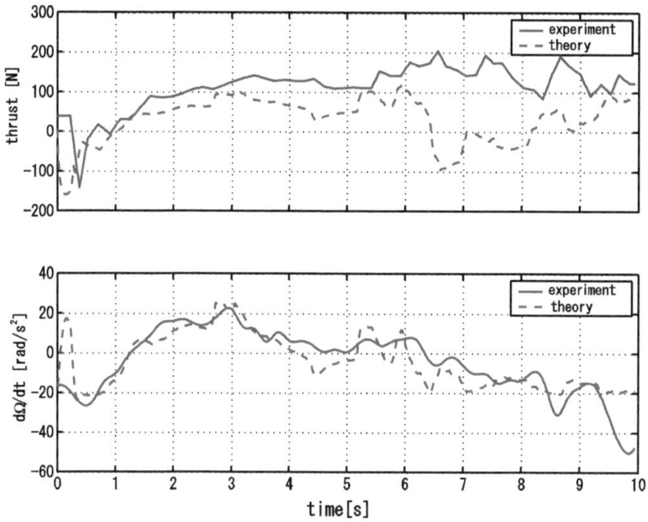

Fig. 6.6 Experimental and calculated thrusts and angular accelerations

Figure 6.6 shows the theoretical and experimental thrust and angular accelerations. The two angular accelerations are more or less equal, but the two thrusts increase after 6 s.

6.4.3 Improvement in Induced Velocity Approximation

Figure 6.7 is the induced velocity used for the calculations in Fig. 6.6. The equation number used for the calculation is also shown. The relationship between the number of equations and the region is shown in Table 6.1. The operation point reaches the vortex ring state after 6 s, where the thrust error in Fig. 6.6 becomes large. From this analysis, it is assumed that there is a problem with the approximation of induced velocity. Therefore, we have introduced a new approximation of the induced velocity. Using (6.9), the induced velocity is calculated by the following inverse equation.

$$v_d = 2\Omega R\left(\frac{-2T}{\rho c b a \Omega^2 R^3} + \frac{1}{3}\theta_0 - \frac{1}{4}tw\right) - V_z \tag{6.18}$$

We approximate the calculated induced velocity as follows:

$$v_d = k_1 V_z + k_2\theta_0 + k_3 V_h + k_4\Omega \tag{6.19}$$

Here

$$V_h = \sqrt{V_x^2 + V_y^2} \tag{6.20}$$

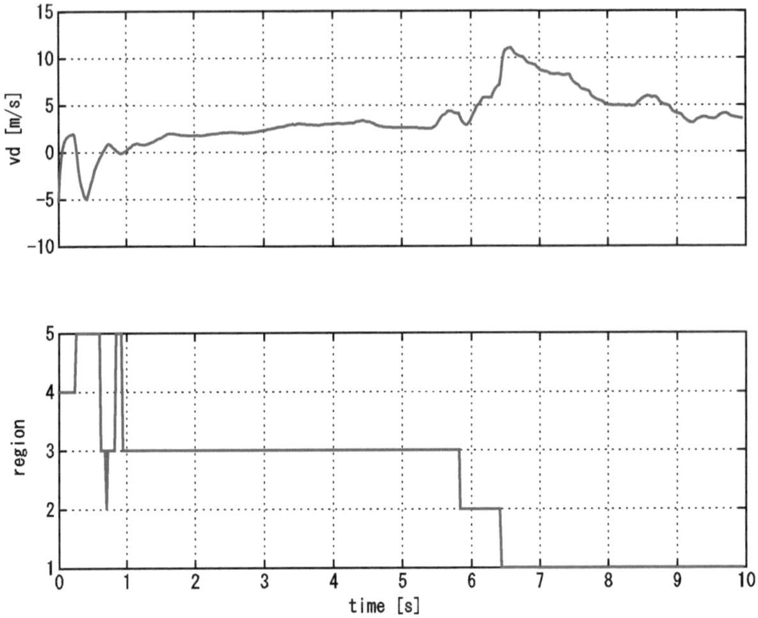

Fig. 6.7 Induced velocity at several operating points

Table 6.1 Equation number and the region where it is applied

Number	Equation No.	Region
1	(6.16)	$1.5v_h < V_Z < 0$
2	(6.17)	$-2.0v_h < V_Z < -1.5v_h$
3	(6.18)	$V_Z < -2.0v_h$
4	(6.14)	$V_Z > 0$
5	(6.14)	$T < 0$

The horizontal velocity changes the amount of induced velocity. Figure 6.8 shows the induced velocity calculated by (6.18) (experiment), (6.19) (linear approximation), and the momentum theory. The approximation by (6.19) accurately approximates the experimental induced velocity.

Figure 6.9 shows the theoretical thrust and angular acceleration using the new approximation by (6.19). By comparing this with Fig. 6.6, the error decreases.

6.4.4 *Validity of Approximated Induced Velocity*

In the previous section, a new approximation of the induced velocity was proposed. However, this was calculated based on the assumption that (6.9) is correct. In order to validate our approximation, we introduced a suitable induced velocity for (6.9) and approximated this value using (6.19). Our second approximation of the induced

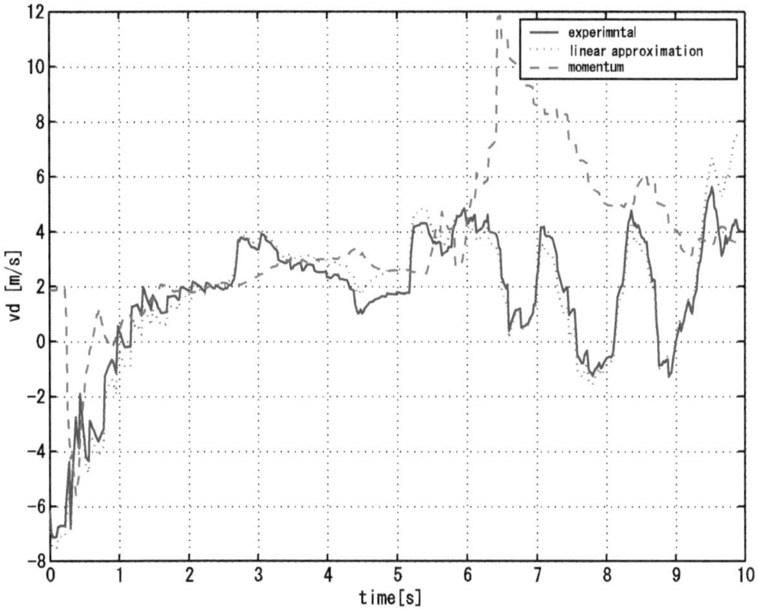

Fig. 6.8 Induced velocity (experimental, approximated, momentum theory)

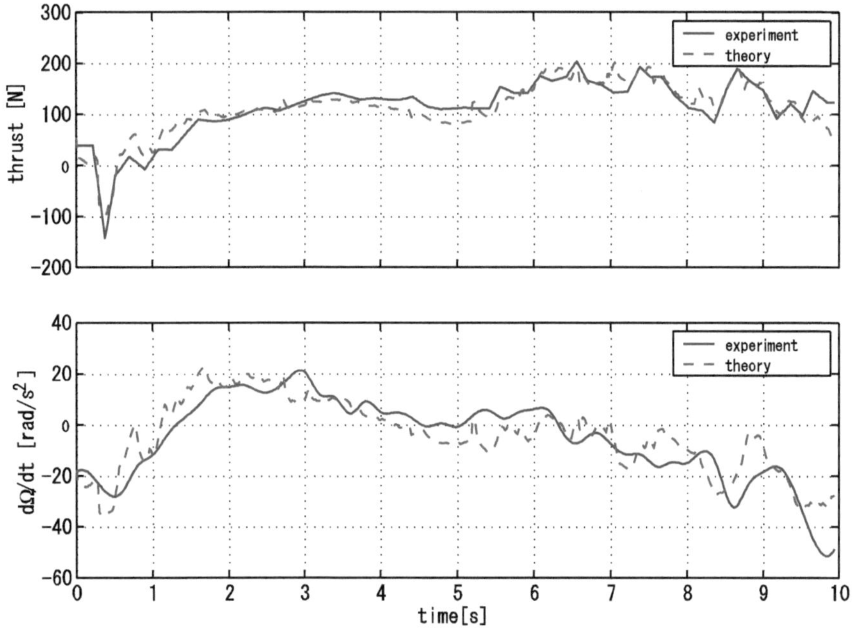

Fig. 6.9 Experimental and calculated thrusts and angular accelerations by using the approximated induced velocity

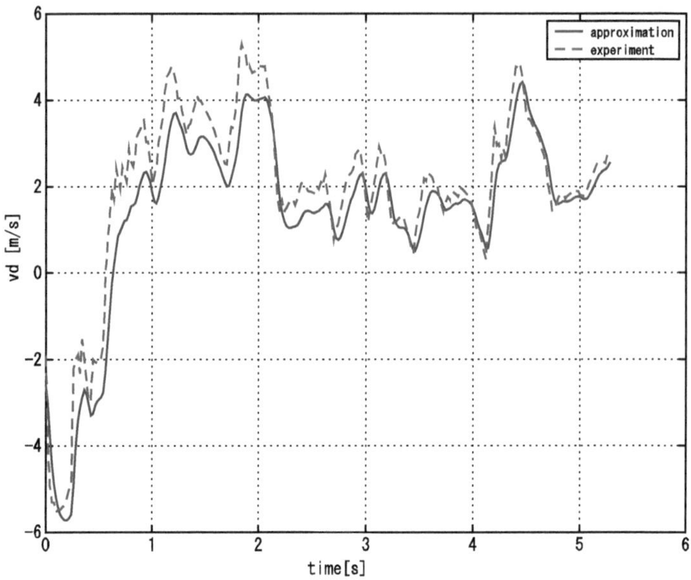

Fig. 6.10 Experimental and approximated induced velocities

velocity also improved the accuracy of the theoretical value. However, if we approximate the induced velocity by using (19) in several experiments, we can definitely say that the proposed induced velocity and approximation are valid.

Figure 6.10 shows the induced velocity calculated by (6.18) in an experiment that was not used for the determination of each parameter of (6.19). Equation (6.19) accurately approximates the induced velocity. We think that our proposed induced velocity is valid.

6.4.5 Simulation

Figure 6.11 shows the simulated results using the derived nonlinear model and proposed induced velocity. We input the same collective pitch and horizontal velocity as in the experimental data, shown in Fig. 6.5, into the simulation model and compare the simulated output with the measured experimental output.

Although the simulation was performed as an open loop, we obtained good simulation results. This implies that the derived nonlinear model and induced velocity are valid.

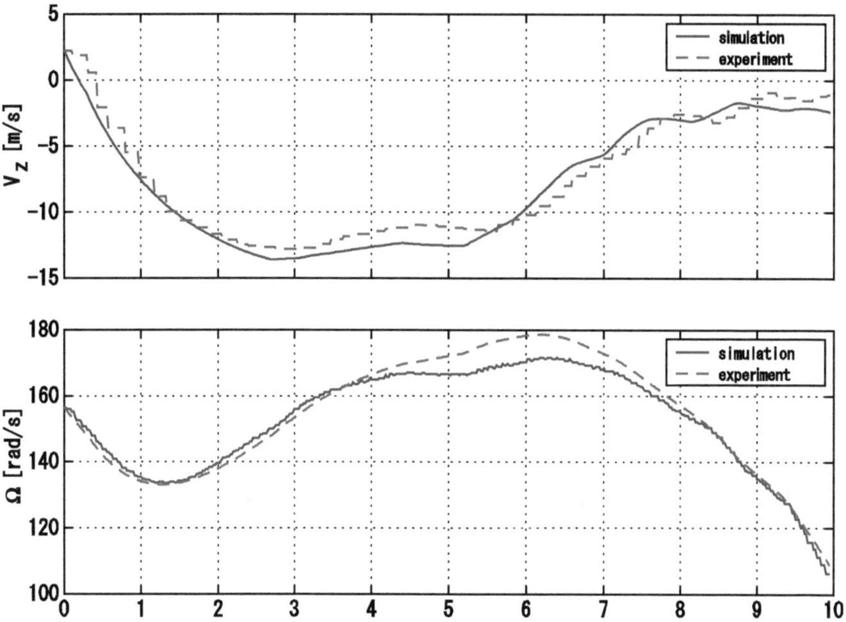

Fig. 6.11 Simulated (nonlinear model) and experimental velocity data and rotor

6.5 Experiment

To achieve autorotation landing, the final vertical velocity must be 0 m/s. To reach this, the revolution energy of the main rotor must be retained while the fuselage is descending.

In this section, as a first step of autorotation landing, we attempt to control the main rotor revolution and vertical velocity.

6.5.1 Autorotation Landing Control

Figure 6.12 shows the control structure for this experiment. After stopping the engine, the rotor revolution is controlled to be 160 rad/s. When the rotor revolution becomes larger than the reference, the controller is changed to the velocity controller. The final vertical velocity is controlled with the reference of 0 m/s.

Both controllers for the revolution and velocity are proportional integral(PI) controllers and the gains are tuned using the derived nonlinear model simulation. Figure 6.13 shows the experimental results. The final vertical velocity is -1.5 m/s. We believe this indicates that we are closer to an autonomous autorotation landing.

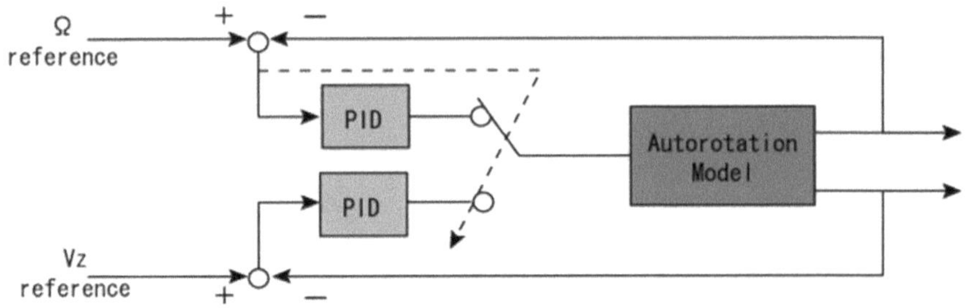

Fig. 6.12 Controller structure for autorotation landing

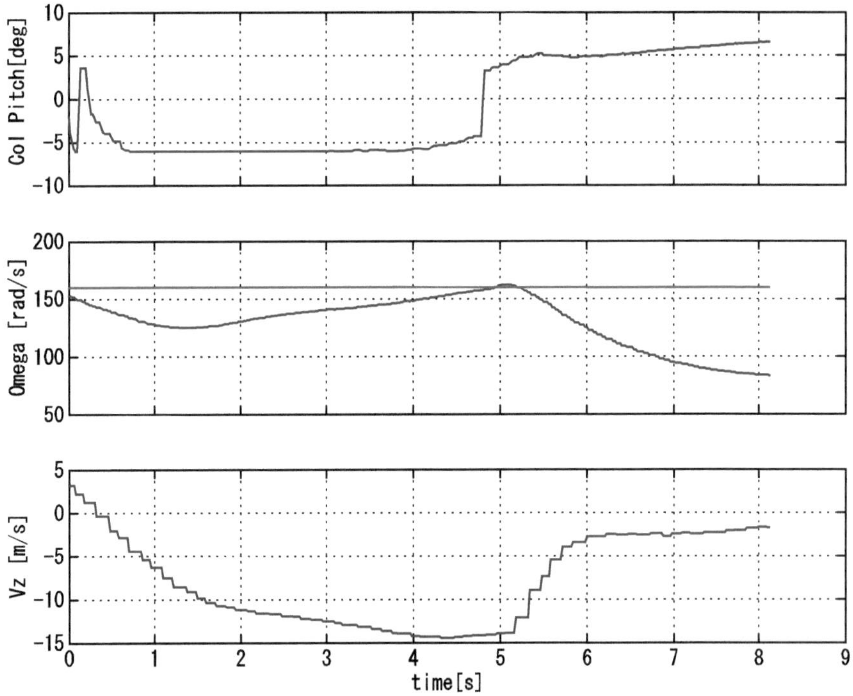

Fig. 6.13 Experimental result of autorotation landing

6.5.2 Vertical Velocity Control

Another important control subject is the control of vertical velocity during descent. We applied the same velocity controller used in the previous experiment in this experiment.

Figure 6.14 shows the experimental result. The velocity reference was 0.7 m/s. Despite using a simple PI controller, the vertical velocity is accurately controlled.

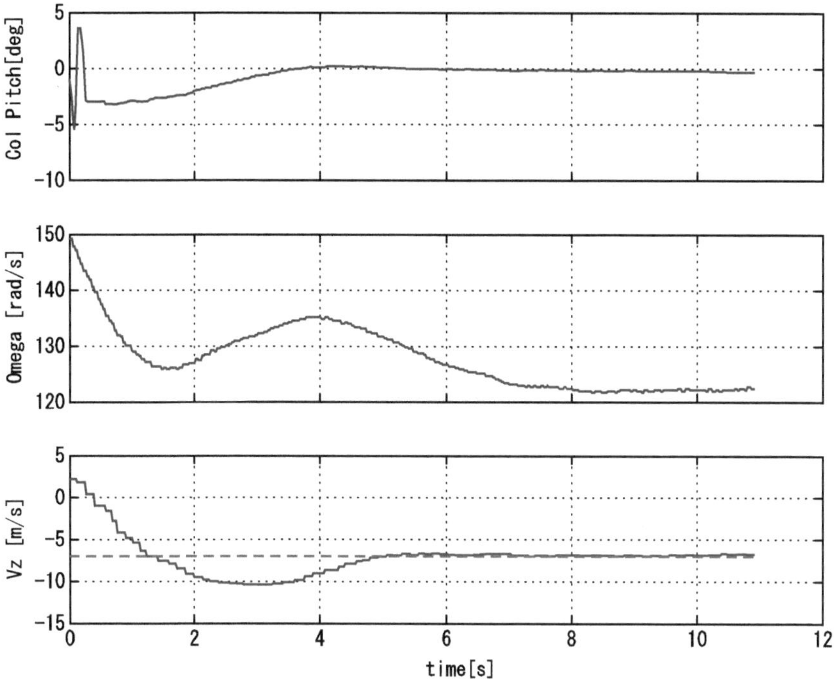

Fig. 6.14 Experimental results of descent velocity control with reference of $-7\,\mathrm{m/s}$

6.6 Linearization

As stated above, our ultimate purpose is autonomous autorotation landing. To achieve this in the near future, it is more advantageous to apply a well-developed linear control method.

Some linear control theories do not require a linear model of the controlled object, but most do. Therefore, in this section, we derive a linearized autorotation model based on the nonlinear model derived in the previous sections.

6.6.1 *Discrete State Space Model*

Any nonlinear system can be expressed by a linear model around an operation point. Therefore, we assume that the linear autorotation model is expressed by the following continuous state space model.

$$\frac{d}{dt}\begin{bmatrix}\Omega \\ V_Z \\ v_d\end{bmatrix} = \begin{bmatrix}X_\Omega & X_{VZ} & X_{vd} \\ Y_\Omega & Y_{VZ} & Y_{vd} \\ Z_\Omega & Z_{VZ} & Z_{vd}\end{bmatrix}\begin{bmatrix}\Omega \\ V_Z \\ v_d\end{bmatrix} + \begin{bmatrix}X_\theta & X_{Vh} \\ Y_\theta & Y_{Vh} \\ Z_\theta & Z_{Vh}\end{bmatrix}\begin{bmatrix}\theta \\ V_h\end{bmatrix} \tag{6.21}$$

As the discrete state space, it becomes

$$\begin{bmatrix} \Omega_{(k+1)} \\ V_{Z(k+1)} \\ v_{d(k+1)} \end{bmatrix} = \begin{bmatrix} dX_{\Omega} & dX_{VZ} & dX_{vd} \\ dY_{\Omega} & dY_{VZ} & dY_{vd} \\ dZ_{\Omega} & dZ_{VZ} & dZ_{vd} \end{bmatrix} \begin{bmatrix} \Omega_{(k)} \\ V_{Z(k)} \\ v_{d(k)} \end{bmatrix} + \begin{bmatrix} dX_{\theta} & dX_{Vh} \\ dY_{\theta} & dY_{Vh} \\ dZ_{\theta} & dZ_{Vh} \end{bmatrix} \begin{bmatrix} \theta_{(k)} \\ V_{h(k)} \end{bmatrix} \tag{6.22}$$

$$\begin{bmatrix} \Delta\Omega_{(k+1)} \\ \Delta V_{Z(k+1)} \\ \Delta v_{d(k+1)} \end{bmatrix} = \left(\begin{bmatrix} dX_{\Omega} & dX_{VZ} & dX_{vd} \\ dY_{\Omega} & dY_{VZ} & dY_{vd} \\ dZ_{\Omega} & dZ_{VZ} & dZ_{vd} \end{bmatrix} - I \right) \begin{bmatrix} \Omega_{(k)} \\ V_{Z(k)} \\ v_{d(k)} \end{bmatrix} + \begin{bmatrix} dX_{\theta} & dX_{Vh} \\ dY_{\theta} & dY_{Vh} \\ dZ_{\theta} & dZ_{Vh} \end{bmatrix} \begin{bmatrix} \theta_{(k)} \\ V_{h(k)} \end{bmatrix} \tag{6.23}$$

or

$$\Delta X_{(k+1)} = \Delta A X_{(k)} + B_d u(k) \tag{6.24}$$

6.6.2 *Determination of Parameters by a Neural Network*

If we can determine the 15 parameters included in (6.23), the discrete linear model can be calculated and linearization is achieved. In aviation, in general, such parameters are determined by using the first-order term of the Taylor expansion. However, it is difficult to know the accurate parameters of the fuselage in this study and to determine the point where the Taylor expansion is implemented, because the most dominant state variable has a wide range. Therefore, we decided to linearize the model by using a neural network without any operation point.

Figure 6.15 shows the neural network structure for system identification. Since the purpose of this section is to derive a linear model, we adopt a linear function such as y = x as a characteristic function. The relationship of the input and output of the neural network is

$$y = Wu \tag{6.25}$$

$$u = \begin{bmatrix} \Omega_{(k)} & \mathrm{V}_{\mathrm{Z}(k)} & v_{\mathrm{d}(k)} & \theta_{(\mathrm{k})} & V_{h(k)} \end{bmatrix}^T \tag{6.26}$$

$$W = \left[\begin{array}{ccc|cc} w_{11} & w_{12} & w_{13} & w_{14} & w_{15} \\ w_{21} & w_{22} & w_{23} & w_{24} & w_{25} \\ w_{31} & w_{32} & w_{33} & w_{34} & w_{35} \end{array} \right] = [\Delta A \quad B_d] \tag{6.27}$$

If the left term of (6.23) is given to the neural network as a training signal, the internal weights are automatically optimized, and they become parameters in (6.23).

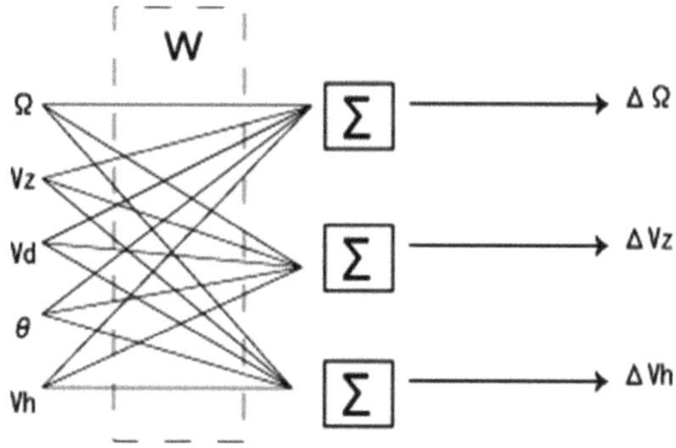

Fig. 6.15 Neural network structure for system identification

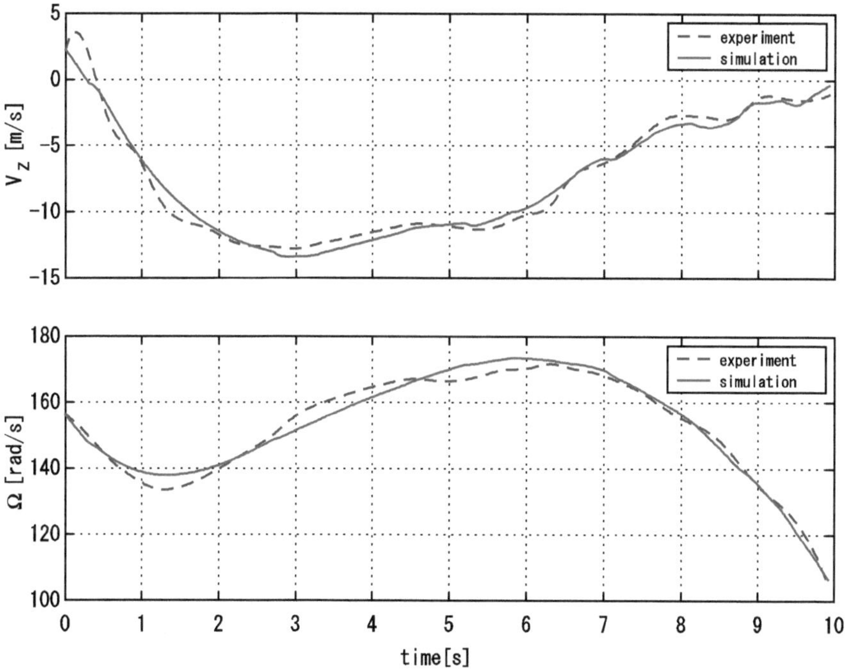

Fig. 6.16 Simulated (linear model) and experimental velocity data and rotor angular velocity

6.6.3 Simulation

Figure 6.16 shows the experimental and simulated data using the derived linear model. If we compare these two signals, the linear model output signal follows the experimental data with almost the same accuracy as Fig. 6.11 (nonlinear model).

6.7 Summary

In this chapter, because our purpose was to achieve autonomous autorotation, we derived a nonlinear model based on BET and a linear model with a neural network.

The nonlinear model was derived with a new approximation of the induced velocity at the vortex ring state. To achieve autonomous autorotation landing, we must include the ground effect in the nonlinear model. Since the ground effect can be explained by changes in the induced velocity, we believe that the derived nonlinear model can be applied to autorotation landing near the ground with a modified induced velocity for the ground effect.

We verified that it is possible to control the rotor revolution and vertical velocity in autorotation with simple PI controllers. With this experiment, we are closer to achieving autonomous autorotation landing. In autorotation landing, the vertical velocity must be 0 m/s at the ground surface. However, in this experiment, we neglected the altitude information. In the future, we will attempt to control the vertical velocity at 0 m/s at the ground surface using altitude information.

Unlike altitude information, the attitude and translational velocity controls in the autorotation maneuver are important. We developed a control system for normal flight in our previous work. However, we are not certain that the same controller works well in autorotation. Next, we will apply the controllers to actual autorotation and verify their performance.

References

1. Tsutsui Y (1986) Helicopter engineering and its operation. Kantousya, Tokyo
2. Leishman JG (2000) Principles of helicopter aerodynamics. Cambridge University Press, Cambridge
3. Stepniewski WZ, Keys CN (1984) Rotary-Wing Aerodynamics Dover Publications, Inc.

Chapter 7
Autonomous Acrobatic Flight Based on Feedforward Sequence Control for Small Unmanned Helicopter

Abstract In this chapter, we describe the simple logic for the execution of an autonomous axial roll of a small unmanned helicopter using a feedforward sequence control method. To develop the feedforward input sequence, we study the maneuver logic and strategies developed by expert human pilots to execute acrobatic maneuvers. We first explain the hardware structure of our SF 40 helicopter and the avionics system used to overcome hardware limitations for executing aggressive maneuvers and then explain the simple feedforward control logic based on recorded pilot control command sequences. This approach may be useful in executing smooth high-speed maneuvers that are less time consuming as it does not require complicated nonlinear control methods.

Video Link:

Fully autonomous acrobatic flight control
http://mec2.tm.chiba-u.jp/monograph/Videos/Chapter7/1.wmv

7.1 Introduction

Small unmanned helicopters have become an increasingly important research topic due to their advantages over other types of flying vehicles, such as their vertical takeoff and landing capability, ability to hover, low cost, and easy handling. Due to these advantages, small unmanned helicopters can be applied in many tasks such as the inspection of power lines, disaster management, and aerial mapping.

In a recent study, our research group successfully demonstrated fully autonomous hovering control, trajectory tracking control, and height control based landing, and takeoff control. These abilities were achieved using simple model based LQ control theory. Because of the closed loop structure and hardware limitations, helicopters can perform only slow maneuvers. Therefore, these maneuvers are very modest as compared to the aggressive maneuvers executed by experienced pilots such as barrel rolls, hammerheads, spilt-S, and inverted flight.

If autonomous aggressive control could be achieved, there could be much wider application of autonomous helicopters, such as faster disaster management and

K. Nonami et al., *Autonomous Flying Robots: Unmanned Aerial Vehicles and Micro Aerial Vehicles*, DOI 10.1007/978-4-431-53856-1_7, © Springer 2010

military applications. MIT has carried out a research on the autonomous acrobatic flight control of small unmanned helicopters [1], using a complicated nonlinear helicopter dynamic model for angular velocity based tracking control to execute autonomous acrobatic maneuvers.

In this chapter, we first explain the hardware structure of our SF 40 helicopter and the avionics system used to overcome hardware limitations and then explain the simple control logic based on recorded pilot control command sequences. Here, we first consider executing an autonomous axial roll maneuver as our first step toward achieving autonomous acrobatic flight. We describe all flight test data recorded during repeated flight tests of an axial roll maneuver conducted by an expert pilot. From this, we illustrate the repeatability of the pilot's input sequences. Since these control sequences were developed by experienced pilots, the process of learning control strategies from expert pilots may be taken as a closed loop system identification problem. Using the recorded flight test data, we determine the helicopter aileron, collective, elevator, and rudder control input sequences to execute an axial roll. Then, we explain our simple execution logic for executing an autonomous axial roll and present simulation results In addition, we describe our executed autonomous axial roll flight. Finally, we compare the simulated autonomous roll with the axial roll manually executed by an experienced pilot to demonstrate the accuracy of our method.

7.2 Hardware Setup

The fully instrumented HIROBO SF40 helicopter is shown in Fig. 7.1. The SF40 is a hobby-class helicopter that has a 40 cc gasoline powered two-stroke engine, a 1.8 m diameter main rotor with a tuned muffler to enhance the engine output, and a full payload capacity of approximately 17.5 kg. To enhance the helicopter

Fig. 7.1 Instrumented SF40 for acrobatic flight

agility, we limited the mass of the developed avionics system to be less than 2 kg and the fully instrumented SF40 to be less than 12 kg. Our avionics system includes a Crossbow three-axis angular sensor, an onboard Hitachi SH2 computer, a modem, a power supply unit, and a GPS (only for data logging purposes). By distributing the avionics system into two parts, we could reduce the size of the skids and reduce the air resistance during high-speed flight.

Data from the sensors on the helicopter are collected by using the SH2 computer and using a data transfer protocol, sent to a ground computer via a wireless modem. Flight data are simultaneously stored and displayed. Calculated control commands (rotational angles of the servomotors) are converted into pulse position modulation (PPM) signals by using a pulse generator developed by our group and sent to the servomotors through an RC transmitter.

7.3 Manual Maneuver Identification

As the first step toward achieving autonomous acrobatic flight, we asked our pilot to execute several 360° axial rolls with our fully instrumented SF40 and collected a series of input and sensor output data sequences. Two sets of collected data sequences are shown in Figs. 7.2 and 7.3. By analyzing the data, we found that aileron and

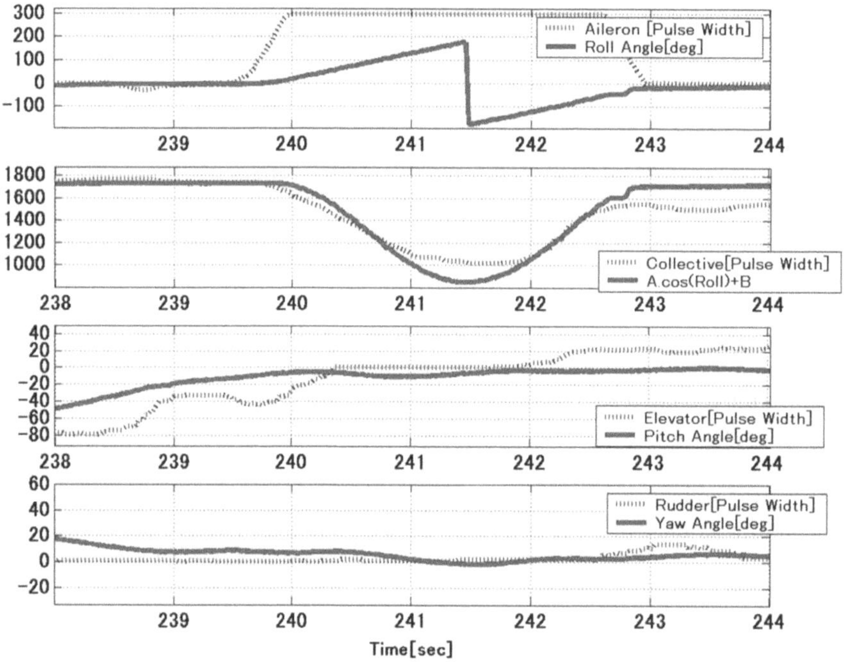

Fig. 7.2 Recorded state trajectories during a manual axial roll (experiment no. 1)

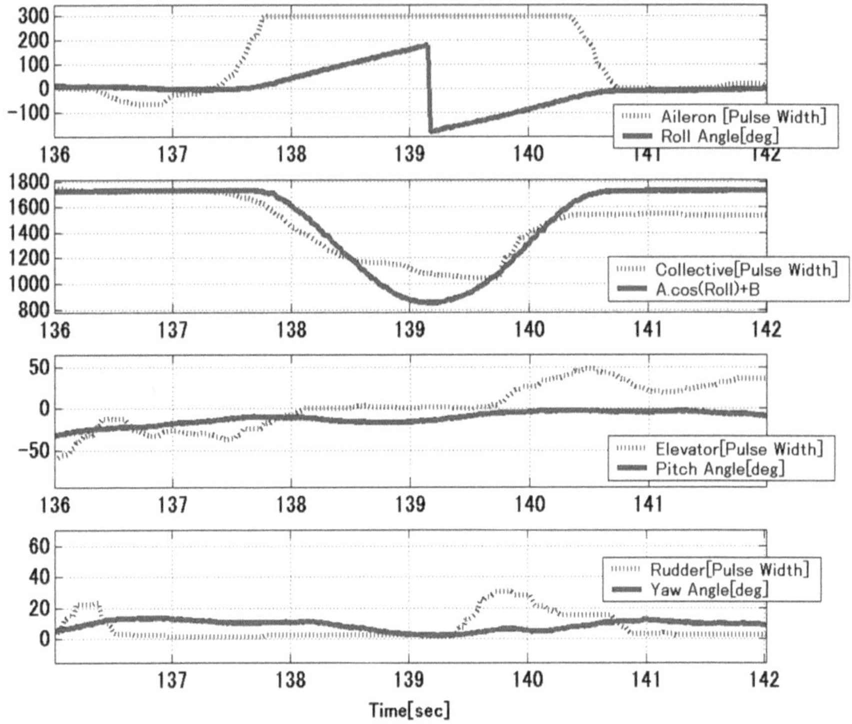

Fig. 7.3 Recorded state trajectories during manual axial roll (experiment no. 2)

collective input sequences were most important for executing an axial roll maneuver. The elevator and rudder inputs were almost constant during the maneuver. In addition, by analyzing different data sets (such as those in Figs. 7.2 and 7.3), we found that the pilot's execution sequence is repeatable mainly in terms of the aileron and collective inputs.

The aileron input sequence is initially zero so that the helicopter roll angle was close to zero, then ramps up to full deflection, remains at full deflection until the helicopter reaches a specific roll angle, and finally ramps down to zero deflection to stabilize the roll angle.

Before entering the roll maneuver, the pilot maintains the collective input higher than that in natural hovering, then ramps it down as the roll angle increases, reaches the inverted collective input when the roll angle reaches 180°, ramps up the collective input with the roll angle reaching 0°, and finally reaches the natural hovering collective input. The collective input sequence illustrates the strong correlation between the collective input and the roll angle during the roll maneuver. We approximate this correlation as (7.1). Here, constants *A and B* were found by matching the collective input data with the equation. This curve matching is shown in the second diagrams of Figs. 7.2 and 7.3.

$$Collective = A\cos(Roll) + B \tag{7.1}$$

On the basis of our understanding of the repeatability of the input sequences and the problems associated with the nonlinearity and input saturation of the attitude model, we decided to implement a fully open loop feedforward sequence control method using only the roll angle as the switching feedback signal for the aileron input sequence and the feedforward collective input signal given by (7.1).

7.4 Trajectory Setting and Simulation

The feedforward aileron input trajectory developed using the recorded flight data is shown in Fig. 7.4. The aileron trajectory has five stages.

1. LQ-based roll attitude control with a roll angle reference of 0° for a period of 0.8 s to stabilize the helicopter before entering the roll (this stage is not shown in Fig. 7.4).
2. Ramping up of the signal to reach full deflection in 0.5 s.

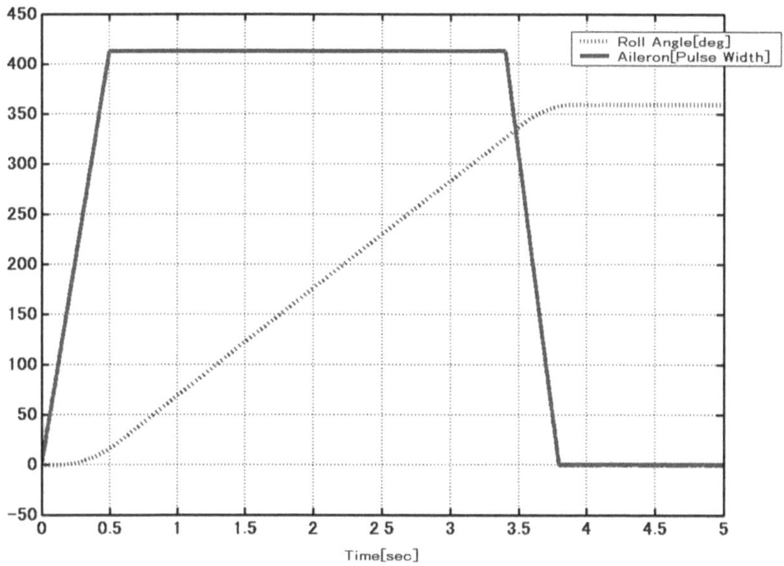

Fig. 7.4 Simulation results for the designed aileron input trajectory

3. Constant full deflection aileron input until the helicopter reaches 326° (as determined by the attitude sensor reading of −34°).
4. Ramping down of the signal to zero aileron input from the full deflection in 0.4 s, at which time the aileron input exits from the feed forward sequence. (If the roll angle reaches 0° within 0.4 s, the aileron input becomes zero and exits from the sequence.)

The roll angle simulated using a fifth-order attitude model for the feedforward aileron trajectory is shown in Fig. 7.4 (indicated by the dashed line). This roll attitude model is described by (7.2) [2]. Here, the aileron pulse width is the input signal and the helicopter roll angle is the output. In this model, we have considered the second-order servomotor approximated dynamics and the time delay of the entire control system.

$$G_{roll} = e^{-Ls} \frac{K_\theta\, \omega_{ns}^2}{(s^2 + 2\varsigma_s\, \omega_{ns}\, S + \omega_{ns}^2)(T_\theta\, S + 1)S} \tag{7.2}$$

From the simulation, we can guarantee the suitability of our trajectory as the roll angle reaches 0° after rotating 360° as expected at the end of the trajectory.

7.5 Execution Logic and Experiment

Feedforward input trajectories for each of the aileron, collective, elevator, and rudder channels and the respective state trajectories during a fully autonomous 360° axial roll are shown in Fig. 7.5. The maneuver is executed in the following steps.

1. The pilot manually brings the helicopter to a constant speed and close to zero attitude at a safe height.
2. Then, he/she switches on the RC transmitter to enter a fully autonomous axial roll maneuver. In the first 0.8 s, the helicopter is stabilized by LQ-based roll attitude control with a roll angle reference of 0°. The feedforward aileron input is then given as described in the previous section. The collective input is given according to (7.1), but this input is saturated at the minimum deflection of the real collective input as shown in the second diagram in Fig. 7.5. The elevator and rudder inputs are kept at zero during the maneuver and are shown in the third and fourth diagrams of Fig. 7.5. When the helicopter completes the 360° roll, the helicopter is again stabilized using LQ-based roll and pitch attitude control with roll and pitch attitude references of 0°.
3. At the completion of the maneuver, the pilot switches off the transmitter and reverts to manual control.

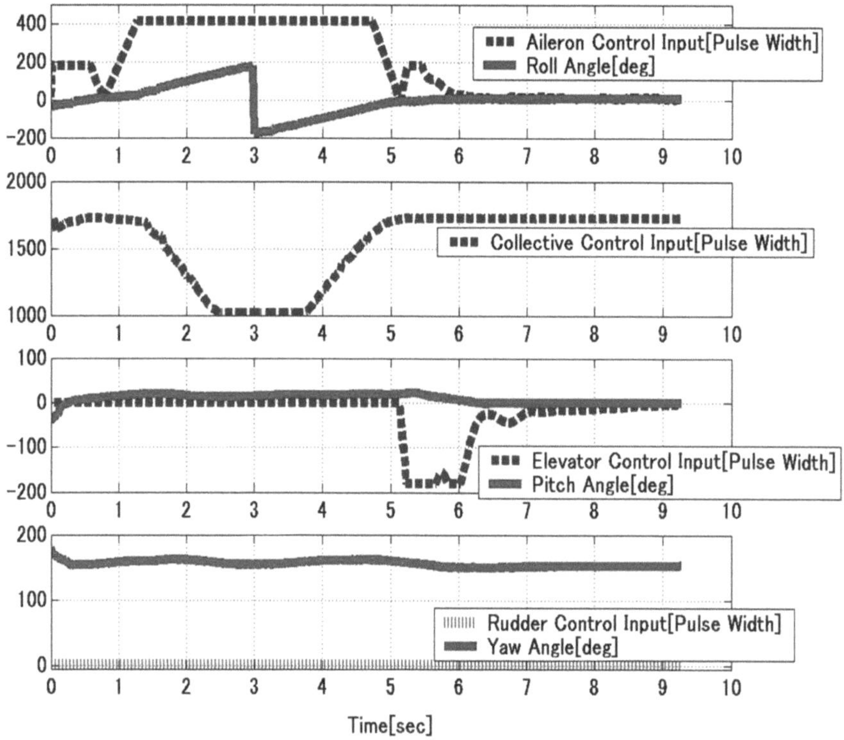

Fig. 7.5 Recorded state trajectories during a fully autonomous axial roll

The state trajectories of roll, pitch, and yaw shown in Fig. 7.5 illustrate that a smooth axial roll could be executed. In addition, entering and exiting the maneuver could also be executed smoothly by using our execution logic.

7.6 Height and Velocity During Maneuver

The height and velocity time histories (measured using the GPS device) of manual and fully autonomous maneuvers are shown in Figs. 7.6 and 7.7, respectively. The status of the GPS data is also shown in the figures. In the case of height data, GPS status of 150 implies that data can be received and GPS status of 155 implies a loss of GPS data. In the case of velocity data, a GPS status of 0 implies that data can be received and a GPS status of 5 implies a loss of GPS data. Data losses are due to the loss of satellite signals when the GPS antenna is bent beyond the allowed range.

As seen from the height data of the manual maneuver in Fig. 7.6, the helicopter falls around 25 m due to its weight and air resistance. From the height data of the fully autonomous flight in Fig. 7.7, we find that there is a height drop of less than 25 m when using our collective input logic.

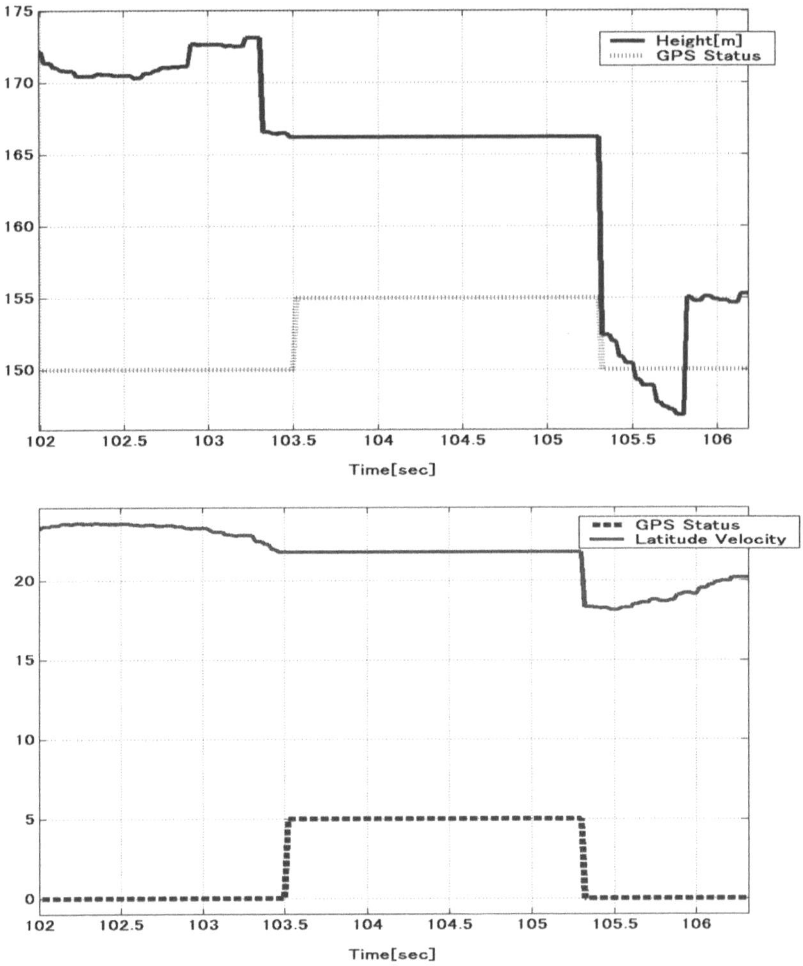

Fig. 7.6 Helicopter altitude and lateral velocity during manual axial roll

The drop in forward velocity in the autonomous maneuver is considerably greater than that in the manual maneuver. This is due to the constant elevator input applied during the entire autonomous maneuver, as the forward velocity due to the elevator is inverted when the helicopter is around its inverted position. Therefore, the constant elevator input reduces the helicopter forward velocity drastically in the inverted attitude range. To keep the forward velocity constant during the autonomous maneuver, it is necessary to define an elevator trajectory that also inverts with the roll angle.

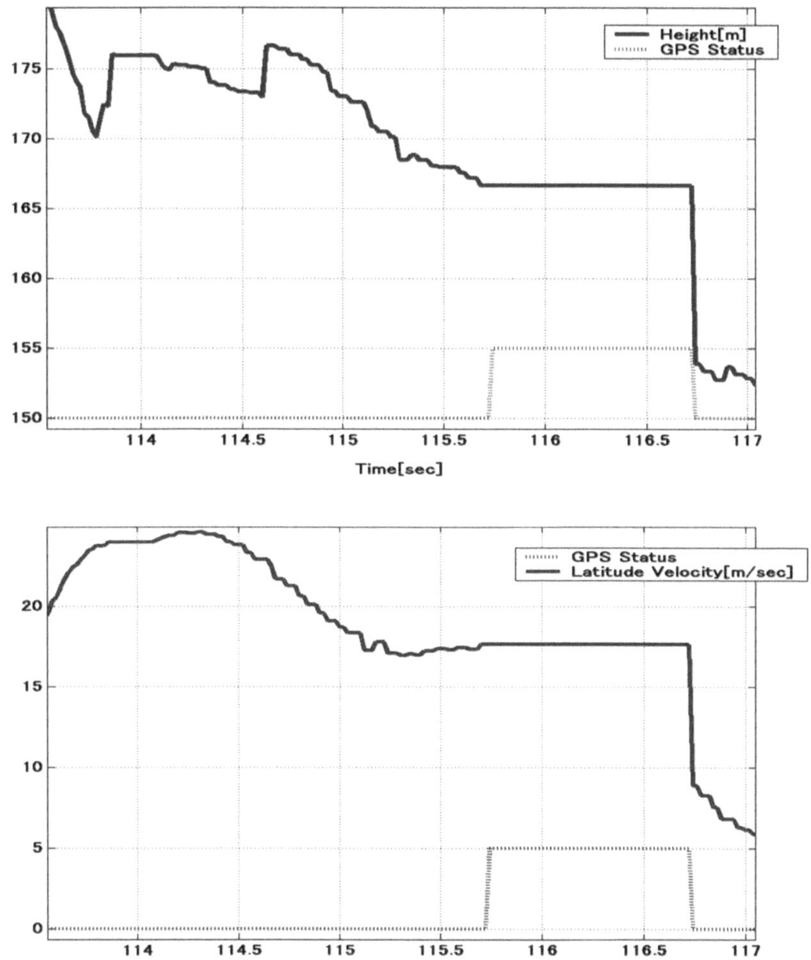

Fig. 7.7 Helicopter altitude and lateral velocity during a fully autonomous axial roll

7.7 Summary

An autonomous acrobatic axial roll maneuver was successfully executed using a complete feedforward sequence control method developed by analyzing manual maneuver data. Robustness of the maneuver is achieved by switching the input trajectory using the roll angle. This simple method is successful since acrobatic maneuvers do not require very complicated control inputs and also such maneuvers are performed in a very short period of time. To achieve complete autonomous acrobatic flight, we have to develop an autonomous high-speed forward-velocity control system and a guaranteed system to recover after a maneuver. In addition, switching between different control methods has to be considered to execute a completely smooth acrobatic maneuver.

References

1. Gavrilets V, Frazzoli E, Mettler B, Feron E (2001) Aggressive maneuver of small autonomous helicopters: a human-centered approach. Int J Robot Res 20:795–807
2. Hazawa K, Shin J, Fujiwara D, Dilshan F, Igarashi K, Nonami K (2003) Autonomous flight control of unmanned small hobby-class helicopter. J Robot Mechatron 15(5):546–554

Chapter 8
Mathematical Modeling and Nonlinear Control of VTOL Aerial Vehicles

Abstract In an effort to make autonomous flight behaviors available to mini and micro rotorcraft, an embedded and inexpensive autopilot was developed. In this chapter, we present the main steps for designing a nonlinear flight controller for mini rotorcraft Unmanned Aerial Vehicles (UAVs). The proposed control system is based on the nonlinear model of rotorcraft UAVs and uses the inner and outer-loop control scheme. It considers system's nonlinearities and coupling and results in a practical controller that is easy to implement and to tune. The asymptotic stability of the complete closed-loop system was proven by exploiting the theories of systems in cascade. In addition to controller design and stability analysis, the chapter provides information about the air vehicle, sensors integration and real-time implementation of guidance, navigation and control algorithms. A mini quadrotor UAV, equipped with the embedded autopilot, has undergone an extensive program of flight tests, resulting in various flight behaviors under autonomous control from takeoff to landing. Experimental results that demonstrate the capabilities of our autonomous UAV are presented.

Video Link:

Waypoint Navigation and Trajectory Tracking
http://mec2.tm.chiba-u.jp/monograph/Videos/Chapter8/1.wmv

8.1 Introduction

Rotorcraft Unmanned Aerial Vehicles (RUAVs) are being widely used in a multitude of applications, mostly military, as well as civilian. Miniature RUAVs offer major advantages when used for aerial surveillance, reconnaissance and inspection in complex and cluttered environments such as urban and indoor environments. Their usefulness results from their low-cost, small size, VTOL (Vertical Take-Off and Landing) capabilities and their ability to fly in very low altitudes, hover, cruise and achieve aggressive maneuvers.

K. Nonami et al., *Autonomous Flying Robots: Unmanned Aerial Vehicles and Micro Aerial Vehicles*, DOI 10.1007/978-4-431-53856-1_8, © Springer 2010

The design of autopilots for mini RUAVs has many theoretical and technical challenges. Indeed, due to their complex dynamics, nonlinearities and high degree of coupling among control inputs and state variables, the design of reliable and robust controllers is a challenge. Furthermore, the limited payload of miniature flyers imposes severe constraint on the selection of navigation sensors and onboard electronics. During the last decade, most research activities on miniature RUAVs have focussed on addressing these issues by deriving mathematical models of their dynamics and synthesizing controllers for basic stability. However, little work has been done on the development of fully autonomous mini RUAVs that can achieve real-world applications. Milestones in UAVs and most reported experimental results about advanced flight behaviors have been achieved using fixed-wing aircraft and small-scale helicopters that weigh several kilograms [5, 13, 26, 27, 29].

This chapter focuses on the development of an embedded autopilot that enables mini and micro RUAVs to achieve several flight behaviors like automatic takeoff and landing, hovering, waypoint navigation, trajectory tracking, etc. This is accomplished by selecting carefully lightweight avionics components that fit the limited payload of miniature vehicles. To compensate for the weaknesses of the low-cost sensors, we put our efforts in designing a reliable model-based nonlinear controller.

Conventional approaches to flight control and most initial attempts to achieve autonomous helicopter flight have been based on linear controllers design. These approaches involve simplified model derivation, model linearization about a set of pre-selected equilibrium conditions or trim points and linear controllers design like decentralized Single Input Single Output (SISO) Proportional Integral Derivative (PID) controllers [2, 10, 19], Multiple Input Multiple Output (MIMO) Linear Quadratic Regulator (LQR) controllers [11, 30], robust controllers [21], etc. The main disadvantage of linear controllers relates to overlooking coupling terms and nonlinearities, resulting in performance degradation when the aircraft moves away from a design trim point or flight mode.

In order to overcome some of the limitations and drawbacks of the previous linear approaches, a variety of nonlinear flight controllers have been developed. Among these, feedback linearization [20], model predictive control [29], dynamic inversion [25], adaptive control [13] have been successfully applied to control small-scale helicopters. Many researchers have also developed nonlinear control systems for mini quadrotor vehicles. One such system, based on the Draganflyer quadrotor airframe, has demonstrated successful hovering using nested saturation-based nonlinear controllers [3, 15]. In the X4-flyer project [7], Backstepping technique was applied to design a nonlinear controller for a miniature quadrotor vehicle. Attitude control was also demonstrated on tethered quadrotor testbeds using sliding mode control [22]. Another tethered testbed [32] used a quaternion-based nonlinear control scheme to perform attitude stabilization. This list is not exhaustive and there are many other interesting works on the control, navigation and visual servoing of mini quadrotor UAVs [1, 9], etc. The main difficulties of nonlinear controllers are related to their implementation on on-board microprocessors with limited computational power and adjustment of their parameters.

Our objective in this research is to design a practical nonlinear flight controller that has the following characteristics:

- It is based on the mathematical model of RUAV and considers system's nonlinearities and coupling between control and state variables.
- It is a multipurpose controller that can handle different flight modes like hovering, flying forward, flying sideward, takeoff and landing, trajectory tracking, etc.
- It guarantees the asymptotic stability of the entire closed-loop system.
- It is simple to implement on on-board microprocessors.
- It is easy to tune and to adjust its parameters on-line even when the plant parameters are not well-known.
- It results in good flight performance.

In order to design a flight controller with such characteristics, we have derived a mathematical model for mini RUAV dynamics and exploited its structural properties to transform it into two cascaded subsystems coupled by an interconnection term. Partially passivation design has been used to synthesize control laws for each subsystem, thereby resulting in an outer-loop with slow dynamics which controls the position and an inner-loop with fast dynamics which controls the orientation. The asymptotic stability of the entire connected system is proven by exploiting the theories of systems in cascade. The resulting nonlinear controller is thus easy to implement and to tune and it guaranties the asymptotic stability of the closed-loop system.

In addition to flight controller design and stability analysis, the chapter provides a comprehensive description of the autopilot hardware, guidance, navigation and control algorithms implementation as well as performance evaluation through real-time experiments using a mini quadrotor platform. Figure 8.1 shows our quadrotor aerial vehicle, equipped with the developed autopilot, during autonomous flight.

8.2 Dynamic Model of Small and Mini VTOL UAVs

Controller design for rotorcraft or VTOL UAVs and Micro Air Vehicles (MAVs) is a complex matter and typically requires the availability of the mathematical model of its dynamics. Indeed, most control design techniques require at least availability of a system model describing plant dynamics. Highly complex models that include actuators and rotors dynamics, fuselage aerodynamics and propellers flapping are not practical for most advanced control design methods. The first step towards dynamic modelling of a rotorcraft is to consider it as a rigid body with six degrees of freedom (DOF) that incorporates a mechanism for generating the necessary forces and moments. The resulted simplified model is sufficient for controller design for mini and micro RUAVs where rotors dynamics and other parasitic forces and moments can be neglected.

Fig. 8.1 Our rotorcraft MAV during autonomous flight

8.2.1 Rigid Body Dynamics

Modeling the rigid body dynamics aims at finding the differential equations that relate system outputs (position and orientation) to its inputs (force and torque vectors). The equations of motion for a rigid body of mass $m \in \mathbb{R}$ and inertia matrix $J \in \mathbb{R}^{3\times 3}$ subject to external force vector $F \in \mathbb{R}^3$ and torque vector $\Gamma^b \in \mathbb{R}^3$ are given by the following Newton–Euler equations, expressed in the body-fixed reference frame $\mathscr{B}$ [6]

$$\begin{cases} m\dot{V} + \Omega \times mV = F \\ J\dot{\Omega} + \Omega \times J\Omega = \Gamma^b \end{cases} \tag{8.1}$$

where $V = (u, v, w)$ and $\Omega = (p, q, r)$ are, respectively, the linear and angular velocities in the body-fixed reference frame. The translational force F combines gravity, main thrust and other body forces components.

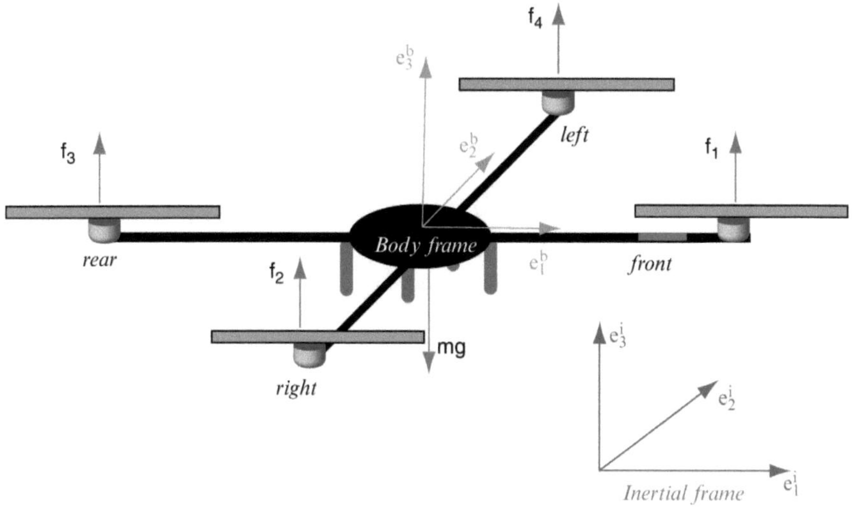

Fig. 8.2 Rigid body dynamics and the associated frames

Using Euler angles parametrization and "ZYX" convention, the airframe orientation in space is given by a rotation matrix R from $\mathscr{B}$ to the inertial reference frame $\mathscr{I}$, where $R \in SO3$ is expressed as follows:

$$\begin{aligned} R &= R_\psi R_\theta R_\phi \\ &= \begin{pmatrix} c\theta\, c\psi & s\phi\, s\theta\, c\psi - c\phi\, s\psi & c\phi\, s\theta\, c\psi + s\phi\, s\psi \\ c\theta\, s\psi & s\phi\, s\theta\, s\psi + c\phi\, c\psi & c\phi\, s\theta\, s\psi - s\phi\, c\psi \\ -s\theta & s\phi\, c\theta & c\phi\, c\theta \end{pmatrix} \end{aligned} \tag{8.2}$$

where $\eta = (\phi, \theta, \psi)$ denotes the vector of three Euler angles and $s.$ and $c.$ are abbreviations for $sin(.)$ and $cos(.)$.

By considering this transformation between the body-fixed reference frame $\mathscr{B}$ and the inertial reference frame $\mathscr{I}$ as seen in Fig. 8.2, it is possible to separate the gravitational force from other forces and write the translational dynamics in $\mathscr{I}$ as follows:

$$\begin{aligned} \dot{\xi} &= \upsilon \\ m\dot{\upsilon} &= RF^b - mge_3^i \end{aligned} \tag{8.3}$$

where $\xi = (x, y, z)$ and $\upsilon = (\dot{x}, \dot{y}, \dot{z})$ are the rotorcraft position and velocity in $\mathscr{I}$. In the above notation, g is the gravitational acceleration and F^b is the resulting force vector in $\mathscr{B}$ (excluding the gravity force) acting on the airframe.

In the literature, different models have been proposed to represent the rotational dynamics like those based on quaternion, rotation matrix R or Euler angles. In the following, we will use Euler angles parametrization to express the rotational dynamics in an appropriate form for control design.

Let us first, write the kinematic relation between Ω and $\dot{\eta}$.

$$\dot{\eta} = \Phi(\eta)\Omega \tag{8.4}$$

where the Euler matrix $\Phi(\eta)$ is given by

$$\Phi(\eta) = \begin{pmatrix} 1 & \sin(\phi)\tan(\theta) & \cos(\phi)\tan(\theta) \\ 0 & \cos(\phi) & -\sin(\theta) \\ 0 & \sin(\phi)\sec(\theta) & \cos(\phi)\sec(\theta) \end{pmatrix} \tag{8.5}$$

It is important to note that the matrix Φ has a singularity at $\theta = \pm\pi/2$, and its inverse matrix $\Psi(\eta) = \Phi^{-1}(\eta)$ is given by

$$\Psi(\eta) = \begin{pmatrix} 1 & 0 & -\sin\theta \\ 0 & \cos\phi & \cos\theta\sin\phi \\ 0 & -\sin\phi & \cos\theta\cos\phi \end{pmatrix} \tag{8.6}$$

By differentiating (8.4) with respect to time, and recalling the second equation of (8.1), we write

$$\ddot{\eta} = \dot{\Phi}\Omega + \Phi\dot{\Omega} = \dot{\Phi}\Psi\dot{\eta} - \Phi J^{-1} sk(\Omega) J\Omega + \Phi J^{-1}\Gamma^b$$

The sk operation is defined here from $\mathbb{R}^3$ to $\mathbb{R}^{3\times3}$ such that $sk(x)$ is a skew-symmetric matrix associated to the vector product $sk(x)y := x \times y$ for any vector $y \in \mathbb{R}^3$.

By multiplying both sides of the last equation by $M(\eta) = \Psi(\eta)^T J \Psi(\eta)$, we obtain

$$M(\eta)\ddot{\eta} + C(\eta, \dot{\eta})\dot{\eta} = \Psi(\eta)\Gamma^b \tag{8.7}$$

with $M(\eta)$ is a well defined positive inertia matrix, provided that $\theta \neq k\pi/2$. The *Coriolis* term $C(\eta, \dot{\eta})$ is given by[1]

$$C(\eta, \dot{\eta}) = -\Psi(\eta)^T J \dot{\Psi(\eta)} + \Psi(\eta)^T sk(\Psi(\eta)\dot{\eta}) J \Psi(\eta)$$

Thus, the nonlinear model of the rotorcraft, used for flight controller design, is

$$\begin{cases} m\ddot{\xi} = RF^b - mge_3^i \\ M(\eta)\ddot{\eta} + C(\eta, \dot{\eta})\dot{\eta} = \Psi(\eta)^T \Gamma^b \end{cases} \tag{8.8}$$

The expressions of F^b and Γ^b are detailed in the next section for some classical VTOL UAVs configurations.

[1] We have exploited the fact that: $\dot{\Psi} = \Psi\dot{\Phi}\Psi$.

8.2.2 Aerodynamics Forces and Torques

An aerial robot should include a mechanism for generating forces and torques that are required to control its motion. For RUAVs, this mechanism is mainly based on rotors and propellers that produce thrust perpendicular to rotors rotation plane. The main thrust is generated along the body vertical axis and it is used to compensate the gravity force and to control the vertical movement. The horizontal movements along the x and y-axes are controlled by directing the force or thrust vector in the appropriate direction (thrust vectoring control), thereby resulting in longitudinal and lateral force components. The control torques are thus used to control the aircraft body orientation which controls the rotorcraft horizontal movement. These torques are generated in different ways according to the rotorcraft configuration and type. In the case of conventional helicopters, control torques are obtained by using a tail rotor and swash-plates to change the angle of the main rotor blades. Two coaxial rotors with cyclic-pitch propellers is also a common mechanism for generating force and torque vectors in small-scale RUAVs. There exist also other multirotor-based mechanisms with fixed-pitch propellers like two tilting rotors, three rotors, four rotors, etc. Each mechanism has its advantages and drawbacks especially in terms of mechanical complexity, maneuverability, aerodynamics and power efficiency, complexity for control design, etc.

However, most of VTOL aerial vehicles are under-actuated mechanical systems with six DOF and four main control inputs. Indeed, many small-scale rotorcraft UAVs can be characterized by three main control torques $\tau = (\tau_\phi, \tau_\theta, \tau_\psi)^T$ and one main control force $F^b = (0, 0, u)^T$. In this representation, we neglect the parasitic forces and moments that are induced by control inputs coupling and other small body forces and moments like aerodynamic effects, rotors dynamics, gyroscopic effects, etc. There is a considerable published research in detailed modeling of different rotorcraft configurations and associated aerodynamics forces and moments. In this chapter, we focus on the standard model that describes the main control forces and moments because:

- For mini RUAVs, the parasitic and small body forces and moments are dominated by the main force and torque vectors, thereby resulting in small and bounded effects on the rotorcraft dynamics.
- The complexity of a model essentially depends on the expression of aerodynamic forces and moments.

In the case of a quadrotor helicopter for example, the control and torque vectors are obtained in an easy way. It features a simple mechanism where the four control inputs $(u, \tau_\theta, \tau_\phi, \tau_\psi)$ are obtained by independently controlling the rotation speed of each motor. The collective lift u is the sum of the thrusts generated by the four propellers

$$u = \sum_{i=1}^{4} f_i \tag{8.9}$$

The airframe torques generated by the rotors are given by [4]

$$\tau_\phi = l(f_2 - f_4) \tag{8.10}$$
$$\tau_\theta = l(f_3 - f_1) \tag{8.11}$$
$$\tau_\psi = Q_1 + Q_3 - Q_2 - Q_4 \tag{8.12}$$

l represents the distance from the rotors to the center of mass of the helicopter and Q_i is the fan torque due to air drag.

Propellers thrust and torque are generally assumed to be proportional to the square of the rotor angular velocity ω. In fact, the relations between the rotor speed w_i and the generated lift f_i and torque Q_i are very complex [23,24]. Therefore, the algebraic model for generating the force and control torques can be written in the following form:

$$\begin{pmatrix} u \\ \tau_\phi \\ \tau_\theta \\ \tau_\psi \end{pmatrix} = \begin{pmatrix} \rho & \rho & \rho & \rho \\ 0 & -l\rho & 0 & l\rho \\ -l\rho & 0 & l\rho & 0 \\ \kappa & -\kappa & \kappa & -\kappa \end{pmatrix} \begin{pmatrix} w_1^2 \\ w_2^2 \\ w_3^2 \\ w_4^2 \end{pmatrix} \tag{8.13}$$

where (μ, κ) are positive constants characterizing the propellers aerodynamics. The expressions in (8.13) are valid approximations that are used in cases of hovering and low-speed displacements.

The dynamical model, considered for control design and that can represent a large range of RUAV configurations is given by the following equations:

$$\begin{cases} m\ddot{\xi} = uRe_3^i - mge_3^i \\ M(\eta)\ddot{\eta} + C(\eta,\dot{\eta})\dot{\eta} = \Psi(\eta)^T \tau \end{cases} \tag{8.14}$$

8.3 Nonlinear Hierarchical Flight Controller: Design and Stability

This section describes the design of a flight controller which is based on the nonlinear model of the rotorcraft, developed in Sect. 8.2. Our objective is to design a 3D flight controller that performs well in practice as well as in theory. Indeed, a control system is required to be easy to implement and to tune while guaranteeing good flight performance. Furthermore, it is important to analyze the stability of the closed-loop system. To reach this goal, we have separated the rotorcraft model into two connected subsystems by exploiting the structural properties of rotorcraft UAVs model. The inner-loop with fast dynamics which performs attitude tracking and generates the required torques. The outer-loop with slow dynamics is used to generate the thrust and the reference angles, required to follow a commanded translational trajectory. The asymptotic stability of the complete connected system is proven by exploiting the theories of systems in cascade.

8.3.1 Flight Controller Design

Controller design for nonlinear systems subject to strong coupling offers both practical significance and theoretical challenges. In this chapter, the control design for miniature rotorcraft UAVs is addressed by transforming the nonlinear model (8.14) into two linear sub-systems coupled by a nonlinear interconnection term.

Since the attitude dynamics in (8.14) is a fully-actuated mechanical system for $\theta \neq k\pi/2$, then it is exact feedback linearizable. In fact, by considering the following change of variables

$$\tau = J\Psi(\eta)\tilde{\tau} + \Psi^{-1}C(\eta, \dot{\eta})\dot{\eta} \tag{8.15}$$

and recalling the expression of R, the system (8.14) can be written in the following form:

$$\begin{cases} \ddot{x} = \dfrac{1}{m}u(\cos\phi \sin\theta \cos\psi + \sin\phi \sin\psi), & \ddot{\phi} = \tilde{\tau}_\phi \\ \ddot{y} = \dfrac{1}{m}u(\cos\phi \sin\theta \sin\psi - \sin\phi \cos\psi), & \ddot{\theta} = \tilde{\tau}_\theta \\ \ddot{z} = \dfrac{1}{m}u\cos\theta\cos\phi - g, & \ddot{\psi} = \tilde{\tau}_\psi \end{cases} \tag{8.16}$$

Now, we apply the backstepping principle in order to transform system (8.16) into two cascaded subsystems. In contrast to the complexity of standard backstepping approaches, the control strategy considered here is simple and easy to implement.

Let us first, define a virtual control vector $\mu \in \mathbb{R}^3$ as follows:

$$\mu = \mathfrak{f}(u, \phi_d, \theta_d, \psi_d) = \frac{1}{m}uR(\phi_d, \theta_d, \psi_d)e_3^i - ge_3^i \tag{8.17}$$

where $\mathfrak{f}(.) : \mathbb{R}^3 \to \mathbb{R}^3$ is a continuous invertible function. Physically, the control vector μ corresponds to the desired force vector. Its magnitude is the total thrust u produced by propellers, and its orientation is defined by the body attitude (ϕ, θ, ψ). ϕ_d, θ_d and ψ_d in (8.17) are thus, the desired roll pitch and yaw angles.

By recalling (8.17), the components of μ are given by

$$\begin{cases} \mu_x = \dfrac{1}{m}u(\cos\phi_d \sin\theta_d \cos\psi_d + \sin\phi_d \sin\psi_d) \\ \mu_y = \dfrac{1}{m}u(\cos\phi_d \sin\theta_d \sin\psi_d - \sin\phi_d \cos\psi_d) \\ \mu_z = \dfrac{1}{m}u\cos\theta_d\cos\phi_d - g \end{cases} \tag{8.18}$$

(μ_x, μ_y, μ_z) are the force vector components along x–y–z axes that are needed for tracking some reference trajectory. These desired control inputs are computed by the outer-loop controller. They are then, used to compute the desired force vector magnitude and desired attitude,[2] $(u, \phi_d, \theta_d) = \mathfrak{f}^{-1}(\mu_x, \mu_y, \mu_z)$, that is

$$\begin{cases} u = m\sqrt{\mu_x^2 + \mu_y^2 + (\mu_z + g)^2} \\ \phi_d = \sin^{-1}\left(m\dfrac{\mu_x \sin\psi_d - \mu_y \cos\psi_d}{u}\right) \\ \theta_d = \tan^{-1}\left(\dfrac{\mu_x \cos\psi_d + \mu_y \sin\psi_d}{\mu_z + g}\right) \end{cases} \tag{8.19}$$

Since the desired angles $(\phi_d, \theta_d, \psi_d)$ are the outputs of the orientation subsystem, they can not be assigned or provided instantaneously. They are thus, considered as reference trajectories for the inner-loop controller. Therefore, we define the following tracking error vector $e = (e_\eta, e_{\dot\eta})^T \in R^6$ such that $e_\eta = \eta - \eta_d$ and $e_{\dot\eta} = \dot\eta - \dot\eta_d$.

Now, by replacing (ϕ, θ, ψ) in (8.16) by $(\phi_d + e_\phi, \theta_d + e_\theta, \psi_d + e_\psi)$, and exploiting some useful relations between trigonometric functions such as

$$\begin{cases} \sin(a+b) = \sin(a) + \sin(b/2)\cos(a+b/2) \\ \cos(a+b) = \cos(a) - \sin(b/2)\sin(a+b/2) \end{cases} \tag{8.20}$$

the rotorcraft translational dynamics (8.16) can be written in the following form:

$$\begin{cases} \ddot x = \dfrac{1}{m}u[(\cos\phi_d \sin\theta_d \cos\psi_d + \sin\phi_d \sin\psi_d) + h_x(\phi_d, \theta_d, \psi_d, e_\phi, e_\theta, e_\psi)] \\ \quad = \mu_x + \dfrac{1}{m}u\, h_x(.) \\ \ddot y = \dfrac{1}{m}u[(\cos\phi_d \sin\theta_d \sin\psi_d - \sin\phi_d \cos\psi_d) + h_y(\phi_d, \theta_d, \psi_d, e_\phi, e_\theta, e_\psi)] \\ \quad = \mu_y + \dfrac{1}{m}u\, h_y(.) \\ \ddot z = \dfrac{1}{m}u[\cos\theta_d \cos\phi_d + h_z(\phi_d, \theta_d, e_\phi, e_\theta)] - g = \mu_z + \dfrac{1}{m}u\, h_z(.) \end{cases} \tag{8.21}$$

The components of the interconnection vector $h(\phi_d, \theta_d, \psi_d, e_\phi, e_\theta, e_\psi) \in \mathbb{R}^3$ are composed of the multiplication and summation of sin(.) and cos(.) functions. Their exact expressions can be obtained in the same manner as in [14].

By defining the position–velocity tracking error $\chi = (\xi - \xi_d, \upsilon - \upsilon_d)^T \in \mathbb{R}^6$, and recalling (8.16) and (8.21), we can write

[2] The desired yaw angle ψ_d is given by the user or by some high-level guidance system.

$$\begin{cases} \dot{\chi} = \underbrace{A_1\chi + B_1(\mu - \ddot{\xi}_d)}_{f(\chi,\mu,\ddot{\xi}_d)} + \underbrace{\frac{1}{m}u\,H(\eta_d, e_\eta)}_{\Delta(u,\eta_d,e_\eta)} \\ \dot{e} = A_2 e + B_2(\tilde{\tau} - \ddot{\eta}_d) \end{cases} \tag{8.22}$$

where $H(\eta_d, e_\eta) = (0, 0, 0, h_x, h_y, h_z)^T$. The matrices $A_1 \in \mathbb{R}^{6\times6}$, $B_1 \in \mathbb{R}^{6\times3}$, $A_2 \in \mathbb{R}^{6\times6}$ and $B_2 \in \mathbb{R}^{6\times3}$ are defined below:

$$A_1 = A_2 = \begin{bmatrix} 0&0&0&1&0&0 \\ 0&0&0&0&1&0 \\ 0&0&0&0&0&1 \\ 0&0&0&0&0&0 \\ 0&0&0&0&0&0 \\ 0&0&0&0&0&0 \end{bmatrix}, \quad B_1 = B_2 = \begin{bmatrix} 0&0&0 \\ 0&0&0 \\ 0&0&0 \\ 1&0&0 \\ 0&1&0 \\ 0&0&1 \end{bmatrix} \tag{8.23}$$

The rotorcraft control problem is thus, formulated as the control of two linear subsystems which are coupled by a nonlinear term $\Delta(u, \eta_d, e_\eta)$. It can be controlled using partially or fully passivation design [28].

In partial-state feedback designs like the one used here, the two linear subsystems are controlled independently ($\mu = \alpha(\chi, \ddot{\xi}_d)$ and $\tilde{\tau} = \beta(e, \ddot{\eta}_d)$). In this case, the interconnection term $\Delta(u, \eta_d, e_\eta)$ acts as a disturbance on the χ-subsystem which must be driven to zero. Such partial-state feedback designs are of interest because of their simplicity especially for implementation on physical systems. However, it is difficult to prove the stability property for the entire connected system.

In full-state feedback designs, the interconnection term plays an active role in controlling the χ-subsystem. In this case, the error vector "e" in (8.22) is treated as an additional input for the χ-subsystem. This results in a relatively simple stability analysis, but at the expense of the control law complexity.

In this research, we consider the partial-state feedback design. The control objective is thus, to synthesize the control laws $\mu(\chi, \ddot{\xi}_d)$ and $\tilde{\tau}(e, \ddot{\eta}_d)$ such that the tracking errors χ and e will asymptotically converge to zero.

To synthesize the control laws $\mu = \alpha(\chi, \ddot{\xi}_d)$ and $\tilde{\tau} = \beta(e, \ddot{\eta}_d)$, we will use the following Theorem [31]:

Theorem 8.1. *If there is a feedback $\mu = \alpha(\chi, \ddot{\xi}_d)$ such that $\chi = 0$ is an asymptotically stable equilibrium of $\dot{\chi} = f(\chi, \alpha(\chi, \ddot{\xi}_d), \ddot{\xi}_d)$, then any partial state feedback control $\tilde{\tau} = \beta(e, \ddot{\eta}_d)$ which renders the e-subsystem equilibrium $e = 0$ asymptotically stable, also achieves asymptotic stability of $(\chi, e) = (0, 0)$. Furthermore, if the two subsystems are both Globally Asymptotically Stable (GAS), then, as $t \to \infty$, every solution $(\chi(t), e(t))$ either converges to $(\chi, e) = (0, 0)$ (GAS) or is unbounded.*

Therefore, the stability of the connected system (8.22) will be ensured if we choose stabilizing feedbacks $\mu = \alpha(\chi, \ddot{\xi}_d)$, $\tilde{\tau} = \beta(e, \ddot{\eta}_d)$ and prove that all the trajectories $(\chi(t), e(t))$ are bounded.

Since the χ- and e-subsystems in (8.22) are linear, we can use simple linear controllers such as PD or PID. Therefore, we synthesize two control laws

$$\begin{cases} \mu = -K_\chi \, \chi + \ddot{\xi}_d, \ K_\chi \in \mathbb{R}^{3\times 6} \\ \tilde{\tau} = -K_e \, e + \ddot{\eta}_d, \ K_e \in \mathbb{R}^{3\times 6} \end{cases} \tag{8.24}$$

such that the matrices $A_\chi = A_1 - B_1 K_\chi$ and $A_e = A_2 - B_2 K_e$ are Hurwitz.

By substituting (8.24) into (8.22), the closed-loop system dynamics is given by

$$\begin{cases} \dot{\chi} = A_\chi \, \chi + \Delta(\chi, e_\eta) \\ \dot{e} = A_e \, e \end{cases} \tag{8.25}$$

Although A_χ and A_e are Hurwitz, the global asymptotic stability of the closed-loop system (8.25) can not be directly deduced because of the interconnection term $\Delta(\chi, e_\eta)$.

8.3.2 *Stability Analysis of the Complete Closed-Loop System*

In [16] and [14], we have proposed a theorem that renders the stability analysis of a class of connected systems less complex. That theorem is recalled below and its proof can be found in [14] and [28].

Theorem 8.2. *Let $\tilde{\tau} = \beta(e, \ddot{\eta}_d)$ be any C^1 partial-state feedback such that the equilibrium point $e = 0$ is GAS and LES. Suppose that there exist a positive constant c_1 and one class-$\mathcal{K}$ function $\gamma(.)$, differentiable at $e = 0$, such that*

$$\|\chi\| \geq c_1 \Rightarrow \|\Delta(\chi, e_\eta)\| \leq \gamma(\|e_\eta\|) \, \|\chi\| \tag{8.26}$$

If there exist a positive semi-definite radially unbounded function $V(\chi)$ and positive constants c_2 and c_3 such that for $\|\chi\| \geq c_2$

$$\begin{cases} \dfrac{\partial V}{\partial \chi} f(\chi, \alpha(\chi, \ddot{\xi}_d), \ddot{\xi}_d) \leq 0 \\ \|\dfrac{\partial V}{\partial \chi}\| \, \|\chi\| \leq c_3 V(\chi) \end{cases} \tag{8.27}$$

then, the feedback $\tilde{\tau} = \beta(e, \ddot{\eta}_d)$ guarantees the boundedness of all the solutions of (8.22). Furthermore, if $\dot{\chi} = f(\chi, \alpha(\chi, \ddot{\xi}_d), \ddot{\xi}_d)$ is GAS, then the equilibrium point $(\chi, e) = (0, 0)$ is GAS.

Here, we apply Theorem 8.2 in order to prove the global asymptotic stability of the connected closed-loop system (8.25).

Since A_χ and A_e are Hurwitz, the χ-subsystem (without the interconnection term) and the e-subsystem are GES which is stronger than the GAS property. The GES of the χ-subsystem implies that there exist a positive definite radially unbounded function $V(\chi)$ and positive constants c_2 and c_3 such that for $\|\chi\| \geq c_2$: $\frac{\partial V}{\partial \chi} A_\chi \chi \leq 0$ and $\|\frac{\partial V}{\partial \chi}\| \|\chi\| \leq c_3 V(\chi)$. Therefore, the condition (8.27) of Theorem 8.2 is satisfied.

Now, it remains to show that the interconnection term $\Delta(\chi, e_\eta)$ satisfies the growth restriction (8.26) of Theorem 8.2, and this is the main difficult part in proving the stability of a such hierarchical controller.

The norm of the interconnection term $\Delta(\chi, e_\eta)$ can be expressed as follows:

$$\|\Delta(\chi, e_\eta)\| = \frac{1}{m}|u(\chi)| \|H(\chi, e_\eta)\| = \frac{1}{m}|u(\chi)| \sqrt{h_x^2 + h_y^2 + h_z^2} \tag{8.28}$$

where

$$|u(\chi)| = m\|\mu(\chi) + g e_3^i\| = m\sqrt{\mu_x^2 + \mu_y^2 + (\mu_z + g)^2}$$

Before proving the boundedness of the interconnection term $\Delta(\chi, e_\eta)$, we need the following two lemmas:

Lemma 8.1. *Assume that the desired trajectories $\xi_d(t)$ and their time-derivatives are bounded and denote $L_d = \|\ddot{\xi}_d\|_\infty$. Then, there exist positive constants r and k_1 such that the collective thrust feedback $u(\chi)$ satisfies the following properties:*

$$|u(\chi)| \leq \begin{cases} k_1\|\chi\|, & \text{for } \|\chi\| \geq r \\ k_1 r, & \text{for } \|\chi\| < r \end{cases} \tag{8.29}$$

Lemma 8.2. *There exists a positive constant k_2 such that the coupling term $H(\chi, e_\eta)$ satisfies the following inequality:*

$$\|H(\chi, e_\eta)\| \leq k_2\|e_\eta\| \tag{8.30}$$

The proofs of Lemmas 8.1 and 8.2 are given in the Appendix.

From Lemmas 8.1 and 8.2, we can write that for $\|\chi\| \geq r$, we have

$$\|u\ H(.)\| \leq k_1\|\chi\|\ k_2\|e_\eta\| = k\|e_\eta\|\|\chi\| \tag{8.31}$$

where $k = k_1\ k_2$ is a positive constant.

Finally, we obtain the following inequality

$$\|\Delta\| = \frac{1}{m}\|u\ H\| \leq \gamma(\|e_\eta\|)\|\chi\|, \quad \text{for } \|\chi\| \geq r \tag{8.32}$$

where $\gamma(.) = \frac{k}{m}\|e_\eta\|$ is a class-$\mathcal{K}$ function differentiable at $e_\eta = 0$.

So, all the conditions of Theorem 8.2 are satisfied and the GAS of the equilibrium point $(\chi, e) = (0, 0)$ is then guaranteed.

Remark 8.1. The final control inputs $(u, \tau_\phi, \tau_\theta, \tau_\psi)$ are computed using (8.15) and (8.19) which are nonlinear and consider system's nonlinearities and coupling between state variables.

$$\begin{cases} u = m\| \mu(\chi, \ddot{\xi}_d) + g\, e_3^i\| = m\| - K_\chi \chi + \ddot{\xi}_d + g\, e_3^i\| \\ \tau = J\Psi(\eta)\tilde{\tau} + \Psi^{-1}C(\eta, \dot{\eta})\dot{\eta} = J\Psi(\eta)(-K_e e + \ddot{\eta}_d) + \Psi^{-1}C(\eta, \dot{\eta})\dot{\eta} \end{cases} \tag{8.33}$$

Note that the nonlinear attitude controller in (8.33) acts as a linear controller at near-hovering condition where the attitude angles are small. However, in contrast to linear controllers that fail to provide good performance at non-nominal conditions, the nonlinear terms in (8.33) become significant when the angles are big and participate positively to control the attitude, yielding higher tracking accuracy and robustness even at relatively big angles (see Fig. 8.7).

Remark 8.2. In practice, the implementation of the controller (8.33) requires some considerations. First, the intermediary control laws (8.24) have been slightly modified to include an integral term, thereby increasing the tracking accuracy without destroying the GAS property of the closed-loop system. Second, the reference angles $(\phi_d(t), \theta_d(t), \psi_d(t))$ are processed using second-order low-pass digital filters to reduce noise and to compute both first and second derivatives. For effective and robust control in most cases, one has also to pay attention to many other things like integral terms resetting when the vehicle is on the ground (before take-off and after landing) or also when a waypoint is reached (for the outer-loop controller).

Remark 8.3. The gain matrices (K_χ, K_e) in (8.33) were tuned empirically based on observing flight properties during experimental tests. Since the controller gains appear in the linear part of the controller, then it was easy to adjust the gains despite the absence of accurate values for the plant parameters. Table 8.1 shows the controller gains used during flight tests.

Table 8.1 shows the controller gains used during flight tests.

Table 8.1 Controller gains

Parameter	Value	Parameter	Value
kp_x, kp_y	0.8	kp_ϕ, kp_θ	28
ki_x, ki_y	0.02	ki_ϕ, ki_θ	0.5
kd_x, kd_y	1	kd_ϕ, kd_θ	1
kp_z	0.8	kp_ψ	3
ki_z	0.02	ki_ψ	0.05
kd_z	1	kd_ψ	0.2

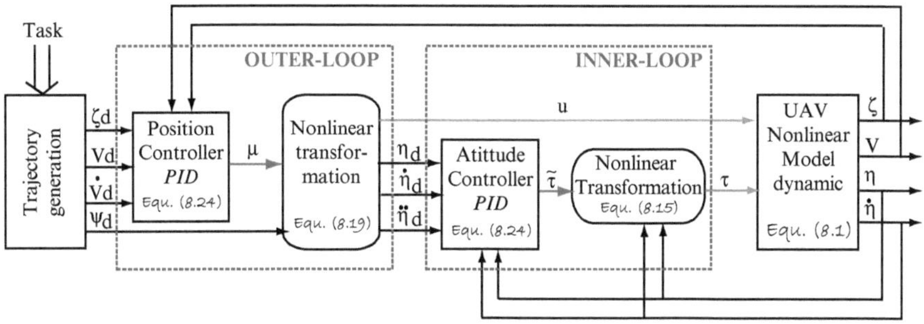

Fig. 8.3 Structure of the inner–outer loop-based controller

The block diagram of the overall controller is shown in Fig. 8.3.

8.4 UAV System Integration: Avionics and Real-Time Software

To demonstrate autonomous flight of mini rotorcraft UAVs, we have developed an experimental aerial platform including autopilot hardware and software. The choice of the air vehicle and avionics is a crucial step towards the development of an autonomous platform. That choice is mainly dependent on the intended application as well as performance, cost and weight. In this section, we will briefly describe the air vehicle and introduce the major subsystems implemented in our RUAV system and shown in Fig. 8.4.

8.4.1 Air Vehicle Description

A versatile hobby quadrotor, the X-3D-BL produced by Ascending Technologies GmbH in Germany is adopted as a vehicle platform [8]. The vehicle is 53 cm rotor-tip to rotor-tip and weights 400 g including battery. Its payload capacity is about 300 g. The propulsion system consists of four brushless motors, powered by a 2,100 mA h three-cell lithium polymer battery. The flight time in hovering is about 12 min with full payload. The complete vehicle is shown in Fig. 8.4.

The original hobby quadrotor, includes a controller board called X-3D which runs three independent PD loops at 1 kHz, one for each rotational axes (roll, pitch and yaw). Angular velocities are obtained from three gyroscopes (Murata ENC-03R) and angles are computed by integrating gyroscopes measurements. More details about the X-3D-BL platform and its internal controller can be found in [8]. In our implementation, the internal attitude controller that comes with the original platform has been disabled except gyro feedback stabilization and yaw angle control, which can not be disabled.

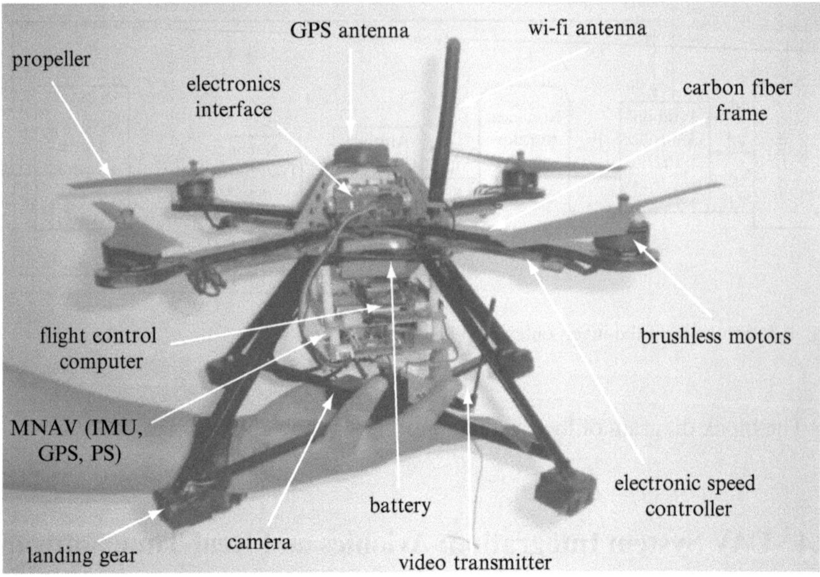

Fig. 8.4 The quadrotor micro air vehicle equipped with the developed autopilot

8.4.2 Navigation Sensors and Real-Time Architecture

Since the maximum payload is limited, it is hard to determine flight hardware with small size and light-weight but high accuracy and reliability. The hardware components that make up the basic flight avionics of our platform include a small micro-controller from *Gumstix* and the MNAV100CA sensor from *Crossbow*.

8.4.2.1 Avionics

The embedded architecture is built around the *Gumstix* (*connex 400*) micro-controller which plays the role of the Flight Control Computer (FCC). A *Gumstix* motherboard is a very small Linux OpenEmbedded computer that features a Marvell PXA255 processor. This computing unit presents many advantages for MAVs applications since it weights only 8 g despite its excellent performances. Indeed, it has a CPU clock of 400 MHz, a flash memory of 16 MB and SDRAM of 64 MB. Two expansion boards (*console-st* and *wifistix*) have been mounted on the motherboard, thereby providing two RS-232 ports and WiFi communication.

After studying and comparing different sensors configuration, we have selected the MNAV100CA sensor from *Crossbow* due to its low-cost and light-weight (about 35 g without GPS antenna). The MNAV100CA is a calibrated digital sensor system with servo drivers, designed for miniature ground and air robotic vehicle navigation and control. All sensors required for complete airframe stabilization and navigation are integrated into one compact module. It includes three-axis accelerometers, three-axis angular rate sensors, and three-axis magnetometers; static pressure (altitude) and dynamic pressure (airspeed) sensors; and a GPS receiver module.

8.4.2.2 Autopilot Architecture

When designing the embedded system, much attention has been paid to simplicity, modularity and safety. A diagram showing the interaction between the different components onboard the vehicle is shown in Fig. 8.5. All sensors data, including IMU measurements (three accelerations, three angular rates and three magnetometers readings), GPS data (latitude, longitude, altitude and three translational velocities in the North–East–Down (NED) frame), pressure sensor data (height) and finally the RC receiver decoded signals (throttle, pitching/rolling/yawing torques, switch, communication status), are sent from MNAV to the *Gumstix* FCC through the serial port RS-232. For wireless communication with the Ground Control Station (GCS), we have mounted the *wifistix* card from *Gumstix* on the 92-pin connector of *Gumstix motherboard*, thereby providing a communication module with high bandwidth (about 50 Mbits/s). The communication range is about 500 m, but it can be increased up to 800 m by reducing the communication bandwidth to 2 Mbits/s.

The FCC has three main tasks: (1) reads data from the MNAV sensor and estimates the vehicle state vector, (2) implements the guidance and control system, and (3) manages the WiFi communication (uplink and downlink or telemetry) with the GCS. In order to interface our autopilot with the X-3D-BL board, we have added a small AVR micro-controller (Atmega32) that receives the control inputs from the FCC and generates a PPM signal which is sent to the *X-base* board, thereby emulating the RC receiver signal. From Fig. 8.5, we can see that the sixth channel of the

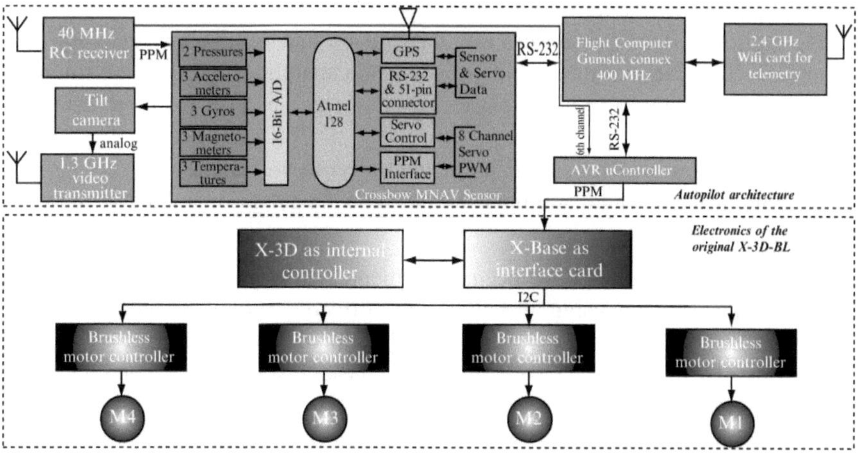

Fig. 8.5 Embedded architecture of the autopilot

RC receiver is also connected to the auxiliary AVR micro-controller. This channel is used to terminate the flight at any moment if the embedded system is not behaving correctly.

Our platform includes also an analog wireless camera which is used as the main mission sensor. The vision module is composed from a camera KX171 from *Range Video*, a 1.3 GHz video transmitter and receiver, and a tilt system which is driven by the FCC through the MNAV servo driver interface.

8.4.3 Guidance, Navigation and Control Systems and Their Real-Time Implementation

The developed autopilot includes onboard guidance and navigation systems which are described in detail in Chap. 10. In this section, we recall briefly the main components of these systems and their real-time implementation onboard the flight control computer.

8.4.3.1 Guidance and Navigation Systems

Guidance, Navigation and Control (GN&C) algorithms are the core of flight software to successfully complete the assigned mission through autonomous flight. It offers to the vehicle the ability to follow waypoints[3] and execute other preprogrammed maneuvers like automatic take-off and landing, hovering and trajectory tracking. As shown in Fig. 8.6, the overall system architecture considered in this chapter consists of six layers:

1. GCS for mission definition and high-level decisions making.
2. Basic guidance system for path planning and trajectory generation.
3. Navigation system for sensors raw data fusion and vehicle's state vector estimation. The standard navigation sensors and capabilities onboard the rotorcraft have been enhanced to include a visual odometer (see [17] for more details on the vision system).
4. Nonlinear controller for stabilization and trajectory tracking.
5. Communication system and interface between the autopilot and the vehicle.
6. The aerial platform.

All GN&C algorithms have been implemented in visual C using Multi-thread programming. In our design, the autopilot software is implemented as process within the Linux OS. It is composed of smaller units of computation, know as tasks (or threads). By using *POSIX* style semantics, theses tasks are called and scheduled separately [12].

[3] Here, waypoint navigation is defined as the process of automatically following a predetermined path defined by a set of geodetic coordinates (GPS coordinates).

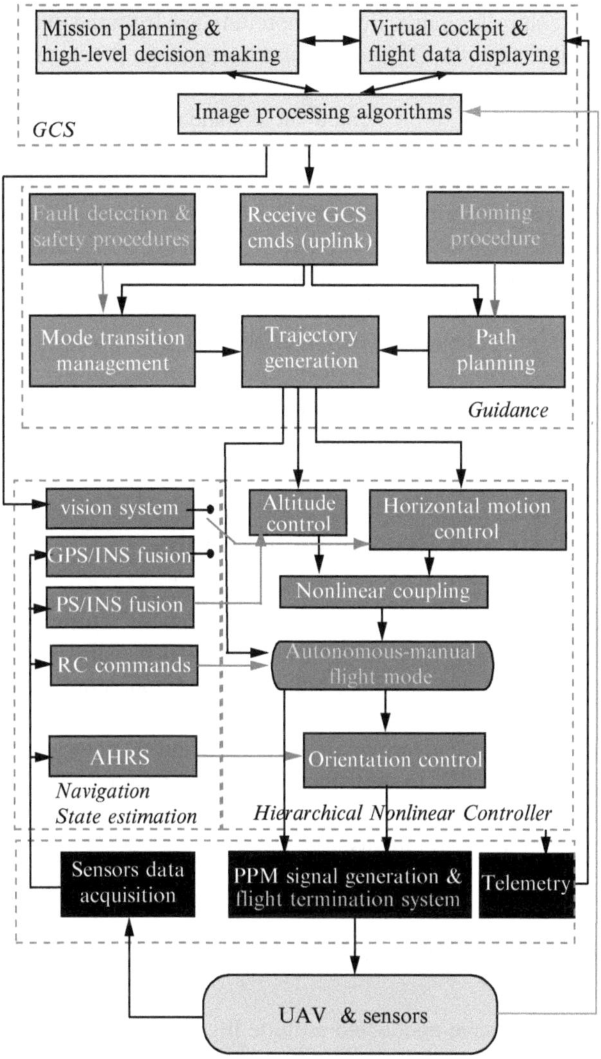

Fig. 8.6 Architecture of the guidance, navigation and control system

More details about the guidance and navigation system as well as the implemented safety procedures can be found in [18].

8.4.3.2 Ground Control Station

The ground control station provides several critical functionalities. The main role of GCS is to display the flight information downloaded from the UAV in real-time and to upload the various operating commands such as operation mode, guidance commands and controller gains. Furthermore, images from the onboard camera are displayed in real-time at the GCS which also hosts image processing algorithms for visual navigation.

The GCS software is implemented on a standard laptop using C++, MFC, OpenGL and OpenCV libraries, and it is based on the Open Source code provided by Crossbow [12]. It includes three different threads that are running independently while sharing memory resources. The first thread runs at 10 Hz and handles wireless communication with the onboard FCC for receiving and decoding flight data (telemetry), and sending visual estimates and other navigation commands. The communication program uses *socket* and UDP protocol. The second thread is running at 5 Hz and implements the GCS interface that allows to display flight data in real-time and to send flight commands by clicking on the appropriate buttons. The third thread, running at 12 Hz, implements image processing algorithms for visual navigation.

8.5 Flight Tests and Experimental Results

In order to evaluate the performance of our autonomous rotorcraft UAV, we have performed real-time flight tests with various mission scenarios. Here, we present experimental results of five tests that highlight the autonomous capabilities of our vehicle including, accurate attitude tracking, hovering with automatic take-off and landing, long-distance flight, waypoint navigation and trajectory tracking. The objective of these flight tests is to show that the designed rotorcraft, equipped with the developed autopilot, is capable to achieve autonomously some real-world applications in obstacles-free environments.

8.5.1 Attitude Trajectory Tracking

In this flight test, we have conducted attitude flight control in order to explore the effectiveness and robustness of the inner-loop nonlinear controller. For best evaluation of the attitude controller performance, we have performed an outdoor test where reference trajectories are generated in the following manner:

1. Between 0 and 25 s, pre-programmed sinusoidal trajectories are generated with 0.5 Hz frequency and a time-varying magnitude. The pitch angle magnitude increases from 0° to 45° while the roll angle magnitude is set to 0° (0–22 s), then to 30° (22–24 s) and finally to 0° (24–25 s).
2. Between 25 and 110 s, the reference trajectories are sent by the operator via the RC transmitter (semi-autonomous control), such that the induced forward velocities are relatively high (about 4 m/s).

The attitude control results are shown in Fig. 8.7 where we can see the rotorcraft accurately tracking the reference commands. Good tracking was obtained even at

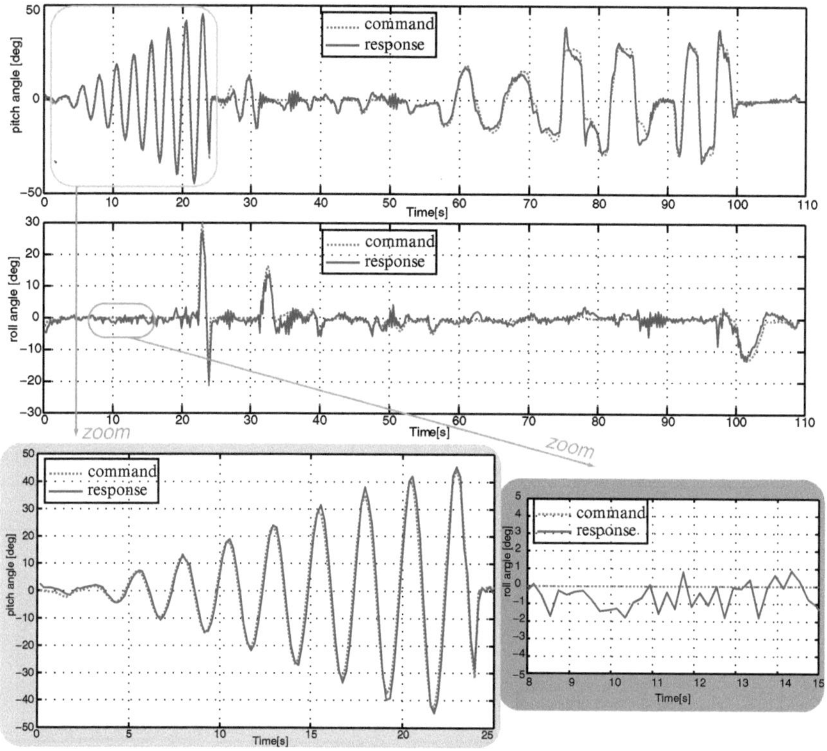

Fig. 8.7 Performance of the inner-loop nonlinear controller during attitude trajectory tracking. The vehicle is able to track attitude commands with high accuracy even at relatively big and rapidly varying angles. In translational flight (70–100 s), the tracking error increased slightly because the pitch and roll dynamics are sensitive to aerodynamic effects like blade flapping

relatively big and rapidly varying angles. Furthermore, the controller handles coupling between pitch and roll axes even when the angles are big.

At relatively high forward speeds, the controller is also able to track reference commands without apparent degradation in performance, as shown in Fig. 8.7 (70–100 s). The small tracking errors are due to aerodynamic effects like blade flapping and pitching-up phenomenon, that occur at relatively high speeds and significant angles of attack.

8.5.2 Automatic Take-off, Hovering and Landing

The nonlinear controller given by (8.33) is used in this flight test to achieve automatic take-off, accurate hovering and precise auto-landing. The experimental results, shown in Fig. 8.8, demonstrate accurate tracking of the height reference command, yielding to effective altitude control and automatic take-off and landing.

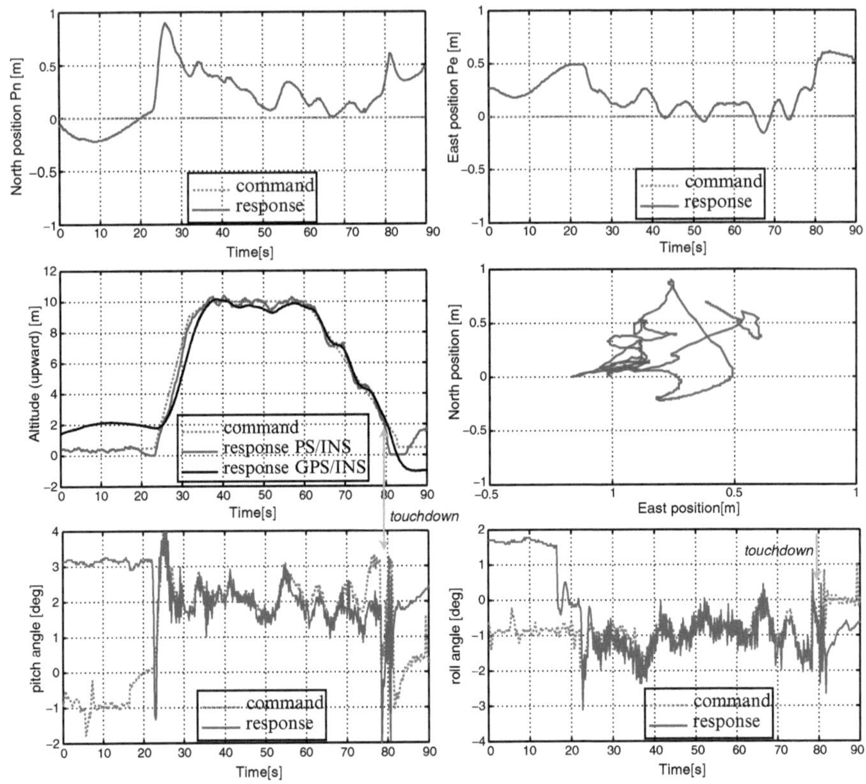

Fig. 8.8 Experimental results from an outdoor flight test where the quadrotor achieved a fully autonomous take-off, hovering and landing. The rotorcraft was able to stay inside of 50-cm-radius circle during hovering. The attitude reference trajectories are accurately tracked with an error of under $0.5°$ during hovering

The rotorcraft achieved also a stable hovering flight and was able to stay inside of 50-cm-radius circle. The horizontal motion is also accurately controlled during take-off and landing maneuvers with less than 1 m error, which is a good performance for this scale rotorcraft flying outdoor and subject to external disturbances like wind.

8.5.3 *Long-Distance Flight*

This flight test has been performed on March 2008 at Agra, India during the US–Asian MAV competition. The objectives of this test were:

1. Demonstrate the capability of our MAV to fly autonomously until the zone of interest, located at about 1 km from the launching point

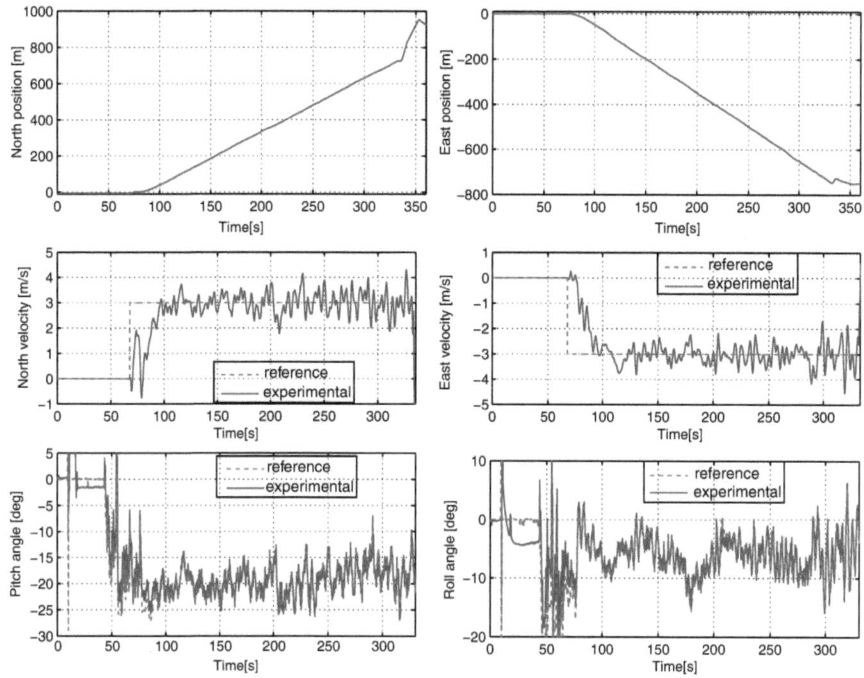

Fig. 8.9 Autonomous forward flight (relatively long-range flight)

2. Check the quality and range of the wireless communication as well as the video transmission

The rotorcraft was thus, tasked to achieve an autonomous forward flight at a translational velocity[4] $V = 4.25$ m/s while transmitting images to the GCS.

The obtained results are shown in Fig. 8.9 where we can see that the velocity and attitude reference trajectories are well tracked. Position trajectories show that the MAV flied a relatively long distance autonomously which is about 1.2 km ($D = \sqrt{x^2 + y^2}$). When the MAV reached the boundaries of the fly zone, the safety pilot switched to manual flight and recovered the vehicle. The test showed also that the range of the WiFi wireless communication was about 600 m while the quality of the video transmission is acceptable until 1,000 m. In this test, the safety procedure related to communication lost was disabled, thereby allowing the vehicle to continue mission execution even when the communication link is lost.

We would like to note that the used quadrotor MAV can achieve a total flight of about 3 km at a forward speed of 10 m/s.

[4] $V = \sqrt{V_x^2 + V_y^2} = \sqrt{3^2 + 3^2} = 4.25.$

8.5.4 Fully Autonomous Waypoint Navigation

Here, we demonstrate the ability of the GN&C system to achieve accurate waypoint navigation and to perform hovering, automatic take-off and landing. In this test, a set of four waypoints were chosen by just clicking the desired locations on the 2D map of the GCS interface (see Fig. 8.10). The MAV should then, pass the assigned waypoints in a given sequence.

The mission is started by clicking the "take-off" button. The MAV performs then, an automatic take-off and hovers at 10 m. The altitude can be changed at any time from the GCS. When the "upload waypoint" button is pushed, the MAV starts waypoint navigation.[5] When the mission is finished, the "land" button can be pushed for automatic landing.

In Fig. 8.10, we can see the selected waypoints as well as the MAV trajectory, plotted in real-time. We can clearly observe that the MAV passed successfully through all the waypoints. The accuracy of this waypoint navigation is confirmed by Figs. 8.11 and 8.12. The reference trajectories are tracked with high accuracy. From the third graph of Fig. 8.11, we can also see that the height, estimated by the pressure sensor, is more accurate than the one estimated by GPS. Despite low-cost sensors, we have achieved an accurate height control (± 1.5 m) even during relatively high-speed flight (5 m/s). Furthermore, the take-off and landing maneuvers have been achieved autonomously. It is interesting however, to note that altitude control is more accurate at low forward speeds as shown in Fig. 8.8 (about 50 cm maximum error). Indeed, the thrust variation created by different angles of attack

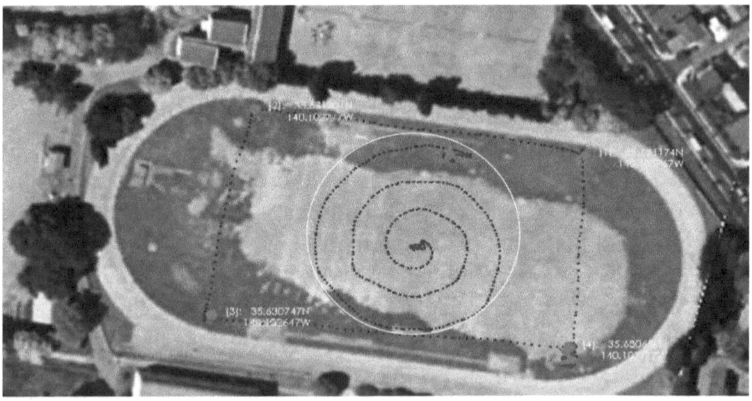

Fig. 8.10 Part of the GCS interface showing a 2D map of test field and MAV trajectories during waypoint navigation and spiral trajectory tracking

[5] Waypoints can be updated and/or modified in real-time.

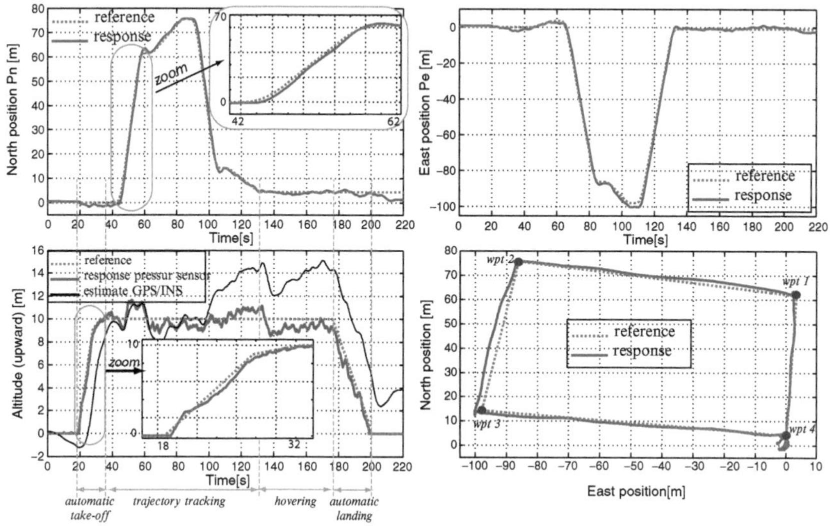

Fig. 8.11 MAV position trajectories during waypoint navigation

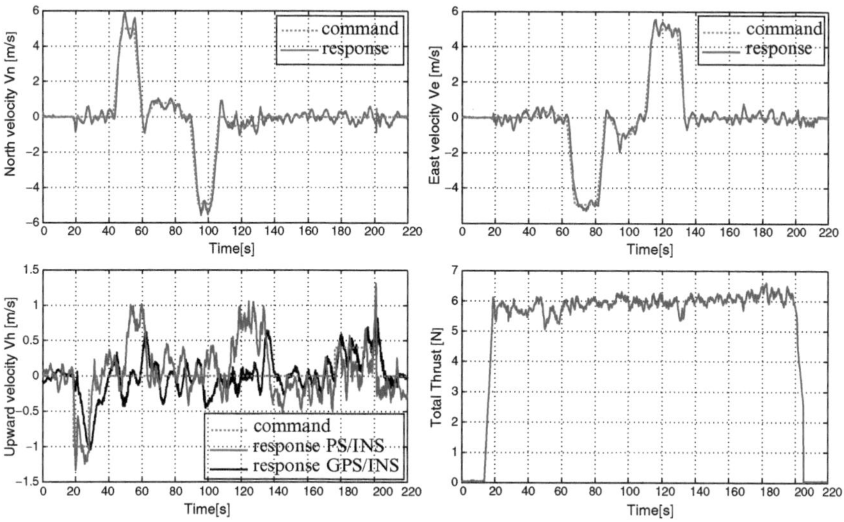

Fig. 8.12 MAV translational velocities during waypoint navigation

at varying forward speeds and wind conditions causes a disturbance that pushes the vehicle above the reference height. The controller takes some time to reject these disturbances because of the time delay in thrust.[6]

[6] We believe that this time delay is introduced voluntary in the original platform to facilitate manual flight.

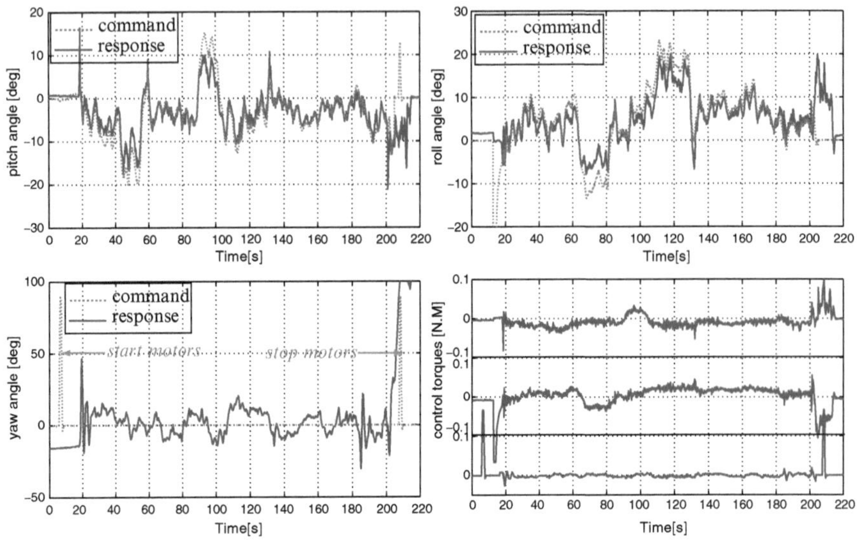

Fig. 8.13 MAV attitude during waypoint navigation

In this test, the desired horizontal velocity was set to 5 m/s (see Fig. 8.12). In fact, our autonomous platform can fly autonomously at higher velocities (10 m/s). However, it was difficult to fly at this high speed because of the test area which is small.

Figure 8.13 shows that the inner-loop performs well and the desired roll and pitch angles are tracked with small errors. As discussed in Sect. 8.5.1, degradation in the attitude tracking capabilities as forward speed increases are observed in this flight test.

A video clip of this flight can be found at http://mec2.tm.chiba-u.jp/monograph/Videos/Chapter8/1.wmv.

8.5.5 Arbitrary Trajectory Tracking

The performance of the autopilot for trajectory tracking is an important evaluation. The final objective of the rotorcraft MAV is to perform autonomously search, rescue and surveillance missions. Thus, trajectory tracking capability is very useful since a spiral trajectory following for example allows the MAV to explore some area of interest. Furthermore, the waypoint navigation is based on straight-line trajectories tracking.

In this flight test, a spiral trajectory was implemented to demonstrate the tracking performance of the nonlinear controller. The reference trajectory was generated using a kinematic model of a modified *Archimedean spiral* in order to obtain a spiral with constant separation distance (10 m) between successive turnings, but also with a constant tangential speed (3 m/s).

Results from a flight test in which the MAV autonomously executes a spiral trajectory tracking are shown in Figs. 8.14 and 8.15 (see also Fig. 8.10). One can see that the MAV tracked successfully the reference trajectories including the height trajectory for automatic take-off and altitude holding. Figure 8.14 shows also the

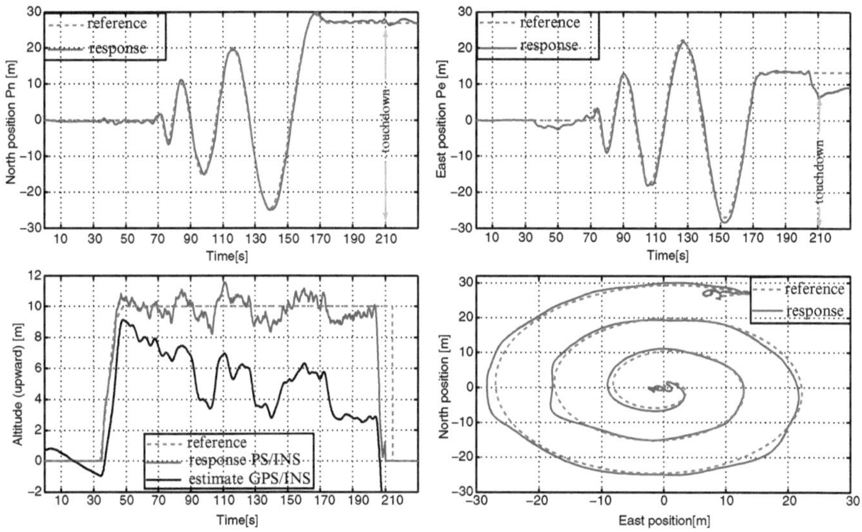

Fig. 8.14 MAV position trajectories during spiral trajectory tracking

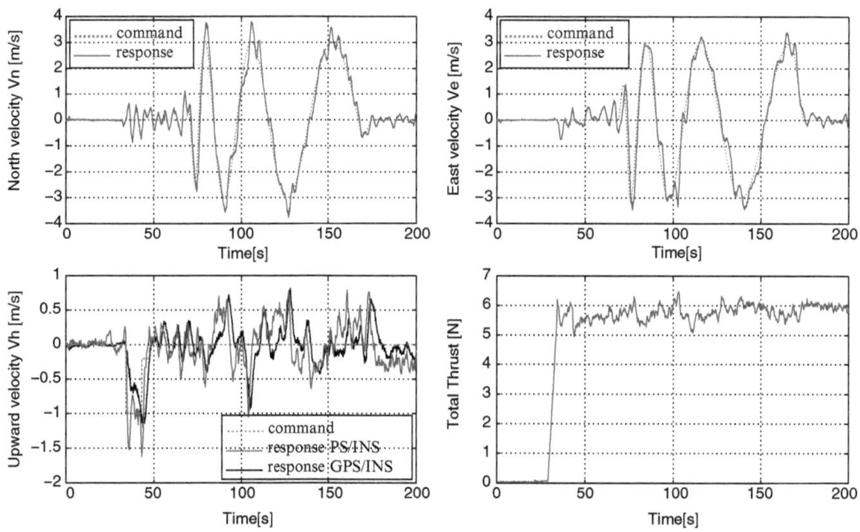

Fig. 8.15 MAV translational velocities during spiral trajectory tracking

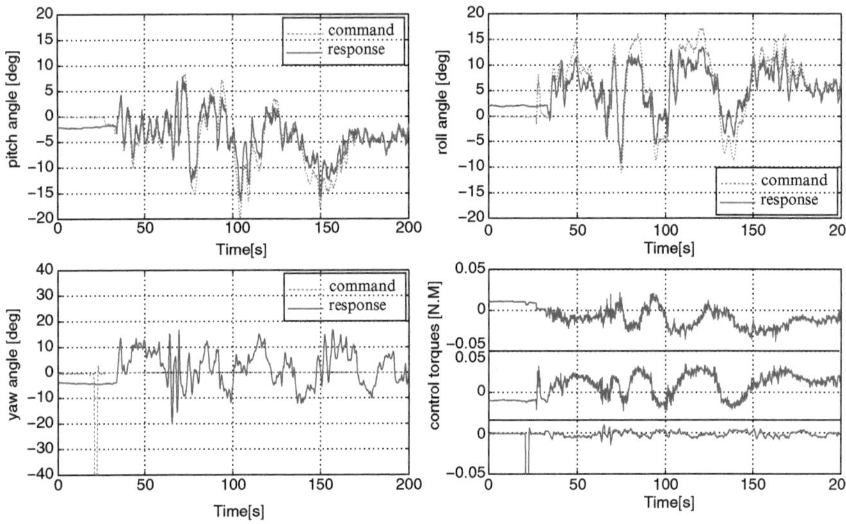

Fig. 8.16 MAV attitude during spiral trajectory tracking

effectiveness and importance of relying on PS/INS rather than GPS/INS for height control, especially for MAVs flying at low altitude where accurate height control is necessary. This is a good tracking performance for this category of RUAVs.

Attitude control results, shown in Fig. 8.16, confirm the good performance of the inner-loop controller even in this complicated motion pattern. One thing to note in the attitude control results, shown in Figs. 8.16 and 8.13, is the significant errors in yaw control compared to pitch and roll control. This is mainly due to the internal yaw controller implemented in the X-3D board which affects the performance of our yaw controller.

In this flight test, we have also demonstrated the flight termination system. The MAV achieved a soft emergency landing without damaging the vehicle, as it can be seen on Fig. 8.14 and the associated video clip, http://mec2.tm.chiba-u.jp/monograph/Videos/Chapter10/1.wmv.

The flight results, presented in this chapter, demonstrate a significant improvement in capability over previous mini rotorcraft and quadrotor MAVs in particular. From the obtained experimental results according to six different mission scenarios, we conclude that the implemented autopilot and GN&C algorithms present good performance for autonomous flight. In contrast to standard linear controllers that fail to provide good performance at non-nominal conditions, the designed nonlinear control system allowed a mini quadrotor to accurately track reference trajectories even at relatively high speeds, big angles, coupled motions and windy conditions. These results demonstrate also that rotorcraft MAVs, weighing less than 0.7 kg, can achieve advanced autonomous flight despite their small size and the weaknesses of the low-cost sensors.

8.6 Summary

In this chapter, we have described the design of a mini rotorcraft UAV with onboard intelligence capabilities that can be used for search and rescue missions. This research dealt with both theoretical and practical issues of autonomous RUAVs ranging from controller design and stability analysis to experimental flight tests. Indeed, we have proposed a nonlinear flight controller which is practical for real-time implementation while guaranteing the global asymptotic stability of the closed-loop system. We have also demonstrated the ability to provide effective waypoint navigation and trajectory tracking capabilities to small low-cost systems such as the mini quadrotor helicopter. This proves the relevance of the developed autopilot which relies solely on low-cost sensors and simple but efficient nonlinear control system. From the flight test results, we conclude that the developed RUAV shows good and reliable performance for autonomous flight, which is an important step towards our goal of developing fully autonomous MAVs capable of achieving real-world applications.

In future work, it would be interesting to improve the control system by considering the aerodynamic effects at higher speeds and aggressive maneuvers. It may also be useful to extend the guidance system capabilities by incorporating additional flight behaviors and a high-level system for decisions making and optimized planning.

Appendix

Proof of Lemma 1: Boundedness of $|u(\chi)|$

Let us recall the expressions of the thrust u and the feedback μ.

$$\begin{cases} |u(\chi)| = m\| \mu(\chi, \ddot{\xi}_d) + g e_3^i \| \\ \mu(\chi, \ddot{\xi}_d) = -K_\xi \chi_\xi - K_v \chi_v + \ddot{\xi}_d \end{cases} \tag{8.34}$$

Let $\lambda_\xi > 0$ and $\lambda_v > 0$ be the maximum values of K_ξ and K_v eigenvalues respectively. Thus, we write

$$\begin{aligned} |u(\chi)| = m\|g e_3^i + \ddot{\xi}_d - K_\xi \chi_\xi - K_v \chi_v\| &\le m(g + \|\ddot{\xi}_d\| + \lambda_\xi \|\chi_\xi\| + \lambda_v \|\chi_v\|) \\ &\le m(g + L_d + \ \max\ (\lambda_\xi, \lambda_v)(\|\chi_\xi\| + \|\chi_v\|)) \\ &\le m(g + L_d) + m\ \max\ (\lambda_\xi, \lambda_v)\sqrt{2}\|\chi\| \end{aligned}$$

because

$$(\|\chi_\xi\| + \|\chi_v\|)^2 = \|\chi_\xi\|^2 + \|\chi_v\|^2 + 2\|\chi_\xi\|\|\chi_v\| \le 2(\|\chi_\xi\|^2 + \|\chi_v\|^2)$$

which implies that

$$\|\chi_\xi\| + \|\chi_\upsilon\| \leq \sqrt{2}\sqrt{\|\chi_\xi\|^2 + \|\chi_\upsilon\|^2} = \sqrt{2}\|\chi\|$$

Setting $c \triangleq m\sqrt{2}\max(\lambda_\xi, \lambda_\upsilon)$, we get

$$|u(\chi)| \leq m(g + L_d) + c\|\chi\| \leq c(\frac{mg + mL_d}{c} + \|\chi\|) \leq c(r + \|\chi\|)$$

$$\text{where } r = \frac{mg + mL_d}{c}$$

From the above inequalities, we deduce that

$$|u(\chi)| \leq \begin{cases} k_1\|\chi\|, & \text{for all} \|\chi\| \geq r \\ k_1 r, & \text{for all} \|\chi\| < r, \quad \text{where } k_1 = 2c \end{cases} \tag{8.35}$$

Proof of Lemma 2: Boundedness of $\|H(.)\|$

We had $\|H(\chi, e_\eta)\| = \sqrt{h_x^2 + h_y^2 + h_z^2}$ where the components (h_x, h_y, h_z) are functions of $\sin(arg)$ and $\cos(arg)$. The argument "*arg*" may be ϕ_d, $e_\phi/2$, $\phi_d + e_\phi/2$, θ_d, $e_\theta/2$, $\theta_d + e_\theta/2$, ψ_d, $e_\psi/2$, $\psi_d + e_\psi/2$.

Let us recall some trivial inequalities that are exploited to prove the boundedness of $\|H(\chi, e_\eta)\|$.

$$\begin{aligned} &|\sin a| \leq |a|\,, \quad |\sin a| \leq 1 \text{ and } |\cos a| \leq 1 \\ &|a||b| \leq \frac{1}{2}(|a| + |b|), \quad \text{for } |a| \leq 1 \text{ and } |b| \leq 1 \\ &|a||b||c| \leq \frac{1}{2}(|a| + |b| + |c|), \quad \text{for } |a| \leq 1,\ |b| \leq 1 \text{ and } |c| \leq 1 \end{aligned} \tag{8.36}$$

By writing the exact expressions of (h_x, h_y, h_z) and doing some mathematical calculations, it is possible to show that:

$$\begin{aligned} |h_x| &\leq c_1(|e_\theta| + |e_\phi| + |e_\psi|), & h_x^2 &\leq c_4(e_\theta^2 + e_\phi^2 + e_\psi^2) \\ |h_y| &\leq c_2(|e_\theta| + |e_\phi| + |e_\psi|), & h_y^2 &\leq c_5(e_\theta^2 + e_\phi^2 + e_\psi^2) \\ |h_z| &\leq c_3(|e_\theta| + |e_\phi|), & h_z^2 &\leq c_6(e_\theta^2 + e_\phi^2) \end{aligned} \tag{8.37}$$

where $c_1, c_2, c_3, c_4, c_5, c_6$ are positive constants.

Proof. In the following, we will show that the component h_z verifies the inequalities in (8.37). By following the same steps, it is easy to show that the other components h_x and h_y satisfy also the inequalities in (8.37).

Let us first write the exact expression of h_z which is given by

$$h_z = \cos\phi_d[-\sin(e_\theta/2)\sin(\theta_d + e_\theta/2)] + \cos\theta_d[-\sin(e_\phi/2)\sin(\phi_d + e_\phi/2)]$$
$$+[-\sin(e_\theta/2)\sin(\theta_d + e_\theta/2)][-\sin(e_\phi/2)\sin(\phi_d + e_\phi/2)] \quad (8.38)$$

By using inequalities in (8.36) and keeping only the $\sin(e_\theta/2)$ and $\sin(e_\phi/2)$ terms, one can write

$$\begin{aligned}|h_z| &\le |\sin(e_\theta/2)| + |\sin(e_\phi/2)| + |\sin(e_\phi/2)||\sin(e_\theta/2)| \\ &\le \frac{3}{2}(|\sin(e_\theta/2)| + |\sin(e_\phi/2)|) \\ &\le \frac{3}{4}(|e_\theta| + |e_\phi|) \\ &\le c_3(|e_\theta| + |e_\phi|), \quad \text{with } c_3 = \frac{3}{4} \end{aligned} \quad (8.39)$$

By computing the square of the previous function and considering that $2|e_\theta||e_\phi| \le e_\theta^2 + e_\phi^2$, one can write

$$\begin{aligned}h_z^2 &\le c_3^2(e_\theta^2 + e_\phi^2 + 2|e_\theta||e_\phi|) \le 2c_3^2(e_\theta^2 + e_\phi^2) \\ &\le c_6(e_\theta^2 + e_\phi^2), \text{ with } c_6 = 2c_3^2 = \frac{9}{8}\end{aligned}$$

After some mathematical developments, we can prove that the norm of the interconnection term $H(.)$ verifies the following inequalities:

$$\|H(\chi, e_\eta)\| = \sqrt{h_x^2 + h_y^2 + h_z^2} \quad (8.40)$$
$$\begin{aligned} &\le \sqrt{c_4(e_\theta^2 + e_\phi^2 + e_\psi^2) + c_5(e_\theta^2 + e_\phi^2 + e_\psi^2) + c_6(e_\theta^2 + e_\phi^2)} \\ &\le \sqrt{(c_4 + c_5 + c_6)e_\theta^2 + (c_4 + c_5 + c_6)e_\phi^2 + (c_4 + c_5)e_\psi^2} \\ &\le \sqrt{(c_4 + c_5 + c_6)e_\theta^2 + (c_4 + c_5 + c_6)e_\phi^2 + (c_4 + c_5 + c_6)e_\psi^2} \\ &\le k\|e_\eta\| \quad \text{with } k = \sqrt{c_4 + c_5 + c_6} \end{aligned} \quad (8.41)$$

and this ends the proof.

References

1. Altug E, Ostrowski JP, Taylor CJ (2005) Control of a quadrotor helicopter using dual camera visual feedback. Int J Robot Res 24(5):329–341
2. Bouabdallah S, Murrieri P, Siegwart R (2005) Towards autonomous indoor micro vtol. Auton Robots 18(2):171–183

3. Castillo P, Dzul A, Lozano R (2004) Real-time stabilization and tracking of a four rotor mini-rotorcraft. IEEE Trans Control Syst Technol 12(4):510–516
4. Castillo P, Lozano R, Dzul A (2005) Modelling and control of mini-flying machines. Advances in industrial control. Springer, Berlin
5. Gavrilets V, Mettler B, Feron E (2004) Human-inspired control logic for automated maneuvering of miniature helicopter. AIAA J Guidance Control Dyn 27(5):752–759
6. Goldstein R (1980) Classical mechanics, 2nd edn. Addison-Wesley, Reading, MA
7. Guenard N, Hamel T, Moreau V (2005) Dynamic modeling and intuitive control strategy for an "x4-flyer". In: Proceedings of 5th international conference on control and automation, Budapest, Hungary, vol 1, pp 141–146
8. Gurdan D, Stumpf J, Achtelik M, Doth KM, Hirzinger G, Rus D (2007) Energy-efficient autonomous four-rotor flying robot controlled at 1 khz. In: Proceedings of the IEEE international conference on robotics and automation, Roma, Italy, pp 361–366
9. He R, Prentice S, Roy N (2008) Planning in information space for a quadrotor helicopter in a GPS-denied environment. In: Proceedings of the IEEE international conference on robotics and automation, California, USA, pp 1814–1820
10. Hoffmann GM, Huang H, Waslander SL, Tomlin CJ (2007) Quadrotor helicopter flight dynamics and control: theory and experiment. In: Proceedings of the AIAA guidance, navigation, and control conference and exhibit, South Carolina, USA, AIAA 2007-6461
11. How JP, Bethke B, Frank A, Dale D, Vian J (2008) Real-time indoor autonomous vehicle test environment. IEEE Control Syst Mag 28(2):51–64
12. Jang JS, Liccardo D (2006) Automation of small UAVS using a low cost mems sensor and embedded computing platform. In: Proceedings of the IEEE/AIAA 25th digital avionics systems conference, Portland, USA, pp 1–9
13. Johnson E, Kannan S (2005) Adaptive trajectory control for autonomous helicopters. AIAA J Guidance Control Dyn 28(3):524–538
14. Kendoul F (2007) Modelling and control of unmanned aerial vehicles, and development of a vision-based autopilot for small rotorcraft navigation. PhD Thesis Report, University of Technology of Compiegne, France
15. Kendoul F, Lara D, Fantoni I, Lozano R (2007) Real-time nonlinear embedded control for an autonomous quad-rotor helicopter. AIAA J Guidance Control Dyn 30(4):1049–1061
16. Kendoul F, Fantoni I, Lozano R (2008) Asymptotic stability of hierarchical inner–outer loop-based flight controllers. In: Proceedings of the 17th IFAC world congress, Seoul, Korea, pp 1741–1746
17. Kendoul F, Fantoni I, Nonami K (2009) Optic flow-based vision system for autonomous 3D localization and control of small aerial vehicles. Rob Auton Syst 57:591–602
18. Kendoul F, Zhenyu Y, Nonami K (2009) Embedded autopilot for accurate waypoint navigation and trajectory tracking: application to miniature rotorcraft UAVS. In: Proceedings of the IEEE international conference on robotics and automation, Kobe, Japan, pp 2884–2890
19. Kim HJ, Shim DH (2003) A flight control system for aerial robots: algorithms and experiments. Control Eng Pract 11(12):1389–1400
20. Koo T, Sastry S (1998) Output tracking control design of a helicopter model based on approximate linearization. In: Proc of the IEEE conf on decision and control, Florida, USA, pp 3635–3640
21. La Civita M, Papageorgiou G, Messner WC, Kanade T (2006) Design and flight testing of an H_∞ controller for a robotic helicopter. AIAA J Guidance Control Dyn 29(2):485–494
22. Madani T, Benallegue A (2006) Backstepping sliding mode control applied to a miniature quadrotor flying robot. In: Proceedings of the 32nd annual conference of the IEEE industrial electronics society, Paris, France, pp 700–705
23. McCormick BW (1995) Aerodynamics, aeronautics and flight mechanics. Wiley, New York
24. Prouty RW (1995) Helicopter performance, stability, and control. Krieger, Malabar, FL
25. Reiner J, Balas G, Garrard W (1995) Robust dynamic inversion for control of highly maneuverable aircraft. AIAA J Guidance Control Dyn 18(1):18–24
26. Saripalli S, Montgomery J, Sukhatme G (2003) Visually-guided landing of an unmanned aerial vehicle. IEEE Trans Rob Autom 19(3):371–381

27. Scherer S, Singh S, Chamberlain L, Elgersma M (2008) Flying fast and low among obstacles: methodology and experiments. Int J Robot Res 27(5):549–574
28. Sepulcre R, Jankovic M, Kokotovic P (1997) Constructive nonlinear control. Communications and control engineering series. Springer, Berlin
29. Shim DH, Kim HJ, Sastry S (2003) A flight control system for aerial robots: algorithms and experiments. Control Eng Pract 11(2):1389–1400
30. Shin J, Fujiwara D, Nonami K, Hazawa K (2005) Model-based optimal attitude and positioning control of small-scale unmanned helicopter. Robotica 23:51–63
31. Sontag E (1988) Smooth stabilization implies coprime factorization. IEEE Trans Automat Contr 34(4):435–443
32. Tayebi A, McGilvray S (2006) Attitude stabilization of a VTOL quadrotor aircraft. IEEE Trans Control Syst Techol 14(3):562–571

Chapter 9
Formation Flight Control of Multiple Small Autonomous Helicopters Using Predictive Control

Abstract In this chapter, we present a model-based formation flight control of multiple small unmanned helicopters as an example of advanced control of unmanned aerial vehicles (UAVs). We design the autonomous formation flight control system as a "leader-following" configuration. In order to achieve good control performance under the system constraints, the "model predictive control" is used for the translational position control of follower helicopters. Position constraints such as moving range and collision avoidance problem are considered in the real-time optimal control calculations. To achieve robustness against disturbance, a minimal-order disturbance observer is used to estimate the unmeasurable state variables and disturbance. The simulation results are presented to show the feasibility of the control strategy. The formation flight control experiment is performed using two helicopters. The experimental results demonstrate an accurate control performance. The position constraint capability is confirmed through the experiments with a single helicopter. Finally, robustness against wind is verified by a windy condition experiment.

Video Links:

ExpMovie_Formation_MFMPC_S_Yaw
http://mec2.tm.chiba-u.jp/monograph/Videos/Chapter9/1.avi
Formation control by model predictive control theory
http://mec2.tm.chiba-u.jp/monograph/Videos/Chapter9/2.avi

9.1 Introduction

In recent years, interest in the autonomous control of unmanned aerial vehicles (UAVs) has been increased because no human operations are required, cost performance is good, and missions can be conducted beyond a visual range. In particular, unmanned small-scale helicopters have received considerable attention because it can perform hovering and vertical takeoff operations [1–8]. In practical missions such as the monitoring of a power transmission lines, surveillance from a skies and

K. Nonami et al., *Autonomous Flying Robots: Unmanned Aerial Vehicles and Micro Aerial Vehicles*, DOI 10.1007/978-4-431-53856-1_9, © Springer 2010

civilian requirements, if we can operate and fly multiple helicopters, we would be able to gather information more efficiently.

Formation control inspired by birds and fishes is a popular topics in multiple vehicle systems [1, 4, 6–19]. Dimarogonas et al. [10] presented a distributed control strategy, which is based on an artificial potential method, for convergence to the desired formation configuration and the collision avoidance of multiple holonomic agents. However, in the autonomous flight control of UAVs, a high control performance is required along with stability. Schouwenaars et al. [8] proposed an approach called mixed-integer linear programming (MILP) to the optimal path planning of multiple UAVs. However, in this method, control calculations are complicated. In order to use MILP, commercially available expensive optimizing software is necessary.

In this chapter, we present a formation flight control of multiple autonomous small unmanned helicopters. We design an autonomous formation flight control system as the "leader-following" configuration. In order to simultaneously achieve accurate trajectory following performance and handle system constraints a distributed "model predictive tracking controller" is used for the guidance control of the helicopters. First, we model the helicopter dynamics as a linear state space equation. Secondly, we design the position controller of each follower by using a model predictive control (MPC). The cost function is formulated in the form of a linear quadratic with a terminal state cost. Further, it is augmented to a form that reflects the collision avoidance maneuver and communication range. Finally, simulation and experimental results are presented to show the feasibility of the control strategy.

9.2 Configuration of Control System

An autonomous flight control system is essential to guide an UAV to intended point. Our guidance-navigation-control (GNC) system of the helicopter SF40 is shown in Fig. 9.1. The system consists of three parts. The navigation system estimates the state of helicopter with respect to its position, velocity, and attitude angles. The guidance system generates attitude commands to fly in designated trajectories by using the estimated position and velocity. The attitude control system calculates

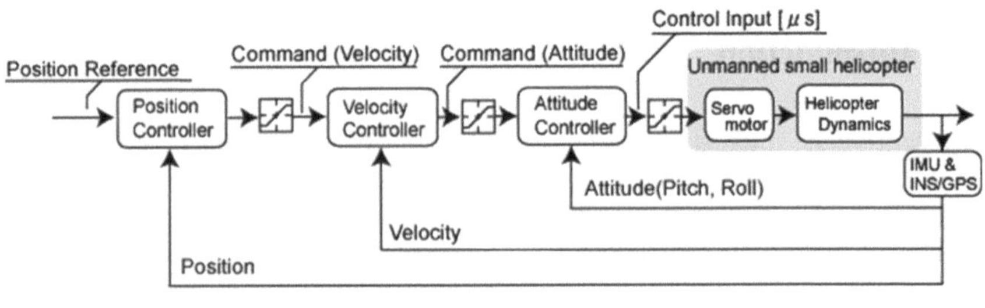

Fig. 9.1 Control system of SF40

control inputs to follow the attitude commands. In the design of the formation flight control described in this chapter, we devised a path planner that determines the shape of formation (position reference of each helicopter) and realizes formation flight control by following the trajectory generated by the path planner. Moreover, the guidance system of each helicopter also has a function for collision avoidance.

9.3 Leader–Follower Path Planner Design

In this section, the design of a path planner for formation control is described. We assign a natural number “p” to the “p”th helicopter and refer to the follower as “p”. The reference trajectory of the follower “p” (X_{rp}, Y_{rp}) is given by

$$\begin{bmatrix} X_{rp} \\ Y_{rp} \end{bmatrix} = \begin{bmatrix} X_l + R_p \cos\psi_l - L_p \sin\psi_l \\ Y_l + R_p \sin\psi_l + L_p \cos\psi_l \end{bmatrix}, \ (p = 1, 2, \cdots, n) \qquad (9.1)$$

Here (X_l, Y_l) is a reference trajectory of a leader and ψ_l is a leader’s angle of direction. R_p, L_p is the offset of the follower “p” from the leader. A shape of the formation is transformable by a value of R_p, L_p. A graphical representation of (9.1) is shown in Fig. 9.2. In this method, whether the leader is an actual helicopter or a virtual one is not important. In the following section, the GNC systems of each follower are designed.

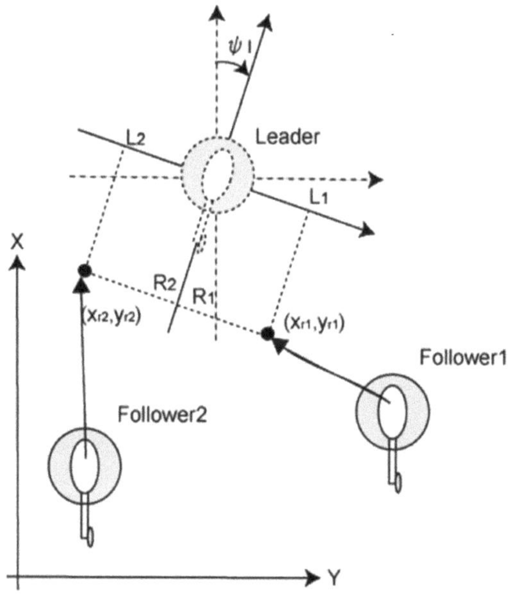

Fig. 9.2 Geometric configuration of the leader following

9.4 Guidance Controller Design by Using Model Predictive Control

In this section, we present a design method of a hierarchical guidance controller that enables each follower to be tracked to the designated reference and construction of the formation. First, we introduce a translational velocity and position model based on (2.16). Then, the guidance controller is derived using the MPC method. In the MPC design, the performance index is described by a quadratic form, and it is augmented to consider the system constraints such as collision avoidance and communication range.

9.4.1 Velocity Control System

In this section, we describe the guidance controller design of each follower. Figure 9.3 shows our coordinate system. Figure 9.3a illustrates the navigation

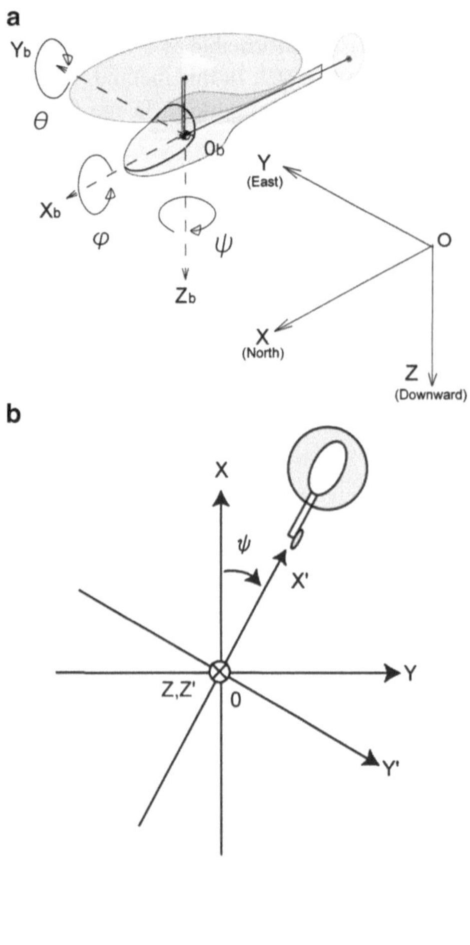

Fig. 9.3 Coordinate systems (**a**) navigation and body fixed frame (**b**) $(X'Y'Z')$ frame

frame (X, Y, Z) and the body fixed frame (X_b, Y_b, Z_b). The attitude angle of the helicopter is expressed by Euler angles. In addition, we apply the $(X'Y'Z')$ frame where in the $X'Y'$ plane is parallel to the XY plane, and we only consider yawing. Figure 9.3b illustrates the $(X'Y'Z')$ frame.

A translational velocity model of a small helicopter that can be adopted up to 5 m/s [5] is given by

$$\dot{x}'_v(t) = A_v x_x(t) + B_v u'(t) \tag{9.2}$$

$$y'(t) = C_v x'_v(t) \tag{9.3}$$

Here

$$A_v = \begin{bmatrix} 0 & 1 & 0 & 0 \\ aT & a-T & 0 & 0 \\ 0 & 0 & 0 & 1 \\ 0 & 0 & aT & a-T \end{bmatrix}, \quad B_v = \begin{bmatrix} 0 & 0 \\ -gT & 0 \\ 0 & 0 \\ 0 & gT \end{bmatrix}, \tag{9.4}$$

$$C_v = \begin{bmatrix} 1 & 0 & 0 & 0 \\ 0 & 0 & 1 & 0 \end{bmatrix}$$

$$x'_v(t) = \begin{bmatrix} \dot{X}' & \ddot{X}' & \dot{Y}' & \ddot{Y}' \end{bmatrix}^T, \quad u'(t) = \begin{bmatrix} \theta(t) & \phi(t) \end{bmatrix}^T \tag{9.5}$$

In these equations, $T = 2.00$, $a = 0.200$, and g is the gravitational constant. Equations (9.2) and (9.3) are the velocity models in a $X'Y'$ coordinate system. However, the velocity data is measured in the XY coordinate system, and hence, we calculate the coordinate transformation as follows:

$$\begin{bmatrix} \dot{X} \\ \dot{Y} \end{bmatrix} = \begin{bmatrix} \cos\psi & -\sin\psi \\ \sin\psi & \cos\psi \end{bmatrix} \begin{bmatrix} \dot{X}' \\ \dot{Y}' \end{bmatrix} \tag{9.6}$$

$$\begin{bmatrix} u_X \\ u_Y \end{bmatrix} = \begin{bmatrix} \cos\psi & -\sin\psi \\ \sin\psi & \cos\psi \end{bmatrix} \begin{bmatrix} \theta \\ \phi \end{bmatrix} \tag{9.7}$$

The coordinate transformed velocity model is described as

$$\dot{x}_v(t) = A_v x_v(t) + B_v u_v(t) \tag{9.8}$$

$$y(t) = C_v x_v(t) \tag{9.9}$$

$$x_v(t) = \begin{bmatrix} \dot{X}(t) & \ddot{X}(t) & \dot{Y}(t) & \ddot{Y}(t) \end{bmatrix}^T, \quad u_v(t) = \begin{bmatrix} u_X(t) & u_Y(t) \end{bmatrix}^T \tag{9.10}$$

(Note that that the suffix "p" represents the "p"th follower.)

$$\dot{x}_{vp}(t) = A_v x_{vp}(t) + B_v u_{vp}(t) \tag{9.11}$$

$$y_{vp}(t) = C_v x_{vp}(t) \tag{9.12}$$

The velocity controller is designed by using the linear quadratic integral (LQI) based on (9.11) and (9.12). The augmented system is described as follows:

$$\dot{x}_{va_p}(t) = A_{va}x_{va_p}(t) + B_{va}u_{vp} + G_{va}r_{v_p} \tag{9.13}$$

$$A_{va} = \begin{bmatrix} A_v & 0_{4\times2} \\ -C_v & 0_{2\times2} \end{bmatrix},\ B_{va} = \begin{bmatrix} B_v \\ 0_{2\times2} \end{bmatrix},\ G_{va} = \begin{bmatrix} 0_{4\times2} \\ I_2 \end{bmatrix} \tag{9.14}$$

$$x_{va_p}(t) = \begin{bmatrix} x_{vp}^T(t) & \varepsilon_{vp}^T(t) \end{bmatrix}^T \tag{9.15}$$

ε_{vp} is the integral of the error between a velocity command and a velocity data estimated by the navigation system. From (9.13) to (9.15), a feedback control input that minimizes the performance index is described as follows:

$$J_{vp} = \int_0^\infty \left\{x_{va_p}^T(\tau)Q_v x_{va_p}(\tau) + u_{vp}^T(\tau)R_v u_{vp}(\tau)\right\}d\tau,$$
$$Q_v \geqslant 0,\ R_v > 0 \tag{9.16}$$

Further,

$$u_{vp}(t) = -F_v x_{va_p}(t) \tag{9.17}$$

$$= -F_{v1}x_{vp}(t) - F_{v2}\varepsilon_{vp}(t) \tag{9.18}$$

$$F_v = \begin{bmatrix} F_{v1} & F_{v2} \end{bmatrix} = R_v^{-1}B_{va}^T P_v$$

$$P_v A_{va} + A_{va}^T P_v - P_v B_{va} R_v^{-1} B_{va}^T P_v + Q_v = 0 \tag{9.19}$$

Equation (9.18) is the attitude angle necessary to follow the designated velocity and the attitude command for the attitude controller. The attitude controller calculates the control inputs based on the Euler angle, and hence, (9.18) is transformed as follows:

$$\begin{bmatrix} \theta_{rp} \\ \phi_{rp} \end{bmatrix} = \begin{bmatrix} \cos\psi_p & \sin\psi_p \\ -\sin\psi_p & \cos\psi_p \end{bmatrix} \begin{bmatrix} u_{Xp} \\ u_{Yp} \end{bmatrix} \tag{9.20}$$

Here θ_{rp} is the command for the pitch angle and ϕ_{rp} is the command for the roll angle. The attitude controller plays the roll of a virtual actuator in the velocity control. The convergence of the attitude controller is sufficiently fast, and hence, we assume that the transfer function of the attitude controller is 1 in the design of the velocity controller.

9.4.2 Position Model

In the design of the position controller, we deal with the inner velocity controller as the position model. The dynamics of the observer of the velocity controller and the attitude controller are sufficiently fast, and we assume that their transfer function is 1. From (9.11) and (9.17), we have the velocity closed loop system as follows:

$$\dot{x}_{va_p}(t) = A_{va}x_{va_p}(t) - B_{va}F_v x_{va_p}(t) + G_{va}r_{vp}(t) \tag{9.21}$$

$$= (A_{va} - B_{va}F_v)x_{va_p}(t) + G_{va}r_{vp}(t) \tag{9.22}$$

By adding an integrator to (9.22), we have a translational position model described as follows:

$$\dot{x}_p(t) = Ax_p(t) + Br_{vp}(t) \tag{9.23}$$

$$y_p(t) = Cx_p(t) \tag{9.24}$$

$$A = \begin{bmatrix} 0 & 0 & 1 & 0 & 0 & 0 & 0 & 0 \\ 0 & 0 & 0 & 0 & 1 & 0 & 0 & 0 \\ 0_{6\times1} & & A_{va} & - & B_{va} & F_v & & \end{bmatrix}, \quad B = \begin{bmatrix} 0_{2\times2} \\ G_{va} \end{bmatrix},$$

$$C = \begin{bmatrix} I_2 \; 0_{2\times6} \end{bmatrix} \tag{9.25}$$

$$x_p(t) = \begin{bmatrix} X_p(t) \; Y_p(t) \; x_{va_p}^T \end{bmatrix}^T \tag{9.26}$$

The input of (9.23) is the velocity command.

Figure 9.4 shows a comparison between an experimental result and a model output. It can be seen that the model can approximate the dynamics of the helicopter.

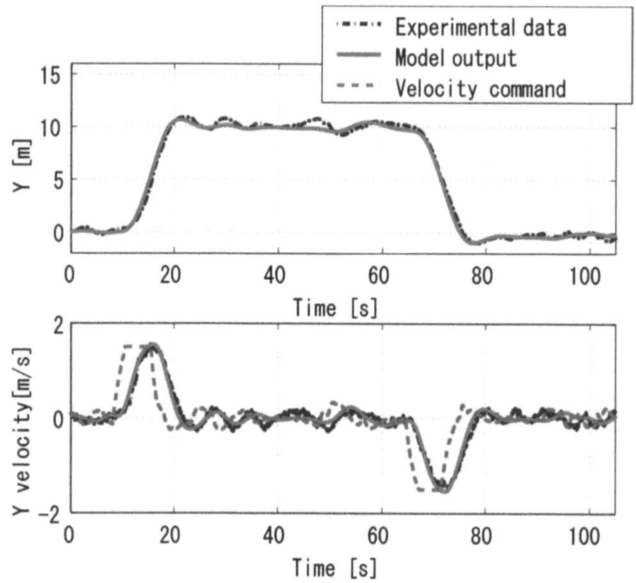

Fig. 9.4 Comparison of the experimental data and model output

9.4.3 Model Predictive Controller Design

In this section, we design the translational position controller for each follower by using the model predictive control theory.

Model predictive control is a model-based control method that optimizes the responses from the present time until a finite time future (horizon) after considering trajectory following or constraints on the system at each time. The horizon of the optimal control decreases with time, and hence, this method is also called the receding horizon control. Figure 9.5 illustrates the concept of model predictive control. In the following discussion, we derive the condition for optimality under the position constraints in the formation flight. Note that we assume that all the helicopters can communicate with each other.

In order to follow the reference position without any tracking error, we design the servo system. The integral of the error ε_p between the reference position

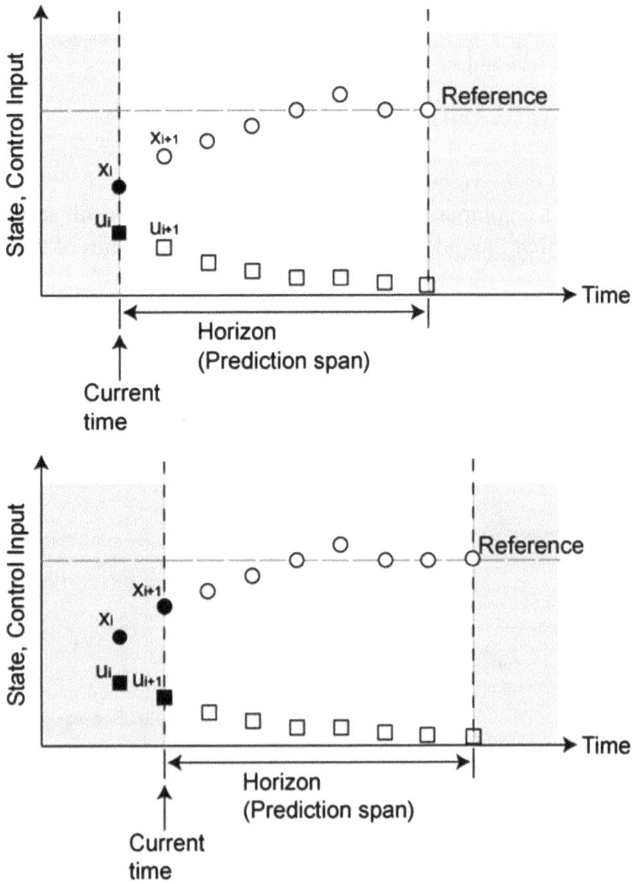

Fig. 9.5 Concept of model predictive control

$r_{vp}(t) = \begin{bmatrix} X_{rp} & Y_{rp} \end{bmatrix}^T$ and the position data estimated by navigation system is included in the state variable, and we have the following augmented state equation as follows:

$$\dot{x}_{ap}(t) = A_a x_{ap}(t) + B_a r_{vp}(t) + G_a r_p(t) \tag{9.27}$$

$$A_a = \begin{bmatrix} A & 0_{8\times 2} \\ -C & 0_{2\times 2} \end{bmatrix}, \; B_a = \begin{bmatrix} B \\ 0_{2\times 2} \end{bmatrix}, \; G_a = \begin{bmatrix} 0_{8\times 2} \\ I_2 \end{bmatrix},$$

$$x_{ap} = \begin{bmatrix} x_p^T & \varepsilon_p^T \end{bmatrix}^T \tag{9.28}$$

Next, we set up the constraints in the flight of multiple helicopters. When we change the shape of the formation by changing (R_p, L_p), considering the distance that each helicopter can communicate, we set the following constraining conditions in order to restrict the position of the follower "p" – $X_{\min} \le (X_p - X_l) \le X_{\max}$ and $Y_{\min} \le (Y_p - Y_l) \le Y_{\max}$ – for changing the shape of the formation.

$$\begin{aligned} g_1(x_{ap}) &= X_{\min} - (X_p - X_l) \le 0 \\ g_2(x_{ap}) &= (X_p - X_l) - X_{\max} \le 0 \\ g_3(x_{ap}) &= Y_{\min} - (Y_p - Y_l) \le 0 \\ g_4(x_{ap}) &= (Y_p - Y_l) - Y_{\max} \le 0 \end{aligned} \tag{9.29}$$

From (9.1), (9.29) can be written as follows:

$$\begin{aligned} g_1(x_{ap}) &= X_{\min} - (X_p - X_{rp} + R_p \cos\psi_l - L_p \sin\psi_l) \le 0 \\ g_2(x_{ap}) &= (X_p - X_{rp} + R_p \cos\psi_l - L_p \sin\psi_l) - X_{\max} \le 0 \\ g_3(x_{ap}) &= Y_{\min} - (Y_p - Y_{rp} + R_p \sin\psi_l + L_p \cos\psi_l) \le 0 \\ g_4(x_{ap}) &= (Y_p - Y_{rp} + R_p \sin\psi_l + L_p \cos\psi_l) - Y_{\max} \le 0 \end{aligned}$$

The following penalty functions are set with respect to these conditions [20].

$$P_j = \begin{cases} (g_j(x_{ap}) - \dfrac{\mu_k}{2} h_k(t))^2 + 3\left(\dfrac{\mu_k h_k(t)}{2}\right)^2, & g_j(x_{ap}) > 0 \\ \dfrac{(\mu_k h_k(t))^3}{g_j(x_{ap}) + \mu_k h_k(t)}, & g_j(x_{ap}) \le 0 \end{cases} \tag{9.30}$$

$(j = 1, 2, 3, 4)$

Here $\mu_k > \mu_{k+1} > 0$, $\lim_{k\to\infty}\mu_k = 0$, and $h_k(t) \geq 0$. When $g_j < 0$, $h_k(t) = 0$ on an optimal trajectory. In this book, $h_k(t)$ are designed as follows:

$$h_k(t) = \begin{cases} k_h g_1(x_{ap})\sqrt{\left|X_{\min} - (X_{rp} - X_l)\right|}, \; g_1(x_{ap}) > 0 \\ k_h g_2(x_{ap})\sqrt{\left|(X_{rp} - X_l) - X_{\min}\right|}, \; g_2(x_{ap}) > 0 \\ k_h g_3(x_{ap})\sqrt{\left|Y_{\min} - (Y_{rp} - Y_l)\right|}, \; g_3(x_{ap}) > 0 \\ k_h g_4(x_{ap})\sqrt{\left|(Y_{rp} - Y_l) - Y_{\min}\right|}, \; g_4(x_{ap}) > 0 \end{cases} \tag{9.31}$$

Moreover, the artificial potential function on the follower "p" for avoiding the collision between the followers "p" and "q" is given as follows:

$$P_c = K \ln\left\{a_p^2(X_p - X_q)^2 + b_p^2(Y_p - Y_q)^2\right\} \tag{9.32}$$

$$K = \begin{cases} K_o, \; \sqrt{(X_p - X_q)^2 + (Y_p - Y_q)^2} < d \\ 0, \; \sqrt{(X_p - X_q)^2 + (Y_p - Y_q)^2} \geq d \end{cases} \tag{9.33}$$

Here K_o, a_p, b_p are the parameters for tuning the performance.

We consider a real-time optimal control problem that minimizes (9.34) under the following constraints:

$$J = \int_t^{t+T} L(x_{ap}(\tau), r_{vp}(\tau))d\tau + \Phi(x_{ap}(t+T)) \tag{9.34}$$

Here

$$L(x_{ap}(\tau), r_{vp}(\tau)) = \frac{1}{2}\{(x_{ap}(\tau) - x_{rp}(\tau))^T Q(x_{ap}(\tau) - x_{rp}(\tau)) + r_{vp}^T R r_{vp}\} \tag{9.35}$$

$$\begin{aligned} \Phi(x_{ap}(t+T)) &= \frac{1}{2}(x_{ap}(t+T) - x_{rp}(t+T))^T S \\ &\quad (x_{ap}(t+T) - x_{rp}(t+T)) \\ &\quad Q, S \geqslant 0, R > 0 \end{aligned} \tag{9.36}$$

$$x_{rp}(t) = \lfloor X_{rp}(t) \quad Y_{rp}(t) \quad 0_{1\times 8} \rfloor \tag{9.37}$$

From (9.27), (9.30), and (9.32), the performance index (9.34) can be augmented as follows:

$$\begin{aligned} J_a = &\int_t^{t+T} L^*(x_{ap}(\tau), r_{vp}(\tau))d\tau + \Phi(x_{ap}(t+T)) \\ &+ \frac{1}{\mu_k}\sum_{j=1}^{4}\int_t^{t+T} P_j dt' + \sum_{p\neq q}^{n}\int_t^{t+T} P_c \tau' \end{aligned} \tag{9.38}$$

Here

$$L^*(x_{ap}(\tau), r_{vp}(\tau)) = L(x_{ap}(\tau), r_{vp}(\tau)) + \lambda_p^T(\tau)(A_a x_{xp}(\tau) + B_a r_{vp}(\tau) + G_a r_p(\tau) - \dot{x}_{ap}(\tau)) \tag{9.39}$$

and λ_p is the Lagrange multiplier. We divide the horizon T into N steps and discretize the optimal control problem. The conditions for the state variable, optimal velocity and multiplier are obtained using the calculus of variations [21] as follows:

$$x_{ap_{i+1}}(t) = A_d x_{ap_i}(t) + B_d r_{vp_i}(t) + G_d r_{p_i}(t) \tag{9.40}$$

$$x_{ap_0}(t) = x_{ap}(t) \tag{9.41}$$

$$R r_{vp_i}(t) + B_d^T \lambda_{p_{i+1}}(t) = 0 \tag{9.42}$$

$$\lambda_{p_i}(t) = Q(x_{ap_i}(t) - x_{rp_i}(t)) + A_d^T \lambda_{p_{i+1}}(t) + \sum_{j=1}^{4} \left[\frac{\partial P_j}{\partial x_{ap}}\right]^T + \sum_{p \neq q}^{n} \left[\frac{\partial P_c}{\partial x_{ap}}\right]^T \tag{9.43}$$

$$\lambda_{p_N}(t) = S(x_{ap_N}(t) - x_{rp_N}(t))(i = 0, 1, 2, \cdots, N-1) \tag{9.44}$$

Here (9.40) denotes the discrete time approximation of (9.27) divided by the sample time T/N. $x_{ap_i}(t)$ denotes the state on the ith step beginning from the initial state $x_{ap}(t)$. The third and fourth terms of (9.43) are zero when a constraint is not violated. By solving (9.40)–(9.44), the sequence of the optimal velocity command is obtained.

First, from (9.40) and (9.41), we predict the sequence of the state variable $\{x_{ap_i}(t)\}_{i=0}^{N}$ for $x_{ap}(t)$ at time t. Secondly, from (9.43), (9.44) and $\{x_{ap_i}(t)\}_{i=0}^{N}$, we have $\{\lambda_{p_i}(t)\}_{i=0}^{N}$. Finally, we have the sequence of the velocity command $\{r_{vp_i}(t)\}_{i=0}^{N-1}$ from (9.42). The velocity command at time t is chosen as $r_{vp}(t) = r_{vp_0}(t)$ from these calculations.

In the control calculations, the calculation of the integrator on the servo system is discontinued because the inputs for the penalties for violating constraints and collision avoidance are cancelled by the integral control. We chose the error state as the state cost in (9.34) in order to operate the guidance system as a regulator when the integrator is stopped.

9.4.4 Observer Design

In this section, we design the observer for the estimation of the unmeasurable state variables is designed. As mentioned in the previous section, the calculation of the integrator of the servo system can be stopped. Hence, it is desirable to compensate the effects of disturbances or modeling errors when the integrator is stopped.

Consequently, we design the disturbance observer to estimate the unobservable state variables and unknown disturbances.

We define w_p as the disturbance that affects the follower "p"; we then have the following state equation.

$$\begin{aligned}\begin{bmatrix}\ddot{X}_p(t)\\ \dddot{X}_p(t)\end{bmatrix} &= \begin{bmatrix}0 & 1\\ aT & a-T\end{bmatrix}\begin{bmatrix}\dot{X}_p(t)\\ \ddot{X}_p(t)\end{bmatrix} + \begin{bmatrix}0\\ -gT\end{bmatrix}u_{X_p}(t) + \begin{bmatrix}D_1\\ D_2\end{bmatrix}w_p(t)\\ &= \begin{bmatrix}A_{11} & A_{12}\\ A_{21} & A_{22}\end{bmatrix}\begin{bmatrix}\dot{X}_p(t)\\ \ddot{X}_p(t)\end{bmatrix} + \begin{bmatrix}B_1\\ B_2\end{bmatrix}u_{X_p}(t) + \begin{bmatrix}D_1\\ D_2\end{bmatrix}w_p(t)\end{aligned} \tag{9.45}$$

Here $\dot{X}_p$ denotes the measurable velocity and $\ddot{X}_p$ is the unobservable state variable. The estimated values of these state variables are common to the velocity and position controllers. We use the estimated values for both the velocity and position controllers.

The augmented system that includes the disturbance as a state variable is constructed as follows:

$$\begin{bmatrix}\ddot{X}_p(t)\\ \dddot{X}_p(t)\\ \dot{w}_p(t)\end{bmatrix} = \begin{bmatrix}A_{11} & A_{12} & D_1\\ A_{21} & A_{22} & D_2\\ 0 & 0 & 0\end{bmatrix}\begin{bmatrix}\dot{X}_p(t)\\ \ddot{X}_p(t)\\ w_p(t)\end{bmatrix} + \begin{bmatrix}B_1\\ B_2\\ 0\end{bmatrix}u_{X_p}(t) \tag{9.46}$$

Now we define

$$\begin{aligned}A'_{12} &= \begin{bmatrix}A_{12} & D_1\end{bmatrix}, A'_{21} = \begin{bmatrix}A_{21}\\ 0\end{bmatrix},\\ A'_{22} &= \begin{bmatrix}A_{22} & D_2\\ 0 & 0\end{bmatrix},\ B'_2 = \begin{bmatrix}B_2\\ 0\end{bmatrix}\end{aligned} \tag{9.47}$$

The minimal order disturbance observer becomes following equation.

$$\begin{aligned}\dot{\hat{z}}_p = {} & A_{21}\dot{X}_p + A'_{22}\hat{z}_p + B'_2 u_{X_p}\\ & + L\{\ddot{X}_p - (A_{11}\dot{X}_p + A'_{12}\hat{z}_p + B_1 u_{X_p})\}\end{aligned} \tag{9.48}$$

Here $\hat{z}_p$ denotes the estimated value of z_p. Now the variable is changed as follows:

$$\tilde{z}_p = \hat{z}_p - L\dot{X}_p$$

And we have

$$\begin{aligned}\dot{\tilde{z}}_p &= (A'_{22} - LA'_{12})\hat{z}_p + (A'_{12} - LA_{11})\dot{X}_p + (B'_2 - LB_1)u_{X_p}\\ &= \hat{A}\tilde{z}_p + \hat{B}\dot{X}_p + \hat{J}u_{X_p}\end{aligned} \tag{9.49}$$

$$\begin{bmatrix} \dot{X}_p \\ \ddot{X}_p \\ \hat{w}_p \end{bmatrix} = \begin{bmatrix} 0 & 0 \\ 1 & 0 \\ 0 & 1 \end{bmatrix} \tilde{z}_p + \begin{bmatrix} 1 \\ L \end{bmatrix} \dot{X}_p \tag{9.50}$$

$$\hat{A} = A'_{22} - LA'_{12}, \ \hat{B} = A'_{21} - LA_{11} + \hat{A}L, \hat{J} = B'_2 - LB_1 \tag{9.51}$$

In these equations, we assume that $D_1 = B_1$ and $D_2 = B_2$. The observer gain L is designed as $L = 4.04 \times 10^2$ by using the duality with LQR. The estimated state variables are used as the initial value for the calculations of model predictive control, and $\hat{w}_p$ is fed back to cancel the unknown disturbance. This observer is discretized in the implementation. Note that the design of the observer on the Y-axis can be achieved by using the same procedure that we have described in this section.

9.5 Simulations and Experiments

The performance of the proposed formation flight control has been demonstrated by simulations and real flight tests using the path planning and MPC methods described in Sects. 9.3 and 9.4.

9.5.1 Simulations

In order to illustrate the efficiency of the proposed method, we will show the simulation results. Figure 9.6 shows the simulation results of the formation flight control with collision avoidance. This figure shows that each follower converges to the reference and constructs formation flight without a collision. Figure 9.6a illustrates the flight trajectory on the XY plane. It is seen that the guidance system operates such that the formation is maintained. Figure 9.6b shows the distance between the followers; it can be seen that the collision avoidance is achieved.

9.5.2 Experiment

Our UAV platform for the formation flight control experiment is based on SF40 (see Fig. 5.17), which is a single-rotor helicopter of 40-cc displacement for aerial monitoring; it was manufactured by Hirobo Limited. It weighs approximately 10 kg with equipments and has approximately 5 kg payload. The diameter of the main rotor is approximately 1.8 m. SF40 is equipped with a gasoline engine and a tuned muffler for payload augmentation.

Figure 9.7 shows the avionics structure of our formation flight control system – it consists of an embedded control system (ECS) and a host-based control system (HCS). The control system of the leader is based on the ECS, while the follower is operated using the HCS.

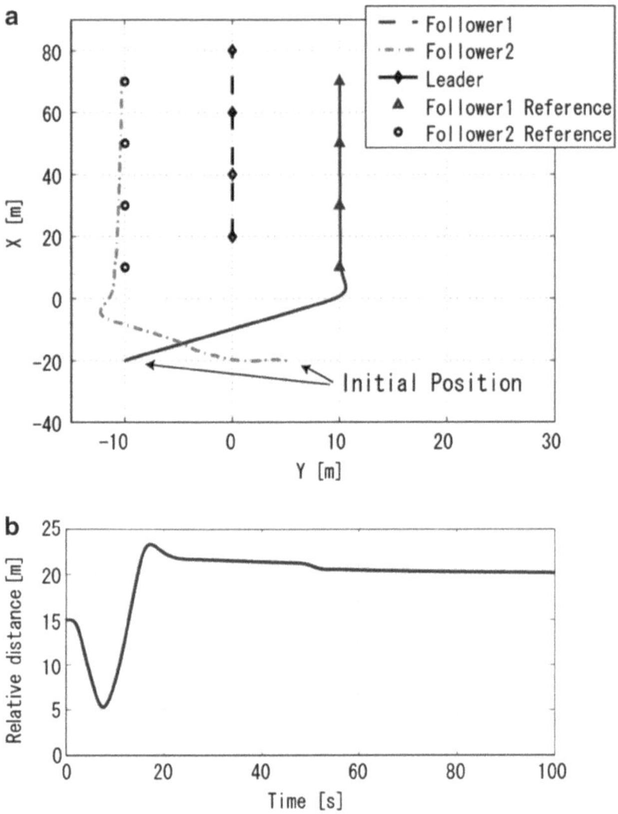

Fig. 9.6 Simulation results of the formation control (**a**) Leader following with collision avoidance. (**b**) Relative distance between followers

The position controller of the leader is a P controller – designed in our previous work [5] – that is implemented on an embedded microcomputer. On the other hand, the position controller of the follower designed by using MPC is implemented on a laptop computer on the ground station using C++. At the ground station, the flight data of each helicopter was monitored, and follower control was calculated. The formation flight control method was applied to experimental helicopter systems. During the experiment, the weather was sunny and calm.

Figure 9.9 shows images of two helicopters in formation flight, while Fig. 9.8 illustrates the experimental results of the flight of two helicopters. In Fig. 9.8, the trajectory on the XY plane, time history of the position, the velocity and pitch angle of the follower, and the relative distance between the leader and follower are shown. In Fig. 9.8a, the helicopter that flies from the origin to the direction of the positive X-axis is the leader. The follower flies ahead of the leader on the right. From Fig. 9.8b, we can see that a precise control performance was achieved, and the

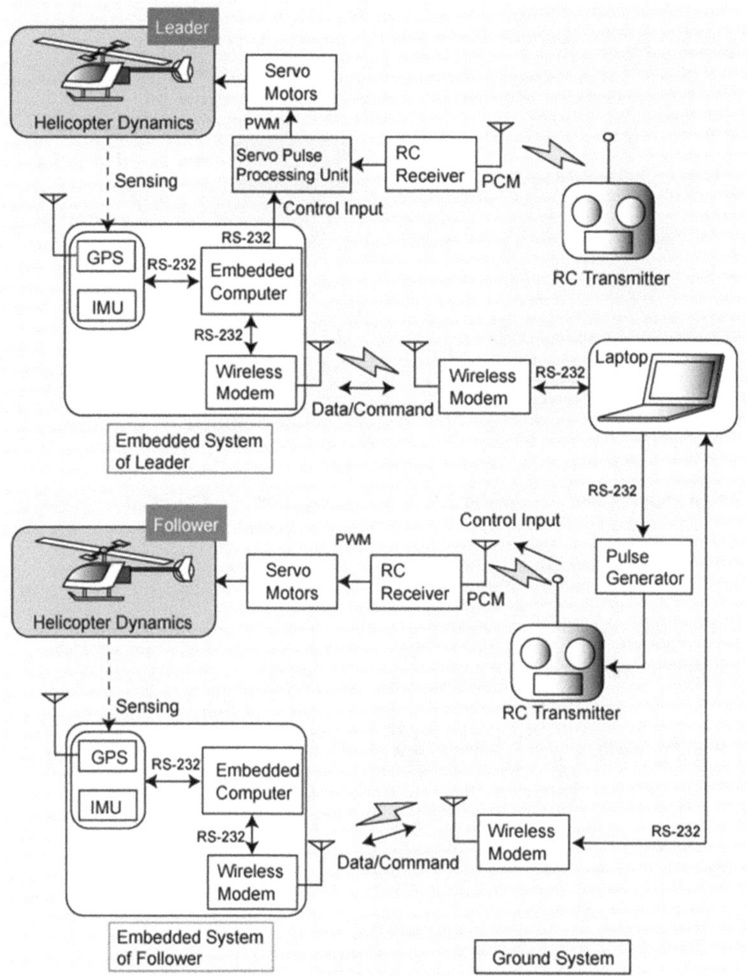

Fig. 9.7 The formation control experimental system

response corresponds to the simulation result. Figure 9.8g shows that the variation in the relative distance is maintained in ± 0.50 m, and the formation is maintained with the desired shape.

9.5.3 Constraint and Collision Avoidance

The experimental results of the position control under the constraints and with the collision avoidance are shown in Figs. 9.10 and 9.11. Experiments in which

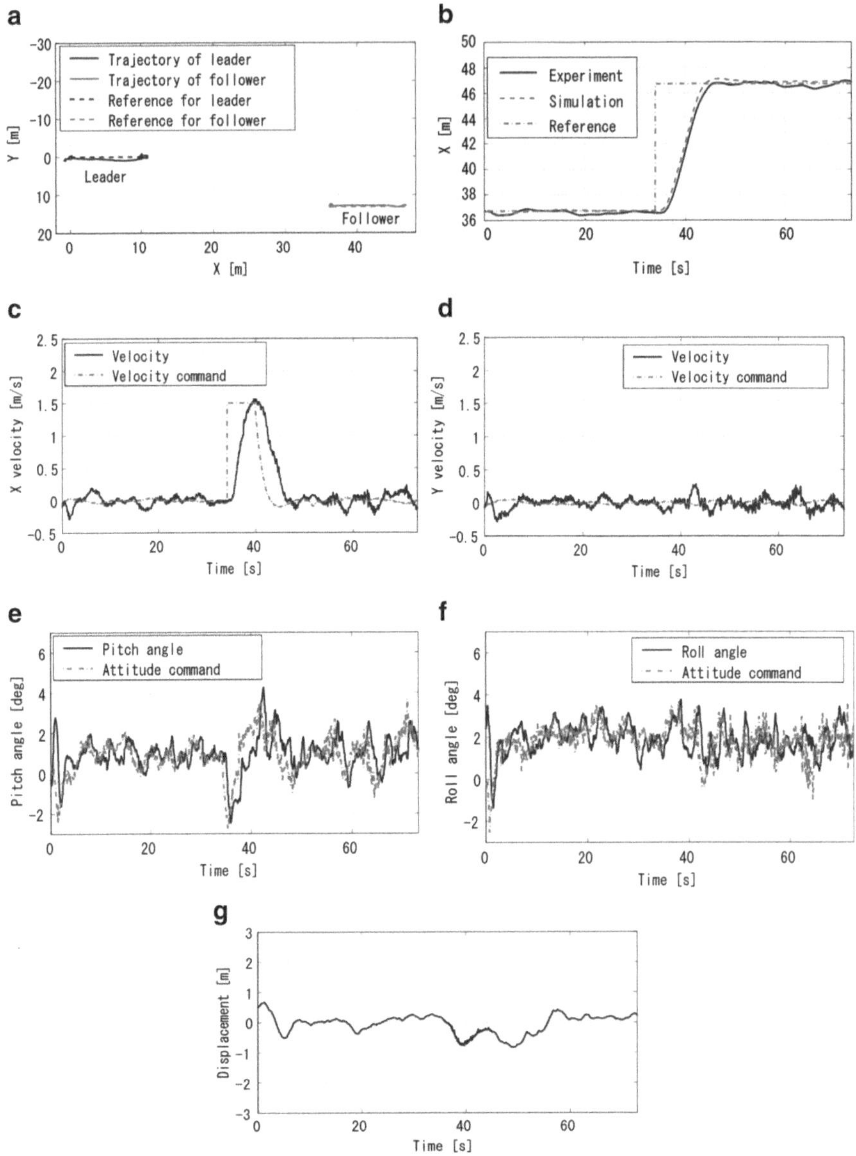

Fig. 9.8 Experimental result of the formation flight control. (**a**) XY plot. (**b**) Position of the follower. (**c**) Velocity of the follower. (**d**) Velocity of the follower. (**e**) Pitch angle of the follower. (**f**) Roll angle of the follower. (**g**) Displacement of the relative distance

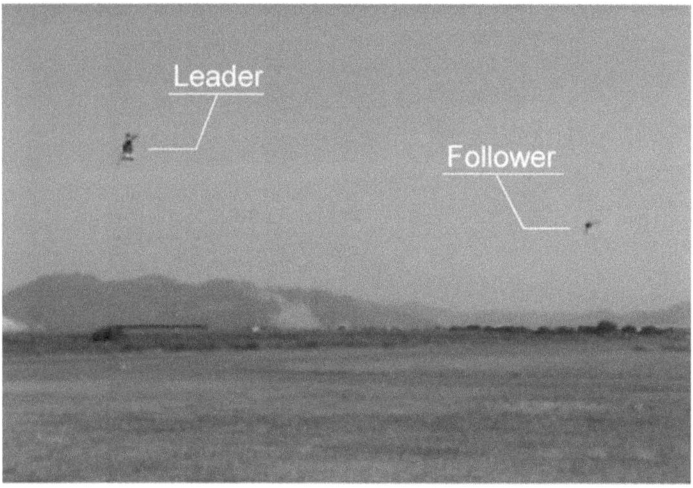

Fig. 9.9 Formation flight control with autonomous helicopters

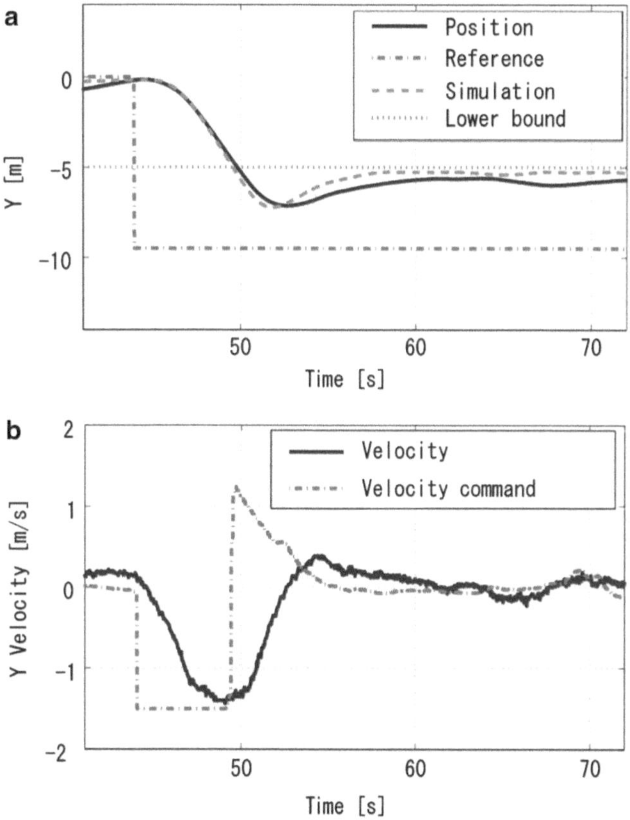

Fig. 9.10 Experimental result of the position control with constraint (**a**) Y. (**b**) Y velocity

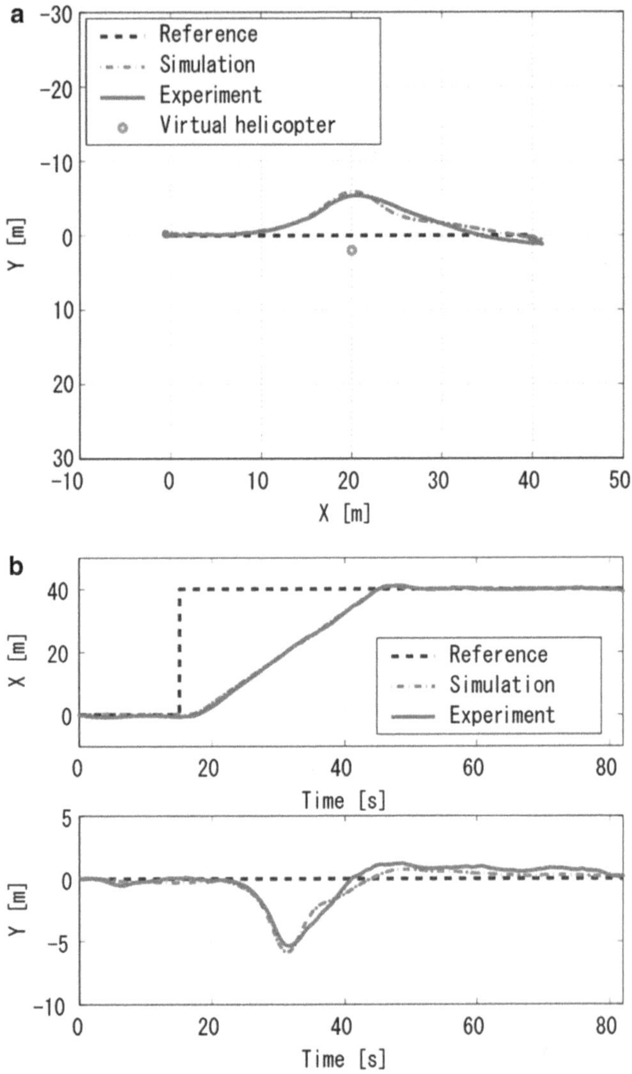

Fig. 9.11 Experimental result of collision avoidance (**a**) XY plot. (**b**) Time domain response

helicopters approach each other are dangerous, and hence, flight tests have been performed by using a single helicopter.

Figure 9.10 shows the flight data under the constraints on the flight area. This figure represents the results under the conditions that the position constraint on the longitude is set as $Y_{\min} = -5$, and the position reference is considered to be violating the constraint. Figure 9.10 demonstrates the handling ability of the designed controller against the constraint. A steady state error exists between the flight data and simulation result; it was caused either by a modeling error or disturbance because when the constraint is violated, the integrator of the position controller stops.

Figure 9.11 shows the results of the collision avoidance. In this experiment, we set a virtual helicopter at $(X, Y) = (20, 2)$, and give the position reference as $(X_{r1}, Y_{r1}) = (40, 0)$ to guide the helicopter. Figure 9.11 demonstrates that the helicopter places the collision avoidance above the trajectory following and achieves collision avoidance. In the flight test, a west wind with wind speeds 3–4 m/s was present. Hence, when the controller stops the integrator to avoid the collision, an error in the experimental result occurs. After the avoidance, the controller begins the integration. Hence, the error converges to 80 s.

9.5.4 Robustness Against Disturbance

Wind disturbance is an important factor for the flight stability. We validated the control performance against the disturbances by conducting experiments under windy conditions.

Figure 9.12a shows the results under the wind speeds 0–3 m/s. On the other hand, Fig. 9.12b shows the results for 5 m/s wind speed. Despite the disturbances, a good performance was achieved. This is sufficient for practical application because the average wind speed is 2–4 m/s in Japan.

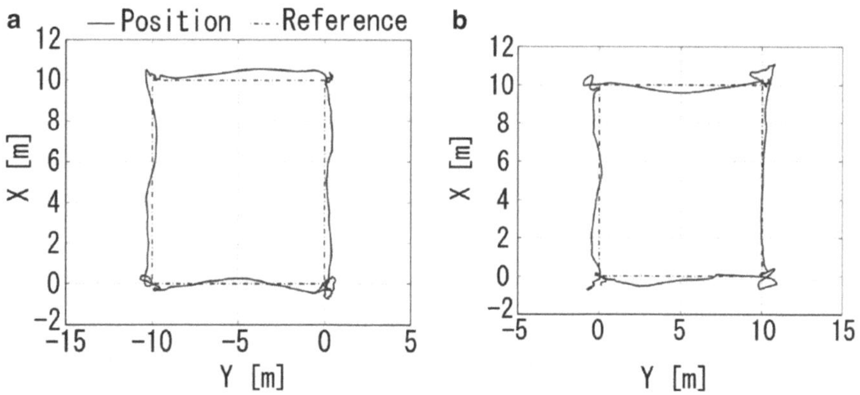

Fig. 9.12 Experimental result of the square trajectory guidance control (**a**) XY plot (weak windy condition). (**b**) XY plot (windy condition)

9.6 Summary

In this chapter, the formation flight control of small-scale unmanned helicopters using predictive control has been presented. In order to construct the formation, we presented the leader–follower strategy. In addition, in order to track the designated formation shape, we designed the position controller of each follower by using the "model predictive control" theory. The optimized performance index was augmented to handle constraints such as collision avoidance and communication range. Simulations and flight tests have demonstrated the effectiveness of our approaches.

References

1. Nakazawa D, Suzuki S, Sakai S, Nonami K (2008) Formation flight control of small unmanned helicopters. Trans Jpn Soc Mech Eng C 74(747):2737–2746
2. Shim DH, Chung H, Kim HJ, Sastry S (2005) Autonomous exploration in unknown urban environments for unmanned aerial vehicles. In: Proceedings of the AIAA guidance, navigation, and control conference and exhibit, pp 1–8
3. Shin J, Fujiwara D, Hazawa K, Nonami K (2002) Attitude control and hovering control of radio-controlled helicopter. Trans Jpn Soc Mech Eng C 68(675):148–155
4. Johnson EN, Calise AJ, Sttigeri R, Watanabe Y, Madyastha V (2004) Approaches to vision-based formation control. In: Proceedings of the IEEE conference on decision and control, pp 1643–1648
5. Hazawa K, Shin J, Fujiwara D, Igarashi K, Fernando D, Nonami K (2003) Autonomous flight control of hobby-class small unmanned helicopter (Report 2: modeling based on experimental identification and autonomous flight control experiments). J Robot Mechatron 15(5):546–554
6. Saffarian M, Fahimi F (2007) A novel leader – follower framework for control of helicopter formation. In: Proceedings of the IEEE aerospace conference, pp 1–6
7. Oh S, Johnson EN (2007) Relative motion estimation for vision-based formation flight using unscented Kalman filter. In: Proceedings of the AIAA guidance, navigation, and control conference, pp 1–17
8. Schouwenaars T, Feron E, How J (2006) Multi-vehicle path planning for non-line of sight communication. In: Proceedings of the 2006 American control conference, pp 5757–5762
9. Galzi D, Shtessel Y (2006) UAV formations control using high order sliding modes. In: Proceedings of the 2006 American control conference, pp 4249–4254
10. Dimarogonas DV, Kyriakopoulos KJ (2006) Distributed cooperative control and collision avoidance for multiple kinematic agents. In: Proceedings of the IEEE conference on decision and control, pp 721–726
11. Casbeer DW, Beard RW, McLain TW, Li S, Mehra RK (2005) Forest fire monitoring with multiple small UAVs. In: Proceedings of the 2005 American control conference, pp 3530–3535
12. Walle D, Fidan B, Sutton A, Yu C, Anderson BDO (2008) Non-hierarchical UAV formation control for surveillance tasks. In: Proceedings of the 2008 American control conference, pp 777–782
13. Borrelli F, Keviczky T, Balas GJ (2004) Collision-free UAV formation flight using decentralized optimization and invariant sets. In: Proceedings of the 43rd IEEE conference on decision and control, pp 1099–1104
14. Shin J, Kim HJ (2009) Nonlinear model predictive formation flight. IEEE Trans Syst Man Cybern A Syst Hum 39(5):1116–1125
15. Moshtagh N, Michael N, Jadbabaie A, Daniilidis K (2009) Vision-based, distributed control laws for motion coordination of nonholonomic robots. IEEE Trans Robot 25(4):851–860

16. Kim S, Kim Y (2007) Three dimensional optimum controller for multiple UAV formation flight using behavior-based decentralized approach. In: Proceedings of the international conference on control, automation and systems 2007, pp 1387–1392
17. Direks T, Jagannathan S (2009) Neural network control of quadrotor UAV formations. In: Proceedings of the 2009 American control conference, pp 2990–2996
18. Ren W (2007) Consensus strategies for cooperative control of vehicle formations. IET Contr Theor Appl 1(2):505–512
19. Wang X, Yadav V, Balakrishnan SN (2007) Cooperative UAV formation flying with obstacle/collision avoidance. IEEE Trans Contr Syst Technol 15(4):672–679
20. Kameyama Y, Sayama H (1974) Penalty method for optimal control problems with state constraints. Trans Soc Instrum Contr Eng 10(3):272–277
21. Kato K (1988) Optimal control – approach to nonlinear control. University of Tokyo Press, Tokyo

Part III
Guidance and Navigation of Short-Range UAVs

Chapter 10
Guidance and Navigation Systems for Small Aerial Robots

Abstract As the capabilities of Unmanned Aerial Vehicles (UAVs) expand, increasing demands are being placed on the hardware and software that comprise their guidance and navigation systems. Guidance, navigation and control algorithms are the core of flight software of UAVs to successfully complete the assigned mission through autonomous flight. This chapter describes some guidance and navigation systems that we have designed and successfully applied to the autonomous flight of a mini rotorcraft UAV that weighs less than 0.7 kg. The real-time flight test results show that the vehicle can perform autonomous flight reliably in indoor and outdoor environments.

Video Link:

Spiral Trajectory Tracking
http://mec2.tm.chiba-u.jp/monograph/Videos/Chapter10/1.wmv

10.1 Introduction

Unmanned aerial vehicles are encountered in an increasing number of civilian and military applications like inspecting and exploring dangerous and complex environments (disaster areas, battlefields, etc.). Autonomous guidance, navigation and control (GNC) are required to achieve these missions without direct or continuous human control. Guidance and navigation systems were originally developed for rockets. The systems entered more widespread use with the advent of spacecraft, guided missiles, and UAVs. Figure 10.1 shows a simple block diagram of guidance, navigation and control systems in a UAV. GNC consists of three essential parts: (1) guidance system which steers the UAV toward a desired state (location, target, etc.); (2) navigation system which tracks current location of the UAV and estimates its state; and (3) control loop (autopilot) which accepts guidance commands and uses navigation data to generate the actual actuator signals to follow the guidance objectives as well as to stabilize the vehicle attitude.

K. Nonami et al., *Autonomous Flying Robots: Unmanned Aerial Vehicles and Micro Aerial Vehicles*, DOI 10.1007/978-4-431-53856-1_10, © Springer 2010

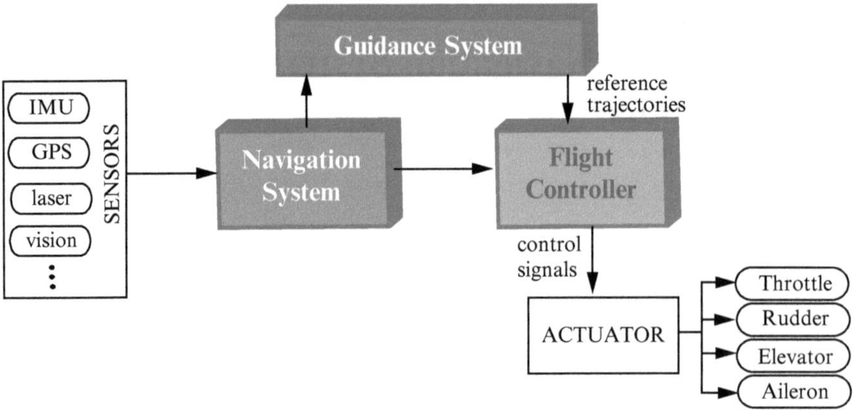

Fig. 10.1 The overall structure of guidance, navigation and control systems in UAV

Guidance systems: A guidance system can be defined as the "driver" of a vehicle to achieve some mission (reach a destination for example) while guaranteeing safety (avoid obstacles for example). It takes inputs from the navigation system (where am I) and uses targeting information (where do I want to go) to send signals to the flight control system that will allow the vehicle to reach its destination within the operating constraints of the vehicle. A guidance system has thus the role of planning the flight path and generating the desired state vectors (position, velocity, heading and height) that must be tracked by the vehicle.

Current operating UAVs are operated remotely by a human operator from a ground control station, or use rudimentary guidance systems, such as following pre-planned or manually provided waypoints or reference trajectories. This is sufficient for UAVs operating at high altitude for reconnaissance and exploration. However, UAVs will require more advanced guidance capabilities to achieve more complex tasks and missions. Furthermore, small and micro UAVs are generally designed to operate at low altitude in cluttered and dynamic environments. Therefore, advanced guidance and navigation capabilities are necessary to detect obstacles, map the environment and plan the flight path.

Planing the UAV flight path or trajectory is the chief problem in autonomous UAV deployment. The basic difficulties include partially known and changing environments, extraneous factors such as threats, and computation complexity because the trajectory planing problem is generally solved using mathematical optimization methods which are computationally too demanding. To address these issue, guidance problems can be simplified according to the intended application and the required vehicle autonomy. Among the approaches applied for UAVs path planing, one can cite belief roadmap [7], MILP [20], evolutionary algorithms [8], receding horizon control [13], occupancy grid [19], etc.

Currently, there is a growing interest in increasing the vehicle autonomy by developing guidance systems that are able to tackle a number of operational events without operator intervention like for example trajectory replanning following the detection of a threat or changes in the mission or environment configuration. Guidance capabilities can be also increased to allow integration with other UAVs and manned vehicles to accomplish coordinated mission. In order to develop appropriate guidance technologies, one needs to understand and evaluate the required autonomy level. This latter can be characterized based on the complexity of operational objectives, vehicle capabilities, environmental elements, and the a priori knowledge about these elements. Generally, as the autonomy increases, the UAV can operate in more complex environments, execute more complex tasks with less a priori knowledge and less amount of operator interaction. The reader can refer to [16] which describes key autonomy capabilities that are critical for future UAV operation.

Navigation systems: In the broad sense, navigation is the art of determining where you are. For autonomous vehicles such as robots and UAVs, navigation can be defined as the process of extracting information about the vehicle's self-motion and also the surrounding environment. This information can be metric like distances, topological like landmarks and/or any other cues that are useful to achieve the task and to guarantee safety like for example using optic flow to avoid obstacles.

The navigation system plays a key role in aircraft system. Typically, standard navigation algorithms fuse information from many sources (multiple sensors, preregistered maps, etc.) to estimate the vehicle's attitude, angular rates, height, velocity and position with respect to some fixed point of reference or relative to some target. Advanced navigation systems include also mapping, obstacles detection, localization, etc. Navigation outputs are used in guidance and control and affect the performance of flight and mission achievement. Much work with UAVs has been done using both Inertial Measurement Unit (IMU) and Global Positioning System (GPS) to achieve autonomous flight and execute the mission [5]. The most used navigation systems are Attitude Heading Reference System (AHRS) and GPS/INS. AHRS fuses data from three gyros, three accelerometers and three magnetometers to estimate the vehicle's attitude and angular rates. The Inertial Navigation System (INS) uses also the same sensors to estimate the vehicle's velocity and position relative to the starting point. All inertial navigation systems suffer from "integration drift" which results in growing errors in velocity and position estimates. Therefore, inertial navigation is usually used to supplement other navigation systems, providing a higher degree of accuracy than is possible with the use of any single system. Indeed, UAVs normally rely on INS and Global Navigation Satellite Systems (GNSS) such as GPS to provide position and velocity information for navigation. For GNSS-denied environments such as indoor spaces, vision is commonly used to provide the navigation solution. Furthermore, most robotic missions are defined within the environment. Terrain relative navigation is thus necessary to achieve these missions. Mapping and obstacles detection can be done by using active sensors (radar, lidar, laser range finder, ultrasonic sensors and infrared sensors) or passive imaging sensors (single camera, stereo camera, photoreceptor, bio-inspired vision ships, etc.).

Today's complex navigation systems typically have multiple redundancy to address drift, improve accuracy (ex. relative to a target) and address isolated system failure. Navigation systems therefore take multiple inputs from many different sensors, both internal to the system and/or external. Kalman filter provides the most common approach to combining navigation data (from multiple sensors) to resolve current position. However, there are multiple approaches to fuse sensors data like particle filters, complementary filters, etc.

In this chapter, we present a simple guidance system that allows small rotorcraft UAV to achieve some missions autonomously in obstacle-free environments. The designed guidance system includes also some safety procedures that handle some events such as sensors failure. A GPS/INS/AHRS navigation system is also described, which provides the guidance loop and the flight controller with all the required data to achieve autonomous flight. In addition to this standard navigation system, four vision-based navigation systems are also presented and demonstrated in real-time experiments.

10.2 Embedded Guidance System for Miniature Rotorcraft UAVs

In this section, we present a guidance system that gives some autonomy for miniature rotorcraft. It offers to the vehicle the ability to follow waypoints[1] and to execute other pre-programmed maneuvers like automatic take-off and landing, hovering and trajectory tracking. The overall system architecture considered here consists of six layers: (1) GCS for mission definition and high-level decisions making, (2) guidance for path planning and trajectory generation, (3) navigation for vehicle's state vector estimation, (4) nonlinear controller for stabilization and trajectory tracking, (5) communication with the GCS and interface between the autopilot and the vehicle, and (6) the MAV platform. All GN&C algorithms have been implemented in visual C using Multi-thread programming. In our design, the autopilot software is implemented as process within the Linux OS. It is composed of smaller units of computation, know as tasks (or threads). By using *POSIX* style semantics, theses tasks are called and scheduled separately.

The proposed guidance strategy is simple and has the role of selecting the appropriate flight mode and generating the corresponding reference trajectories. It is composed of four main parts: (1) path planning, (2) mode transition management, (3) fault detection and fail-safes, and (4) trajectory generation. Although this approach is simple, it provides sufficient capabilities for the tasks required of the rotorcraft MAV in most applications.

[1] Here waypoint navigation is defined as the process of automatically following a path defined by a set of geodetic coordinates (GPS coordinates).

10.2.1 Mission Definition and Path Planning

The flight plan or mission is defined by the human operator using the GCS interface. This plan is expressed in terms of desired 3D waypoints and other maneuvers like hovering, landing, trajectory tracking, target tracking, et. During mission execution, the flight plan can be modified or updated in real-time. These commands are transmitted from the GCS to the processor of the onboard avionics unit via a 2.4 GHz WiFi link.

The proposed guidance strategy is simple and has the role of selecting the appropriate flight mode and generating the corresponding reference trajectories. It is composed of three main parts: (1) path planning, (2) mode transition management and (3) trajectory generation. Although this approach is simple, it provides sufficient capabilities for the tasks required of the rotorcraft MAV in most applications.

The transmitted waypoints from the GCS are expressed in the Geodetic LLA (Latitude–Longitude–Altitude) coordinates. They are then converted into the NED (North–East–Down) local frame where they are expressed in meter according to the starting point. A path planning routine constructs a path, which is defined by a sequence of N desired waypoints and desired speeds of travel along path segment S_i connecting waypoint wpt_i to wpt_{i+1}. The path planner is also required to have some strategy to detect when the waypoint wpt_i is reached by the rotorcraft in order to switch to the next one (wpt_{i+1}). The proposed strategy attempts to achieve the desired position initially with 10 m accuracy in both lateral and longitudinal directions. It takes then, an additional 20 s to achieve the position with 2 m accuracy, and finally takes additional 5 s to achieve higher accuracy. If the waypoint is not achieved to within 2 m in the 20 s, the path planner assumes that external factors (i.e., wind) are interfering and ends the attempt. Flight paths that passed within these thresholds were considered to have reached the waypoint so the guidance law then directs the UAV to reach the next waypoint. In this way, we are guaranteed some baseline level of performance (10 m), and the vehicle will attempt to achieve a higher level of accuracy without excessive time delays.

10.2.2 Flight Mode Management

The mode transition manager has the role of managing three autonomous flight modes (GPS-based flight, vision-based flight and target tracking mode) and various flight behaviors (landing, hovering, trajectory tracking, emergency landing, etc.). These different modes and maneuvers are either selected by the human operator through the GCS interface or by the *Fail-Safe* system, described in Sect. 10.2.3. In order to minimize errors and reduce risk, each flight mode and maneuver has pre-conditions that must be satisfied, otherwise the task will be rejected. For example, when the GPS signal is weak or lost, the system switches automatically to vision-based flight and the landing procedure is activated after some waiting-time to achieve a safe vision-based landing. Furthermore, some maneuvers can not be executed in parallel such as emergency landing and take-off for example. If two non-compatible tasks are selected, then, the maneuver with higher priority will be executed and the other one will be blocked.

10.2.3 Safety Procedures and Flight Termination System

When developing autonomous UAVs and MAVs for civilian applications, some level of safety must be guaranteed. Therefore, we have implemented several safety procedures on our rotorcraft to keep single failures from having catastrophic effects on both people and the vehicle itself.

10.2.3.1 Pre-programmed Fail-Safes

The developed autopilot has a number of pre-programmed fail-safes instructing the vehicle to perform a certain task when some unexpected problem occurs. Some of those are discussed here:

- WiFi communication: if the vehicle loses signal with the GCS for more than 5 s, the hovering mode is automatically enabled. If the communication is not established in 20 s, then the vehicle will switch to the "homing mode", and it will navigate to a pre-assigned location (home).
- Fly zone: when the vehicle travels out of the fly zone, the hovering mode is activated, waiting for new commands from the GCS to correct the trajectory or to recover the vehicle.
- GPS unit failure: in the case of GPS outages or failure, the rotorcraft will achieve hovering using visual estimates. Depending on the situation, the operator may continue the mission using vision system or activate the emergency landing procedure.
- IMU unit failure: if IMU data are not available, emergency landing is achieved by incrementally reducing the thrust.
- Undesired behavior: if the aircraft experiences some undesired behavior, which may be caused by erroneous sensors data, controller problems, mechanical problem, or other, then the safety pilot can switch to manual flight and re-gain control of the vehicle when this later is in the RC radio communication range. Otherwise, the operator at the GCS can terminate the flight.

10.2.3.2 Flight Termination System

The safety procedures discussed above are implemented in the main FCC. A malfunction of the flight computer may occur due to some software bug or hardware problem. Thus, all the programmed safety procedures can not be executed. To provide more safety, a simple independent flight termination system (FTS) is used to reduce and decrement the thrust from its nominal value to zero, thereby resulting in a soft emergency landing.

A secondary micro-controller is used as the primary interface to the physical vehicle. Moreover, it forms a FTS since it implements a flight termination program that can be activated in three different ways:

1. At any point of the mission, the operator can terminate the flight by pushing the "emergency" button on the GCS interface.
2. If the main FCC stops working for any reason, then, the flight termination program is automatically activated.
3. The sixth channel of the RC receiver is directly connected to the FTS. Therefore, the safety pilot can use this channel to terminate the flight at any time, provided that the vehicle is in the radio communication range.

These safety precautions are very important especially during the autopilot development and testing stages where the system is vulnerable due to programming errors, bad controller tuning, etc. We have successfully tested all the listed safety procedures without damaging the vehicle.

10.2.4 Real-Time Generation of Reference Trajectories

The goal of this routine is to generate a physically feasible trajectory according to the flight behavior selected by the mode transition manager. In the case of waypoint navigation for example, the trajectory generation routine takes straight-line path segments and produces feasible time-parameterized desired trajectories which are the desired 3D position $(x_d(t), y_d(t), z_d(t))$, the desired heading $\psi_d(t)$ and their time derivatives $(\dot{x}_d(t), \dot{y}_d(t), \dot{z}_d(t), \dot{\psi}_d(t))$. The kinematic model used for trajectory generation uses specifiable limits on the maximum speed and acceleration the rotorcraft may have during a maneuver. In fact, reference trajectories are generated in two steps: first, the desired velocity trajectory is computed from the specified accelerations and the desired maximum speed during forward flight. Second, the position reference trajectory is obtained by integrating the velocity trajectory with respect to time. When the vehicle approaches a desired waypoint, it starts decelerating in order to reach the waypoint by zero velocity. This simple but effective strategy allows accurate and robust transition from one waypoint to the next without overshoot on sharp corners. Furthermore, the obtained position and velocity trajectories are differentiable with respect to time, which is needed for some control designs.

To achieve non-aggressive take-off and soft landing, smooth reference trajectories $z_d(t)$ along the vertical axis are generated in real-time based on the desired ascent/descent velocity and the actual height of the vehicle. For search operations, a spiral trajectory is generated in real-time, thereby allowing the rotorcraft to explore some target area. Furthermore, this spiral trajectory will allow to evaluate the performance of the developed flight controller for tracking non-straight and arbitrary trajectories. Our implementation allows also to incorporate easily many other trajectories.

Although the resulting trajectories are not necessarily optimal, the lower computation burden of this trajectory generator ensures its implementation on small onboard microprocessors.

The kinematic model used for trajectory generation uses specifiable limits on the maximum speed and acceleration the aircraft may have during a maneuver. Let the trajectory generator have the following form:

$$\begin{cases} \dot{x}_d(t) = V_d(t)\cos(\psi_d(t)) \\ \dot{y}_d(t) = V_d(t)\sin(\psi_d(t)) \\ \dot{z}_d(t) = Vz_d(t) \end{cases}, \quad \dot{V}_d(t) = \begin{cases} a_d & \text{when starting} \\ 0 & \text{when } V_{\max} \text{ is reached} \\ -a_d & \text{when stopping} \end{cases} \tag{10.1}$$

where $V_d(t)$ is the commanded linear speed along the path, $Vz_d(t)$ is the desired vertical velocity and a_d is the desired acceleration.

In the case of spiral trajectory generation, we have modified the *Archimedean spiral* in order to obtain a spiral with constant separation distance between successive turnings, but also with a constant tangential speed (implies time-varying angular speed contrary to standard *Archimedean spiral*). Equivalently, in polar coordinates (r, ψ) it can be described by the equation

$$\begin{cases} r_d(t) = c_1 + c_2\psi_d \\ \dot{\psi}_d(t) = \dfrac{V_d(t)}{r_d(t)} \end{cases}, \quad \dot{V}_d(t) = \begin{cases} a_d & \text{when starting} \\ 0 & \text{when } V_{\max} \text{ is reached} \\ -a_d & \text{when stopping} \end{cases} \tag{10.2}$$

In cartesian coordinates, we obtain

$$\begin{cases} x_d(t) = r_d(t)\cos(\psi_d(t)) \\ y_d(t) = r_d(t)\sin(\psi_d(t)) \\ Vx_d(t) = \dfrac{c_2 V_d(t)}{r_d(t)}\cos(\psi_d(t)) - V_d(t)\sin(\psi_d(t)) \\ Vy_d(t) = \dfrac{c_2 V_d(t)}{r_d(t)}\sin(\psi_d(t)) + V_d(t)\cos(\psi_d(t)) \end{cases} \tag{10.3}$$

This guidance system has been implemented on a small quadrotor UAV and demonstrated in real-time flight tests. When combined with navigation systems described in Sect. 10.3 and Chap. 12, and the flight controller described in Chap. 8, it resulted in an embedded flight software that enabled a quadrotor UAV to achieve autonomously several maneuvers and tasks (see Chaps. 8 and 12 for experimental results and flight video clips).

10.3 Conventional Navigation Systems for Aerial Vehicles

The design of sensing and navigation systems is a crucial step in the development of autonomous flying machines. The main objective of a navigation system is to provide information about the vehicle's self-motion (orientation, position, etc.) and its surrounding environment (obstacles, target, etc.). This information is then exploited by the flight controller for motion control and by the guidance system for mission achievement. Position, altitude and orientation measurements are usually sufficient for the control of UAVs operating at high altitudes or in obstacles-free environments. Therefore, conventional avionics that include GPS IMU and barometric altimeter provide the required information for flight control and waypoint navigation.

The navigation system that we have developed and implemented on our quadrotor platform includes Attitude Heading Reference System (AHRS) for attitude estimation, GPS/INS for inertial position and velocity estimation and Pressure Sensor/Inertial Navigation System (PS/INS) for height computation. The experimental results from autonomous flight, presented in Chap. 8, are obtained using the navigation system described in this section.

10.3.1 Attitude and Heading Reference System

An Extended Kalman Filter (EKF) is used to fuse IMU raw data (three gyro., three accelero., three magneto.) in order to provide estimates about the vehicle's attitude, heading and angular rates at an updating rate of 50 Hz.

it is based on the following nonlinear model:

$$\begin{cases} \dot{X}_1 = f_1(X_1, \Omega) + W_1 \\ Y_1 = h_1(X_1) + V_1 \end{cases} \tag{10.4}$$

where $X_1 = [q \quad b_\Omega]^T \in \mathbb{R}^{7\times 1}$ represents the state of the system that is composed of the quaternion, $q = [q_0 \quad \mathbf{q}]^T \in \mathbb{R}^{4\times 1}$, and rate gyro biases, $b_\Omega \in \mathbb{R}^{3\times 1}$, $\Omega \in \mathbb{R}^{3\times 1}$ is the known angular rates, and $Y_1 = [a \quad \psi]^T \in \mathbb{R}^{4\times 1}$ comprises the acceleration measurement, $a \in \mathbb{R}^{3\times 1}$, in the vehicle's body axes and the magnetic heading angle ψ. (W_1, V_1) are the process and the measurement noises, respectively. The nonlinear functions $f_1(X_1, \Omega)$ and $h_1(X_1)$ are given by

$$f_1(X_1, \Omega) = \begin{bmatrix} \frac{1}{2} \begin{bmatrix} -q_1 & -q_2 & -q_3 \\ q_0 & -q_3 & q_2 \\ q_3 & q_0 & -q_1 \\ -q_2 & q_1 & q_0 \end{bmatrix} \begin{bmatrix} p - b_p \\ q - b_q \\ r - b_r \end{bmatrix} \\ 0^{3\times 1} \end{bmatrix} \tag{10.5}$$

$$h_1(X_1) = \begin{bmatrix} 2g(q_1q_3 - q_0q_2) \\ 2g(q_2q_3 + q_0q_1) \\ g(q_0^2 - q_1^2 - q_2^2 + q_3^2) \\ tan^{-1}\left(\dfrac{2(q_1q_2 + q_0q_3)}{q_0^2 + q_1^2 - q_2^2 - q_3^2}\right) \end{bmatrix} \tag{10.6}$$

10.3.2 GPS/INS for Position and Velocity Estimation

GPS measurements (3D position in LLA[2] coordinates and 3D velocity vector in NED[3] frame) are fused with the INS data using an EKF. The filtered and propagated position and velocity estimates are provided at an updating rate of 10 Hz.

The EKF for the INS uses the following nonlinear model:

$$\begin{cases} \dot{X}_2 = f_2(X_2, a') + W_2 \\ Y_2 = h_2(X_2) + V_2 \end{cases} \tag{10.7}$$

The state vector $X_2 = [p \quad v \quad b_a]^T \in \mathbb{R}^{9\times1}$, is composed of the position $p = [\lambda \quad \Phi \quad h]^T$ in the LLA coordinate, velocity $v = [v_N \quad v_E \quad v_D]^T$ in the local NED frame, and accelerometer biases b_a. $a' = [a'_N \quad a'_E \quad a'_D]^T$ are the accelerations in the NED coordinate. The measurement vector includes GPS data, that is, $Y_2 = [p \quad v]^T \in \mathbb{R}^{6\times1}$. The nonlinear functions $f_2(X_2, a')$ and $h_2(X_2)$ are given by

$$f_2(X_2, a') = \begin{bmatrix} \dfrac{v_N}{R_\lambda + h} \\ \dfrac{v_E}{(R_\Phi + h)\cos\lambda} \\ \dfrac{-v_E^2 \sin\lambda}{(R_\Phi + h)\cos\lambda} + \dfrac{v_E v_D}{R_\lambda + h} + a_N - b_{a_N} \\ \dfrac{v_E v_N \sin\lambda}{(R_\Phi + h)\cos\lambda} + \dfrac{v_E v_D}{R_\lambda + h} + a_E - b_{a_E} \\ -\dfrac{v_E^2}{(R_\Phi + h)} - \dfrac{v_N^2}{R_\lambda + h} + g + a_D - b_{a_D} \\ 0 \in \mathbb{R}^{3\times1} \end{bmatrix} \tag{10.8}$$

and $h_2(X_2) = [I^{6\times6} | 0^{6\times3}] X_2$.

---

[2] Latitude–Longitude–Altitude.

[3] North–East–Down.

10.3.3 Altitude Estimation Using Pressure Sensor and INS

The height estimated by the GPS/INS is not accurate enough (5–10 m error) to allow good altitude control. The altitude controller is thus based on the height estimated by the static pressure sensor after fusing it with the vertical acceleration using a kinematic Kalman filter. We have then improved the height estimation accuracy to ± 1.5 m in normal flight conditions. However, the used low-cost barometric pressure sensor is sensitive to weather conditions like wind and temperature.

10.4 Visual Navigation in GPS-Denied Environments

Milestones in manned and unmanned aircraft have been achieved using conventional navigation sensors such as standard IMUs for orientation, GPS for position, pressure sensors for altitude sensing, radar, ultrasound and laser range-finder for detection of obstacles. Our particular interest, however, involves small and micro UAVs flying close to the ground in cluttered environments like urban and indoor environments. Therefore, GPS information may not be available. Furthermore, the substantial weight and energy constraints imposed by small and micro UAVs preclude the use of conventional sensors. On the other hand, visual sensors are passive, lightweight and can provide rich information about the aircraft's self-motion and surroundings structure. Therefore, computer vision can be used for autonomous localization, which is a crucial step for small aerial robot's control and guidance. However, the design of a reliable vision system for aerial vehicles has many unsolved problems, ranging from hardware and software development to pure theoretical issues, which are even more complicated when applied to small flying machines operating in unstructured environments. Moreover, the difficulty found when using imaging sensors is the high bandwidth of data, and the resulting heavy computational burden.

In recent years, there is an active research in developing and applying vision systems for UAVs guidance and navigation. Visual SLAM [4, 14], stereo vision [1], [9] and Structure-From-Motion [11, 12] techniques are generally used as visual navigation systems to localize and estimate the UAV ego-motion. Target relative navigation systems have been also successfully employed to land a UAV on some ground target [18] or to track an aerial target [10]. Many researchers have been interested in developing bio-inspired optic flow-based systems for reactive navigation of small aerial vehicles [17, 24].

In this section, we present four different vision systems that we have developed in our laboratory and demonstrated in real-time using a mini quadrotor UAV:

1. Optic flow-based vision system for flight control
2. Visual odometer for autonomous flight
3. Color-based vision system for target tracking
4. Stereo vision-based system for accurate positioning and landing of micro air vehicles

10.4.1 Flight Control Using Optic Flow

Recently, there is a growing interest in applying visual navigation principles of insects for UAVs control and guidance. Indeed, recent experimental research in biology has discovered a number of different ways in which insects use optic flow in order to deal with the 3D flight control problem [6, 21]. We thus take inspiration from insects in order to develop a navigation system which is based on optic flow obtained from a minimum sensor suite that includes a single onboard camera and a low-cost IMU. Since the current technologies do not allow to imitate the visual system and neuronal circuits of insects, we use standard computer vision techniques/algorithms for computing and interpreting optic flow. The developed vision system is composed from three subsystems for (1) optic flow computation, (2) fusion of visual measurements and angular rate data, and (3) motion and structure parameters estimation. The proposed computational framework is based on Three Nested Kalman Filters (3NKF) which allowed to combine the three subsystems in a favorable manner, Fig. 10.2. The resulted 3NKF-based algorithm is fast, accurate and robust, which make it suitable for aerial robotic applications.

Details about this optic flow-based vision system can be found in our paper [12] which also contains experimental results about indoor and outdoor autonomous flights. Here, we describe briefly the major components of that system and present some experimental results.

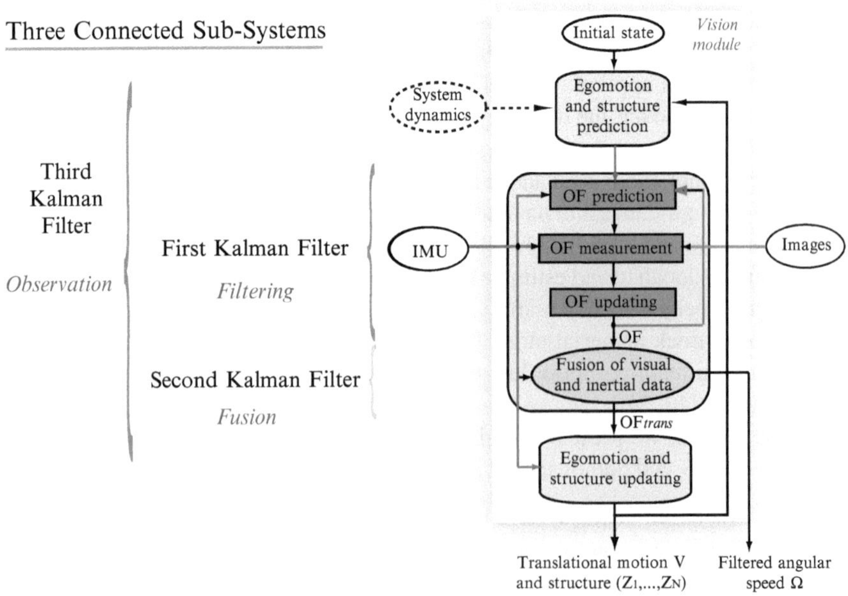

Fig. 10.2 3NKF-based vision algorithm for optic flow computation and interpretation

10.4.1.1 Optic Flow Computation

Much progress has been made in optic flow computation and yet, its efficient and accurate estimation in real-time remains difficult, especially when it is used for aerial robotics applications. The proposed algorithm combines matching and differential techniques for efficient and accurate measurement of large and sub-pixel image displacements. The proposed method takes advantage of UAV dynamics constraining the camera motion to be continuous and smooth. Therefore, based on the 3NKF framework, IMU measurements (angular rates and angles) and SFM-module outputs (velocity and range estimates) are exploited to predict the displacement of a given patch (block of pixels) in the subsequent images. A block matching technique is then applied to compute the nominal image displacement with adapting the patch/template shape to encounter deformations caused by vehicle rotation. Therefore, large image displacements are efficiently and robustly computed without suffering from the main issues of standard block matching methods like quadratic complexity and sensibility to deformations.

Once the nominal displacement $d_n \in \mathbb{Z}^2$ is computed, the new captured image is virtually translated by this vector (i.e., d_n). Now, the displacement between the reference image and the new translated image does not exceed one pixel in the ideal case, and the well-known Lucas–Kanade [15] differential algorithm may compute easily and accurately the remaining sub-pixel displacement $d_s \in \mathbb{R}^2$. Finally, the total image displacement is obtained by summing d_n and d_s, which is then filtered with a Kalman Filter.

10.4.1.2 Optic Flow and Angular Rate Data Fusion

By using the perspective-central camera model, the optic flow $(\dot{x}_i, \dot{y}_i)$ computed at the image location (x_i, y_i), can be expressed in terms of the rotorcraft translation velocities $V \in \mathbb{R}^3$ and angular rates $\Omega \in \mathbb{R}^3$, and depths Z_i (structure parameters).

$$\begin{bmatrix} \dot{x}_i \\ \dot{y}_i \end{bmatrix} = \begin{bmatrix} \frac{-1}{1+\beta Z_i} & 0 & \frac{\beta x_i}{1+\beta Z_i} \\ 0 & \frac{-1}{1+\beta Z_i} & \frac{\beta y_i}{1+\beta Z_i} \end{bmatrix} \begin{bmatrix} V_x \\ V_y \\ V_z \end{bmatrix} + \begin{bmatrix} \beta x_i y_i & -\left(\frac{1}{\beta} + \beta x_i^2\right) & y_i \\ \left(\frac{1}{\beta} + \beta y_i^2\right) & -\beta x_i y_i & -x_i \end{bmatrix} \begin{bmatrix} \Omega_x \\ \Omega_y \\ \Omega_z \end{bmatrix} \tag{10.9}$$

The fusion strategy that we have proposed aims at simplifying the SFM problem with improvements in accuracy and robustness. Our main idea is to divide the SFM estimation process into two steps. In the first step, we integrate the angular rate data and optic flow in a KF in order to estimate the translational component of the optic flow. In fact, at this stage, we aim at subtracting or cancelling the rotational component of the optic flow using a KF that handles measurements noise. The translational optic flow is the only component that depends on depths Z_i, thereby providing useful information for tasks related to depth perception, such as obstacles detection and collision avoidance. In the second step, the reduced SFM problem is formulated to recover translational motion and structure parameters using the translational optic flow estimated in the previous step.

10.4.1.3 Motion Parameters Estimation by Solving the SFM Problem

The computational framework that we use for recursive estimation of UAV motion and structure is the Extended Kalman Filter (EKF), which has been the subject of much work on image sequences. Using the framework discussed thus far, the initial SFM problem is reduced to estimate translational velocity and structure parameters, considering the previously estimated translational optic flow as the measurement vector. So, our composite state vector consists of three parameters for camera/UAV translational motion, and N variables for structure: $X = (V_x, V_y, V_z, Z_1, \ldots, Z_N)^T$.

Thus, the number of unknowns is $3 + N$. Computing optic flow at N image locations introduces N additional unknowns $(Z_1, \ldots, Z_N)$ while providing $2N$ equations or measurements. Consequently, the system is completely determined for $2N \geq N + 3 \Rightarrow N \geq 3$. For more accuracy and stability, we have chosen $N = 9$. These nine OF vectors are computed at well-chosen image locations.

Thus, the evolution of the state vector X is governed by the following discrete dynamic system [12]:

$$X_{k+1} = AX_k + B\gamma_k + w_k, \quad w_k \tilde{} \mathcal{N}(0, \Sigma_w) \tag{10.10}$$

From (10.9), the observation discrete model can be written in the following form

$$Y_k = g(X_k) + \nu_k, \quad \nu_k \tilde{} \mathcal{N}(0, \Sigma_\nu) \tag{10.11}$$

with Y is the measurement vector in $\mathbb{R}^{2N}$ containing the estimated translational optic flow, and $g(X_k)$ is a nonlinear function that can be deduced from (10.9).

Once the system and measurement/observation models have been specified, then the EKF implementation is straightforward.

10.4.1.4 Experimental Results

In order to demonstrate and evaluate the performance of the developed vision system in natural environments, we have performed several tests using a ground vehicle and a small quadrotor UAV.

- *Test 1 (ground vehicle):* this test was performed at the university of technology of Compiegne (France) using the Heudiasyc STRADA vehicle, Fig. 10.3. STRADA

Fig. 10.3 The Heudiasyc intelligent vehicle (STRADA) in a forest road. The camera is mounted on the vehicle's roof and is looking laterally

is a CITROËN car equipped with onboard computer and navigation sensors (GPS, IMU, laser telemeters, ABS, etc.), and used as experimental platform for research on navigation, mapping and driving assistance.

The 3NKF-based algorithm is implemented in the embedded computer, and runs in real-time at 20 Hz. In this test, the vehicle trajectory was straight and parallel to a flat terrain as shown in Fig. 10.3. The *Basler* camera was fixed on the roof of the vehicle, and oriented laterally towards ground with an angle of about $\alpha = 0.1$ rad. Figure 10.4 shows the two components of the computed optic flow at nine image locations, the estimated translational and rotational velocities, and the recovered nine depths. Firstly, we can notice that the algorithm computes robustly large optic flow vectors (until 28 pixels/frame), and the estimated angular velocities are less noisy than the IMU measurements. Secondly, the estimated velocity V_x is accurate and comparable to the velocity measured by wheels-mounted sensors (ABS), considered here as ground-truth.[4] Concerning the structure recovery, the results are also satisfactory. Since the observed scene (ground) is flat and oriented with an angle $\frac{\pi}{2} - \alpha$ relative to the image plane, then, the depth of each point depends essentially on its vertical image location y_i. The estimated depths shown in Fig. 10.4 correspond to the ground-truth values with small errors.

[4] For comparison purpose, the SFM estimates are scaled and normalized in a post-processing step in order to obtain the absolute camera motion and depths.

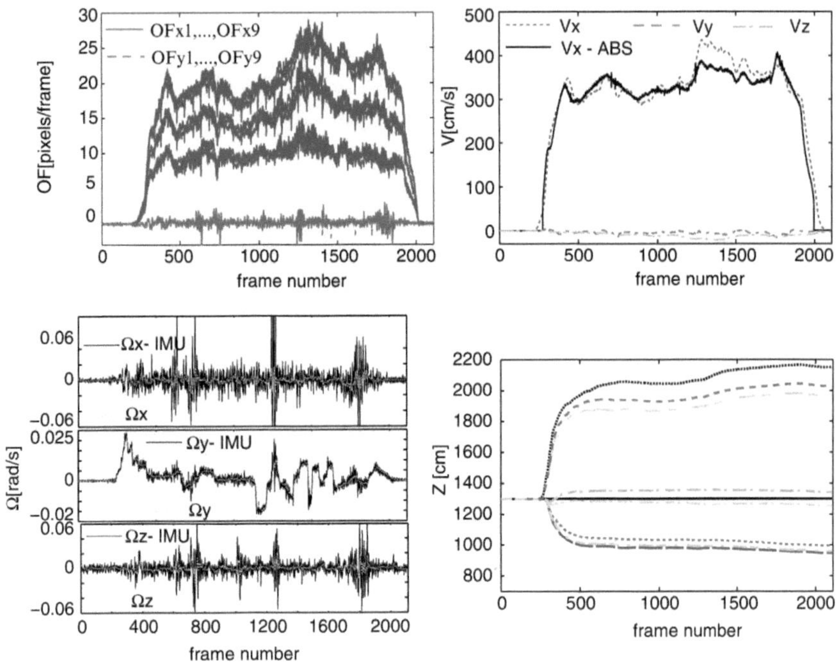

Fig. 10.4 Experimental results from a vehicle-mounted camera. Optic flow is computed robustly despite the poor texture of the scene. Motion and structure parameters are also recovered with good accuracy

The obtained results in outdoor natural environments with poor texture are very promising. They confirm the robustness of the 3NKF-based algorithm and its effectiveness and practicability in real world applications.

• *Test 2 (autonomous flight):* the developed optic flow-based navigation system was also demonstrated indoors and outdoors using a 750-g quadrotor MAV. The objective of these flight tests is to evaluate the performance of the 3NKF-based vision system when used in closed-loop control for autonomous flight.

Detailed description of the aerial platform and flight controller can be found in Chap. 8. Figure 10.5 shows the real-time architecture of vision and control algorithms onboard the quadrotor helicopter.

We have performed several flight tests and the obtained results showed that the developed optic flow-based navigation system allowed a small rotorcraft UAV to achieve autonomous flight. Here, we present experimental results from an indoor flight conducted at *Tokyo Big Sight* during an international exhibition. Since the floor is homogeneous without any texture, we have put some objects on the ground to provide some texture for optic flow computation (see Fig. 10.6). The vehicle was commanded to take off, hover and land autonomously. During the autonomous hovering, we have switched to manual flight for several seconds and then switched back to autonomous flight. The objective of this maneuver was to show to the audience

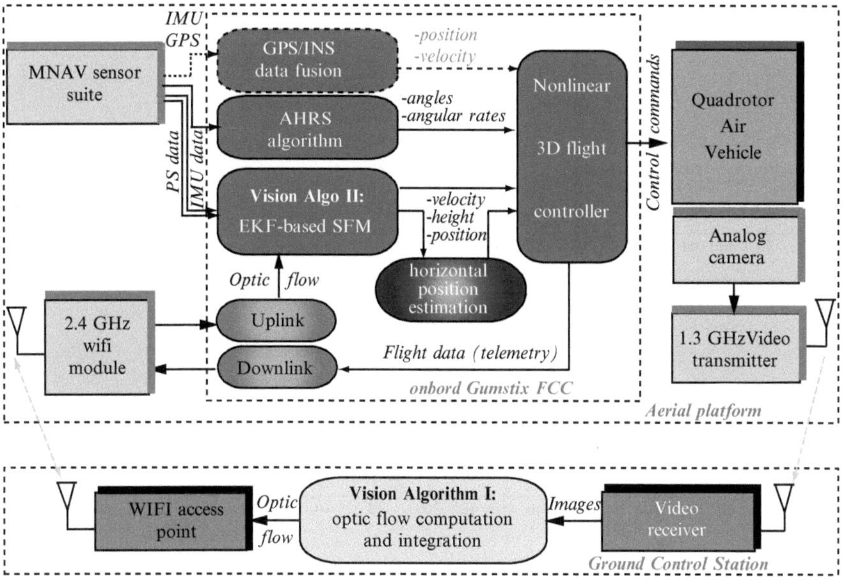

Fig. 10.5 Real-time architecture of the optic flow-based autopilot

Fig. 10.6 The quadrotor UAV in autonomous indoor flight using optic flow

(visitors) that the rotorcraft is naturally unstable and the vision-control system plays the main role in stabilizing the vehicle.

As it can be seen in Figs. 10.6 and 10.7, the rotorcraft achieved autonomously the required task with good performance.

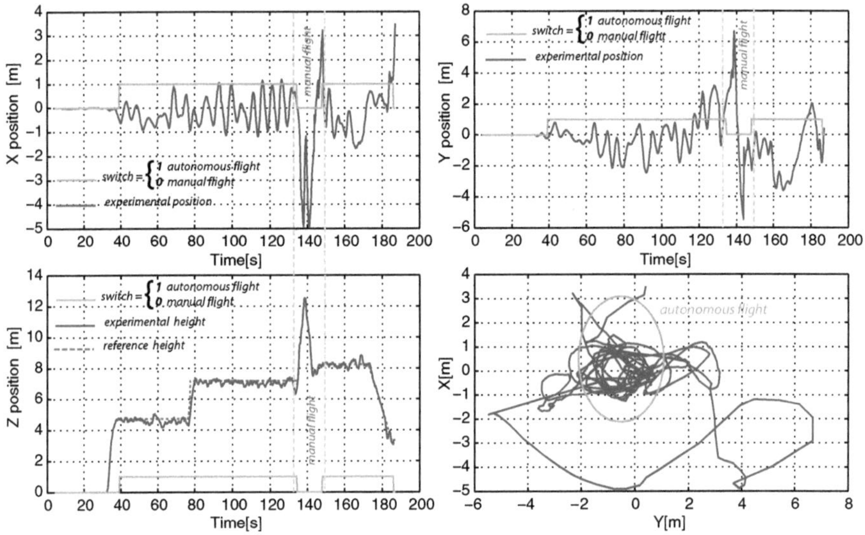

Fig. 10.7 Experimental results of the vision-based autonomous indoor flight

A video clip of this flight test can be found at http://mec2.tm.chiba-u.jp/monograph/Videos/Chapter12/7.wmv.

Experimental results of vision-based outdoor autonomous flights can be found in [12].

10.4.2 Visually-Driven Odometry by Features Tracking

We present here a visual navigation system as an alternative pose estimation method for environments and situations in which GPS is unavailable. The developed visual odometer is an incremental procedure that estimates the vehicle's ego-motion by extracting and tracking visual features in images from an onboard camera. Functionally, a vision algorithm detects and tracks visual features based on monocular images obtained from an on-board single camera looking downwards. By continuously tracking ground objects that appear in the camera field of view and accumulating or integrating their image displacement, it is possible to estimate the travelled flight distance in terms of total image displacement. These visual measurements are then fused with IMU data in order to overcome the non-desired rotation effects. Indeed, it is difficult to sense UAV translation with vision alone since image displacements also occur with aircraft rotation (translation–rotation ambiguity). Another ambiguity is the scale factor or the unknown range that can not be estimated from visual measurements alone. Therefore, the vision algorithm has been augmented by an adaptive mechanism which identifies the scale factor and estimates the range (height) from optic flow, accelerometers data and pressure sensor measurements. the Adaptive Visual Odometer (AVO) generates estimates of the UAV velocity and position with respect to some initial rotorcraft location or some selected target. These visual estimates are then used in the guidance and flight control laws to guide the rotorcraft relative to the initial location or some moving/stationary target.

The proposed system is motivated by problems such as short-range navigation in GPS-denied environments for flight stabilization, accurate landing, target tracking, etc. To enable accurate and robust long-range navigation with the AVO, this latter can be combined with other terrain relative navigation techniques like landmarks recognition.

This visual odometer is described in Chap. 12. Experimental flight test data over various ranges of the flight envelope are also presented in Chap. 12. The obtained results illustrate that the proposed vision-based autopilot performs well and allows a mini rotorcraft UAV to achieve autonomously advanced flight behaviours like automatic take-off, accurate hovering, precise auto-landing, trajectory tracking and moving target tracking.

10.4.3 Color-Based Vision System for Target Tracking

Here, we describe another vision-based navigation system that relies on color to track some selected object and to estimate its position relative to a micro air vehicle. The vision algorithm processes images from a single onboard camera and fuses the visual measurements with inertial data in order to obtain more accurate and robust estimates. An arbitrary target can be selected in real-time from the ground control station, thereby outperforming template and learning-based approaches. The estimated relative position between the vehicle and the target is used by the flight control system thereby resulting in visual servoing of the rotorcraft to track the target.

The use of color for tracking objects is motivated by the fact that the resulting algorithm is relatively simple and can be applied for micro air vehicles.

We recall in this section the basic components of our color-based vision algorithm and show some experimental results when applied to visual servoing of a mini quadrotor vehicle. More details and results can be found in our recent paper [2].

10.4.3.1 Generalities on Integral Image

An integral image (also known as a summed-area table) is a tool that can be used whenever we have a function from pixels to real numbers (for instance, pixel intensity), and we wish to compute the sum of this function over a rectangular region of the image. Integral images, have been applied for example in face detection in images [23] and stereo correspondence [22]. Without an integral image, the sum can be computed in linear time per rectangle by calculating the value of the function for each pixel individually. In our proposed image algorithm, in order to search the target, we need to compute the sum over multiple overlapping rectangular windows,

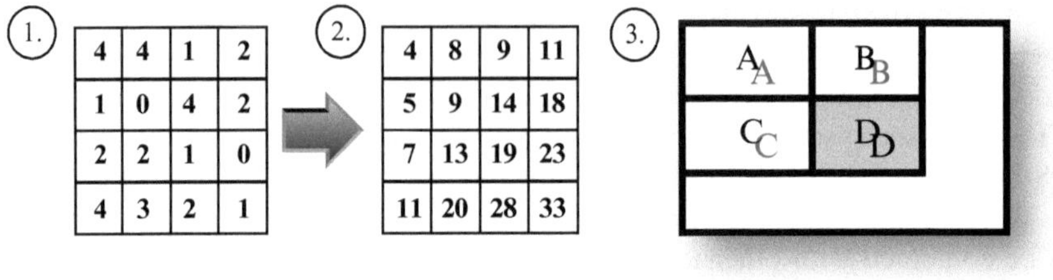

Fig. 10.8 Integral image calculation

so using an integral image would allow us to achieve a constant number of operations per rectangle with only a linear amount of preprocessing. To compute integral image, we store at each location, the sum of all terms to the left and above the pixel (depending on the origin of coordinates). This is accomplished in linear time using the following equation for each pixel (taking into account the border cases). The equation will be as below:

$$I(x, y) = f(x, y) + I(x - 1, y) + I(x, y - 1) - I(x - 1, y - 1) \tag{10.12}$$

An example of computation of an integral image, is shown in Fig. 10.8, (1)–(2). The number of each box in (1) represents for value for each pixel before computing the integral image, and after computing the integral image, value of each pixel is shown in (2). Once we have the integral image, we can compute the sum of the function for any rectangle with upper left corner, and lower right corner, in constant time using the following equation:

$$\sum_{x=x_1}^{x_2} \sum_{y=y_1}^{y_2} = I(x_2, y_2) - I(x_2, y_1 - 1) - I(x_1, y_2) + I(x_1 - 1, y_1 - 1) \tag{10.13}$$

Figure 10.8 (3) shows that calculating the sum of over rectangle D using (10.13) is equivalent to computing the sums over the rectangles

$$(A + B + C + D) - (A + B) - (A + C) + A$$

10.4.3.2 Probability Distributing Chart

Before tracking the intended object using integral-image based algorithm, the image processing algorithm will undergo visual preprocessing stage. After the target object is selected, the image algorithm then extracts the color of the object selected and thresholds it by using the color probability distribution of the whole image. The histogram is computed by counting the number of pixel in a region of interest that have a given color. The colors are quantized into bins. This operation allows similar color values to be clustered as single bin. The quantization into bins reduces the memory and computational requirements. The un-weighted histogram is computed in the following manner:

$$q_u = \sum_{i=1}^{n} \delta[c(x_i) - u] \tag{10.14}$$

where the function $c : \Re^2 \rightarrow 1, \ldots, m$ associates the value of pixel at locationx_i, to the bin number, n is the number of pixels, and δ is the Kronecker delta function. The process of histogram normalization is shown in (10.15).

$$\left\{ p_u = \min \frac{255}{\max(q)} q_u, 255 \right\}_{u=1,\ldots,m} \tag{10.15}$$

According to equation (10.15), value of the histogram and each chroma level is normalized from $[0, \max(q)]$ to the new range $[0, 255]$. Then back-projection is used to associate the pixel values in the image with the value of the corresponding histogram bin. This operation replaces the pixel values of the input image with the value of corresponding bin of the histogram. The value of each pixel in the probability image represents the probability that the pixel belongs to the object of interest.

10.4.3.3 Visual Tracking Using Color and Integral Image

Integral image-based object tracking can be briefly described as follows:

1. When the target is selected by the user, each pixel color intensity in the image frame is filtered based on the color probability distribution. The sum of the intensities (image integral value) of the selected target in the select box is calculated. The ratio of the intensity and object size (density) is then taken as the density reference.
2. The search area for the next image frame would be limited in a radius set by a preset value, that changes according to the current object size (search box size).
3. For each search box location in the search area the intensity ratio is evaluated compared to the reference. The location with the highest score would be the new target location. If there are more than one location of the same score, the nearest to the last detected location would be selected. To prevent miss-detection when the object goes out of the field of view (FOV), a threshold for intensity ratio is set.
4. To ensure that the search box would be always at the target center, the size of the search box would be increased then decreased on its four sides. During the search box increasing and decreasing step, the ratio of intensity value inside the box and the box size ratio is again compared to the reference. The new size of the target will be updated according to the score. The size would not be changed if the highest score does not exceed the intensity ratio threshold.
5. When the object looses its detection for a certain period of time, due to image noise or when it is not in the FOV, the search area would be initialized to search

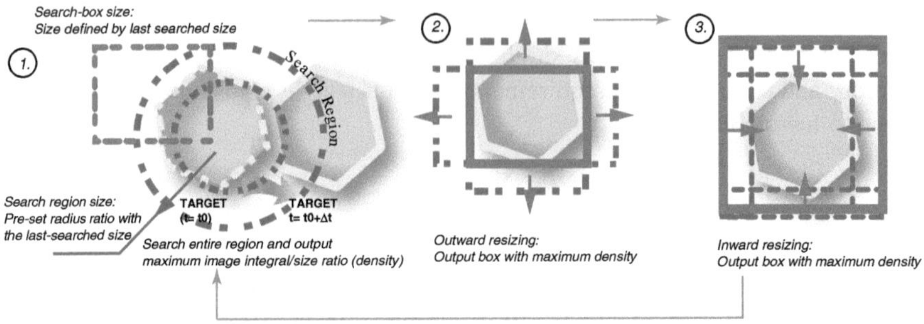

Fig. 10.9 Object tracking strategy. (*1*) Search box is moved around a pre-set area and compare the intensity (image integral: box size) of search box with the template image. (*2*) First step resizing: increase the search area in four sides. (*3*) Second step resizing: decrease the searched area of the second step and output the maximum intensity compared to the template intensity

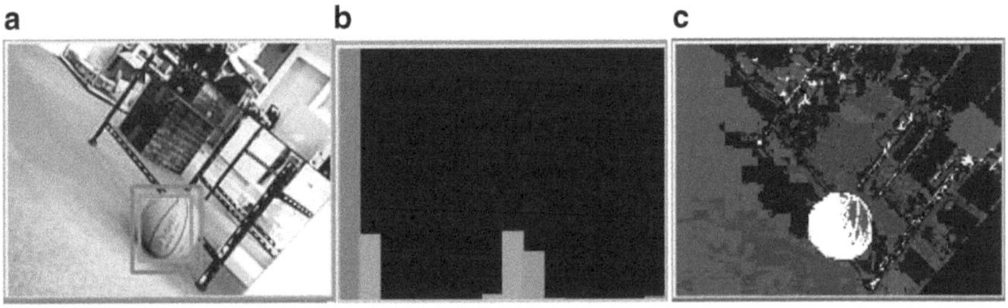

Fig. 10.10 (**a**) Result of the tracking, after rotating the camera, (**b**) Color histogram of the ball, (**c**) The backprojected image based on color probability distribution

the whole image to recover the target. An object missing flag will be raised, if it is not recovered even after searching in a preset number of search-cycles.

6. Rotation is not considered for this application because the MAV is only required to hover on the target.
7. After detecting the center of the target object, the x and y coordinates are input to the Kalman Filter, considering a linear uniform movement.

Figure 10.9 summarizes the object tracking algorithm described above, while Fig. 10.10 shows the image to be tracked and image processing results.

10.4.3.4 Experimental Results

In order to evaluate the performance of the proposed vision algorithm, we have performed outdoor experimental tests using the quadrotor and the nonlinear flight controller described in Chap. 8. These experiments were performed in the Chiba University, outdoor athletic field. The MAV is to be autonomously controlled to hover above a target which is a 0.5 m × 1.0 m red plastic-board on the ground.

Fig. 10.11 *Left*: MAV hovering above target; *Right*: real-time tracking image by the onboard camera

In this test the MAV is controlled manually above a target, and after a target has been chosen, it is switched to vision-based autonomous control. In these experiments GPS data are only used for comparison with the image data, and not for control. The position results obtained for the test are presented in Fig. 10.12. Figure 10.11 (right) shows the result of the real-time object tracking algorithm being displayed on the image streamed to the Ground Control Station (GCS). The target center is the marked by the center of the blue circle, which depicts the search area. The white box and green box are the tracking results of the raw data and after Kalman filtering respectively.

Figure 10.12 shows the MAV position and height obtained during the autonomous hovering mode. After switching to autonomous hovering above the target (at time = 50 s as indicated by the green line on Fig. 10.12), it is observed that the MAV succeeded to hover autonomously above the target for about 7–8 min. This is a good performance for this MAV category which is relying on vision for relative position estimation. The control accuracy is about 2 m. However, higher accuracy can be achieved by adjusting the controller gains. The altitude controller also achieved very good performance since the desired height has been maintained for about 8 min. It is important to note that the hovering time is limited by the battery life and not by the vision and control system.

From Fig. 10.12, it is also observed that the vision/INS estimates are more accurate than the GPS/INS measurements. Moreover, the GPS/INS data have a time delay of about 0.6 s compared to the vision/INS data.

Video recording of the test is available online at http://mec2.tm.chiba-u.jp/monograph/Videos/Chapter12/4.wmv.

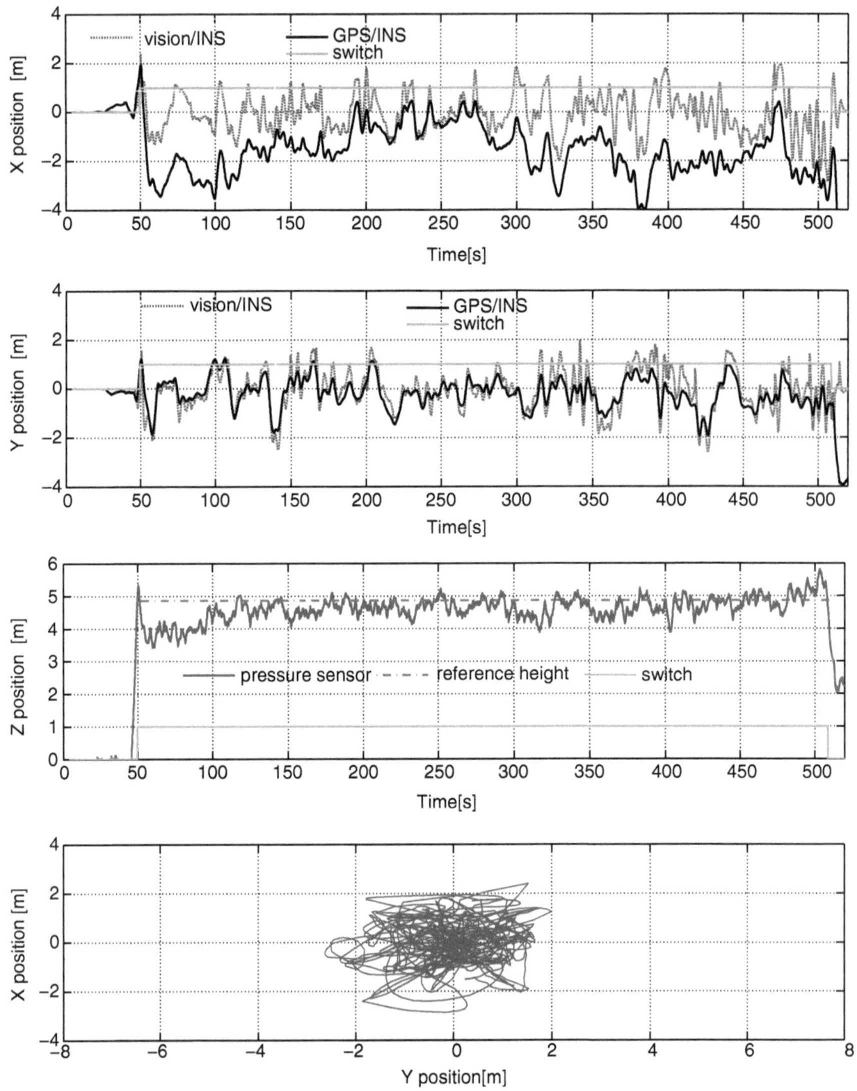

Fig. 10.12 MAV position during vision-based autonomous hovering

10.4.4 Stereo Vision-Based System for Accurate Positioning and Landing of Micro Air Vehicles

The vision systems, previously described are based on a single onboard camera. Here, we investigate the possibility of using stereo vision to control a micro air vehicle. The advantage of using stereo vision is that this latter can estimate the relative 3D position of some target or object without the need of any other sensor.

This characteristic is very useful for UAVs since the estimated pose can be used for accurate positioning and landing, obstacles detection and environment mapping. However, available stereo cameras are relatively heavy and can not be embarked onboard the MAV. Therefore, we propose in this research a practical approach which consists to use an on-ground stereo vision system to detect and estimate the position of a flying vehicle relative to the camera. The data is then transmitted to the onboard flight controller where they are fused with INS and used for vehicle stabilization and accurate landing. This work is motivated by problems such as automatic charging of MAVs, launch and recovering of sub-vehicle MAVs from relatively bigger helicopters that can be considered as mother ships. For these applications, accurate positioning of the MAV is required for performance and safety reasons. Off-board stereo vision is thus an interesting solution.

The developed stereo vision-based navigation system consists of three main parts: (1) computation of the depth image; (2) detection and tracking of the rotorcraft; and (3) fusion of vision and INS data.

10.4.4.1 Depth Image Computation

The depth image is constructed using the *Triclops* software development kit that is provided with the Point Grey Stereo Vision camera. Therefore, depth range images are provided in real-time by exploiting the Triclops^{TM} SDK functions which perform all stereo processing.

10.4.4.2 Object Tracking: CAMSHIFT

Camshift algorithm was firstly developed by Bradski [3]. This algorithm is an application of the mean-shift algorithm for tracking objects in video sequence. The standard mean shift algorithm can only deal with static distributions (i.e., single images) because its search window has a fixed size. Bradski uses a dynamical search window that adapts its size after every video frame, depending on the size of the target object.

Steps of this algorithm can be summarized as follow:

1. Set the region of interest (ROI) of the probability distribution image to the entire image.
2. Select an initial location of the Mean Shift search window. The selected location is the target distribution to be tracked.
3. Calculate a color probability distribution of the region centered at the Mean Shift search window.
4. Iterate Mean Shift algorithm to find the centroid of the probability image. Store the zeroth moment (distribution area) and centroid location.
5. For the following frame, center the search window at the mean location found in Step 4 and set the window size to a function of the zeroth moment. Go to Step 3.

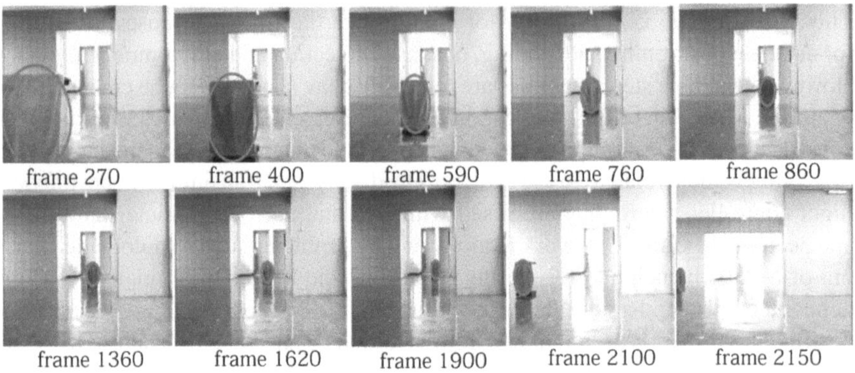

Fig. 10.13 Object tracking by using CAMSHIFT algorithm

For further understanding of CAMSHIFT algorithm, the reader can refer to [3]. Some static results of object tracking by using Camshift algorithm are shown in Fig. 10.13.

10.4.4.3 3D Position Calculation

Stereo camera has the advantage that it can measure the 3D position of an object from camera accurately by applying a simple triangle method. The illustration of the 3D position measurement by using stereo camera is shown in Fig. 10.14.

In order to compute X and Z, one needs first to derive some relations between angles shown in Fig. 10.14

$$\gamma = \tan^{-1}\left(\frac{C_L}{f}\right), \qquad \delta = \tan^{-1}\left(\frac{C_R}{f}\right)$$
$$\alpha = 90^\circ - \gamma = 90^\circ - \tan^{-1}\left(\frac{C_L}{f}\right), \quad \beta = 90^\circ - \delta = 90^\circ - \tan^{-1}\left(\frac{C_R}{f}\right) \tag{10.16}$$

By applying *sine* theorem and using the information of α and β, r can be easily obtained as shown in (10.17)

$$r = \frac{L}{\sin(180^\circ - \alpha - \beta)} \sin(\beta) \tag{10.17}$$

Therefore, X and Z of the object from camera can be expressed as follow:

$$X = r\cos(\alpha) \qquad Z = r\sin(\alpha) \tag{10.18}$$

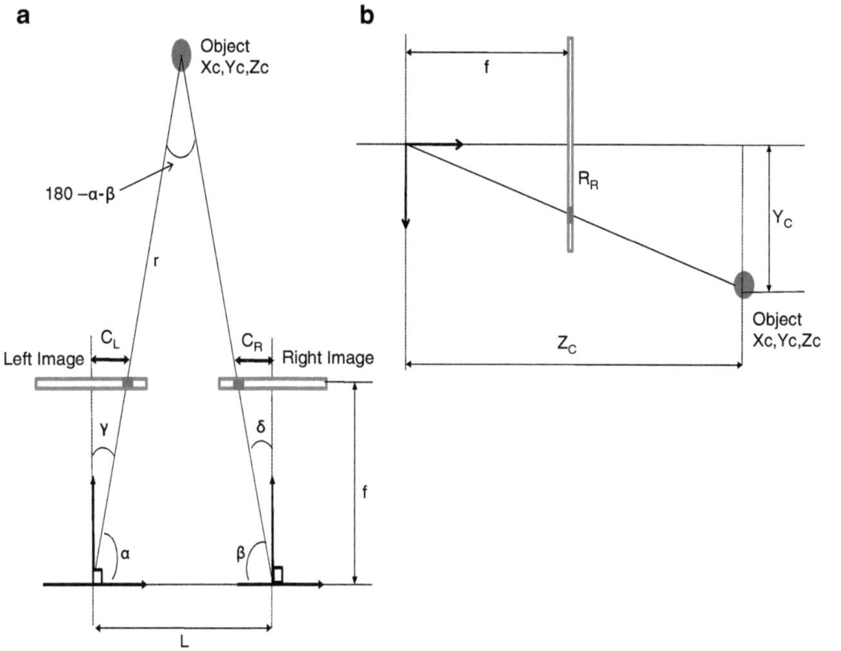

Fig. 10.14 (**a**) Configuration between object and image on X–Z plane. (**b**) Configuration between object and image on Y–Z plane

Fig. 10.15 Experimental setup of autonomous quad-rotor by means of on-ground stereo camera

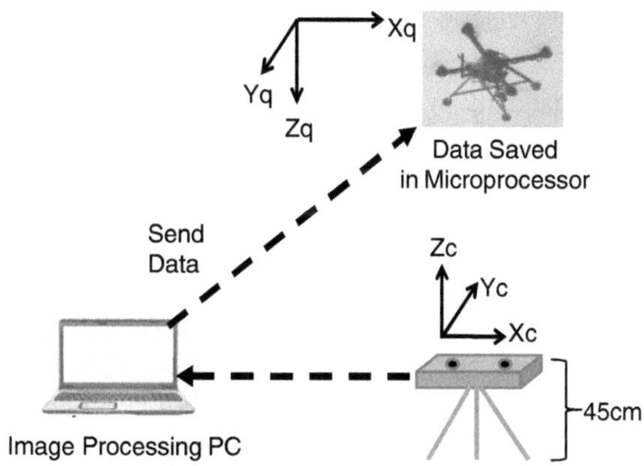

In order to find the distance on the Y direction, Fig. 10.14b is used. The Y distance is calculated in the following manner:

$$Y = Z\frac{R_R}{f} \tag{10.19}$$

The 3D position in the camera coordinate frame is then transformed to the quad-rotor body frame as shown in Fig. 10.15.

Table 10.1 Bumblebee specification

Items	Description
Baseline	12 cm
Resolution	Up to 1,204 × 768
Max. frame rate	15 fps
Horizontal field of view	50°
Weight	375 g

10.4.4.4 Experimental Results

Several indoor and outdoor flight tests have been performed to show the effectiveness and robustness of using on-ground stereo vision system for accurate positioning and landing of mini rotorcraft UAVs. Here, we present experimental results from two autonomous flight tests. The experiments were conducted outdoor using the quadrotor platform and the embedded autopilot described in Chap. 8.

For these experiments, we used the Bumblebee stereo camera which is commercialized with its software. Its specifications are given in Table 10.1.

The experimental setup is as shown in Fig. 10.15. The stereo camera is placed on a tripod and is connected to an image processing computer that will directly send the information about X, Y and Z measurements to the quadrotor through a wireless communication. Further processing is performed onboard the flight computer to fuse visual estimates with INS data. The flight controller implemented onboard the vehicle uses these estimates (3D position and velocity) to achieve the assigned task autonomously.

Autonomous hovering: The procedure of the experiment is first to fly the quadrotor manually. When it is detected by stereo camera (i.e., the vehicle appears in the camera FOV), the operator at the ground station selects the quadrotor as a target and the Camshift vision algorithm will track the aerial vehicle in images. When object tracking is started, the flight mode is changed from manually to autonomous. The stereo vision system provides real-time estimates of the vehicle's position (X, Y, Z) relative to the camera fixed on the ground. The control objective is thus to control the quadrotor motion in order to regulate this relative position to zero, thereby ensuring vision-based hovering above the camera.

Experimental results of the rotorcraft position (X, Y, Z) are shown in Fig. 10.16. The vehicle achieved a stable autonomous hovering using stereo vision estimates. The control accuracy for this vision-based flight is about ± 1 m. This is a good performance for a such small platform flying outdoor and using visual estimates. Higher accuracy can be achieved by adjusting the controller gains and placing the camera in a planar surface such that it has *zero orientation*.

It is also important to note the stereo vision estimates are more accurate than GPS and Pressure Sensor (PS) measurements.

Autonomous landing: The procedure for this experiment is the same as the one used for autonomous hovering. First, the quad-rotor is flied manually. When it is in the field of view of the stereo camera, the quad-rotor is selected as target and the

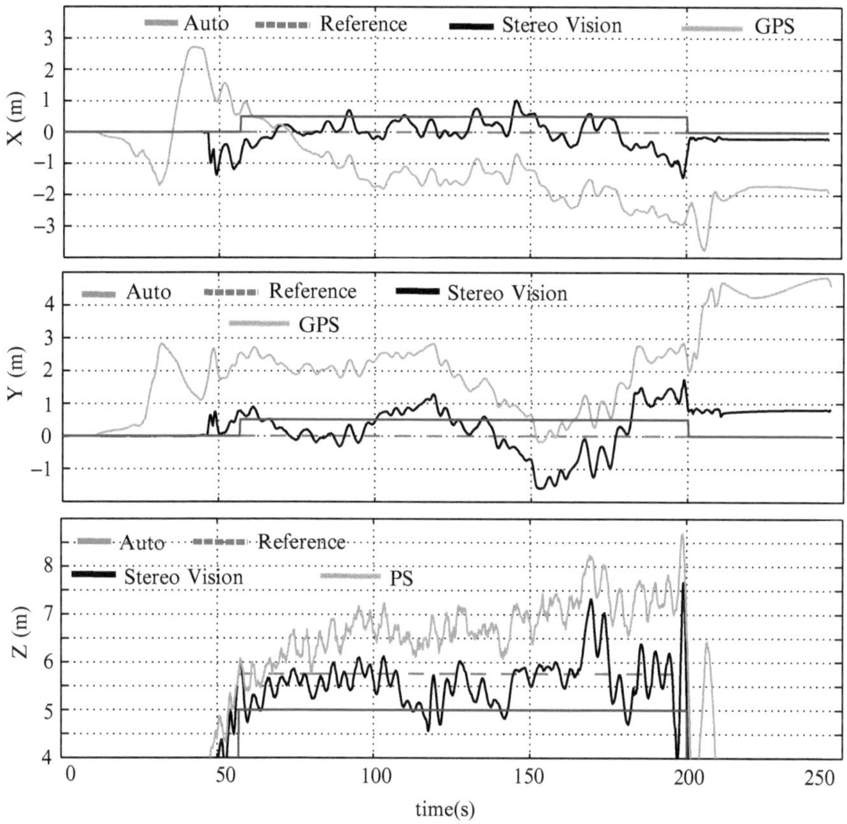

Fig. 10.16 Autonomous hovering using on-ground stereo vision system

object tracking process is conducted. Then, the flight is changed from manually to autonomous. The vehicle is tasked to perform vision-based hovering for about 80 s and then to achieve automatic landing using stereo vision estimates.

Figure 10.17 shows the MAV trajectories during hovering and landing. The task is accomplished successfully with small tracking errors. The hight reference trajectory is accurately tracked using visual estimates, resulting in autonomous and safe landing. As seen in Fig. 10.17 height estimates by stereo vision system are more accurate than the ones provided by pressure sensor because this late is sensitive to weather conditions like wind and temperature. In the third graph of Fig. 10.17, one can see that height reference trajectory is tracked until 2 m altitude. In fact, at about 2 m height, the camera FOV is very small and the vehicle goes out of the field of view and landed at about 1–2 m from the camera. To achieve reliable and precise landing, one approach could consist in achieving descent flight until about 1–2 m height using stereo vision and then perform feed-forward control by incrementally decreasing the thrust to achieve soft landing.

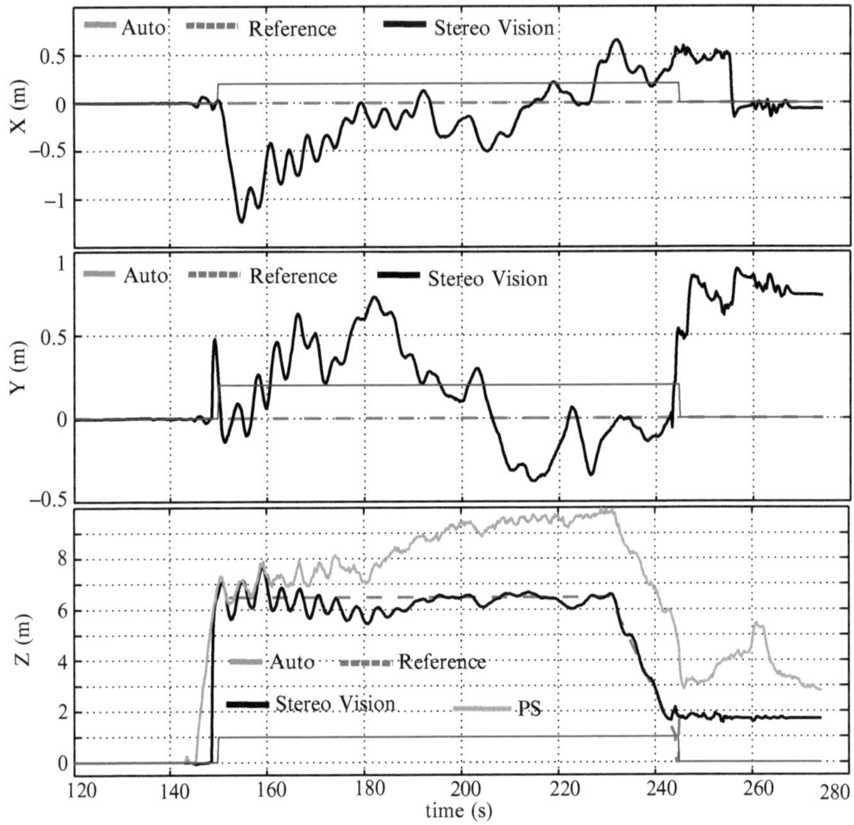

Fig. 10.17 Automatic landing based on stereo vision measurements

From the experimental results, it can be seen that the performance of stereo vision with INS is better than GPS system. This result suggests the possibility of stereo vision system for precise guidance of an MAV. In the future, such kind of system can be used for many applications such as automatic battery recharging system, cooperative control among aerial vehicles or among aerial vehicles and ground vehicles.

The main limitation of this on-ground stereo vision system is the small FOV of the camera especially at low altitudes. One can use cameras with larger FOV or a pan/tilt system.

10.5 Summary

In this chapter, an embedded guidance system and several navigation systems are presented. They are based on low-cost and light-weight sensors to meet the limited payload of miniature aerial platforms. The experimental results from autonomous flights show that the developed systems provide small UAVs with some level of autonomy to achieve tasks and maneuvers such as waypoint navigation, hovering, automatic take-off and landing, vision-based target tracking, etc. In order to achieve more complex tasks, we are conducting active research on the design of guidance, navigation and control systems that can support more advanced functions such as automated decision making, obstacle avoidance, target acquisition, artificial vision, and interaction with other manned and unmanned systems.

References

1. Amidi O, Kanade T, Fujita K (1999) A visual odometer for autonomous helicopter fight. Rob Auton Syst 28(2–3):185–193
2. Azrad S, Kendoul F, Perbrianti D, Nonami K (2009) Visual servoing of an autonomous micro air vehicle for ground object tracking. In: Proceedings of the IEEE/RSJ international conference on intelligent robots and systems, St. Louis, MO, USA
3. Bradski G (1998) Computer vision face tracking for use in a perceptual user interface. Intel Technol J 2
4. Caballero F, Merino L, Ferruz J, Ollero A (2009) Vision-based odometry and slam for medium and high altitude flying UAVS. J Intell Rob Syst 54:137–161
5. Christophersen HB, Pickell RW, Neidhoefer JC et al (2006) A compact guidance, navigation, and control system for unmanned aerial vehicles. AIAA J Aerosp Comput Inform Commun 3:187–213
6. Egelhaaf M, Kern R (2002) Vision in flying insects. Curr Opin Neurobiol 12(6):699–706
7. He R, Prentice S, Roy N (2008) Planning in information space for a quadrotor helicopter in a GPS-denied environment. In: Proceedings of the IEEE international conference on robotics and automation, California, USA, pp 1814–1820
8. Jia D, Vagners J (2004) Parallel evolutionary algorithms for UAV path planning. In: AIAA 1st intelligent systems technical conference, Illinois, USA, AIAA 2004-6230
9. Johnson A, Montgomery J, Matthies L (2005) Vision guided landing of an autonomous helicopter in hazardous terrain. In: Proceedings of the 2005 IEEE international conference on robotics and automation (ICRA), Barcelona, Spain, pp 4470–4475
10. Johnson EN, Calise AJ, Watanabe Y, Ha J, Neidhoefer JC (2007) Real-time vision-based relative aircraft navigation. AIAA J Aerosp Comput Inform Commun 4:707–738
11. Kanade T, Amidi O, Ke Q (2004) Real-time and 3D vision for autonomous small and micro air vehicles. In: Proc. of the 43rd IEEE conference on decision and control, Atlantis, Paradise Island, Bahamas, pp 1655–1662
12. Kendoul F, Fantoni I, Nonami K (2009) Optic flow-based vision system for autonomous 3D localization and control of small aerial vehicles. Rob Auton Syst 57:591–602
13. Keviczky T, Balas G (2005) Software-enabled receding horizon control for autonomous UAV guidance. AIAA J Guidance Control Dyn 29(3):680–694
14. Kima J, Sukkarieh S (2007) Real-time implementation of airborne inertial-slam. Rob Auton Syst 55:62–71
15. Lucas B, Kanade T (1981) An iterative image registration technique with an application to stereo vision. In: Proc. DARPA IU workshop, pp 121–130

16. Mettler B, Valenti M, Schouwenaars T, Kuwata Y, How J, Paunicka J, Feron E (2003) Autonomous UAV guidance build-up: flight-test demonstration and evaluation plan. In: AIAA guidance, navigation, and control conference (AIAA-2003-5744)
17. Ruffier F, Franceschini N (2005) Optic flow regulation: the key to aircraft automatic guidance. Rob Auton Syst 50(4):177–194
18. Saripalli S, Montgomery J, Sukhatme G (2003) Visually-guided landing of an unmanned aerial vehicle. IEEE Trans Rob Autom 19(3):371–381
19. Scherer S, Singh S, Chamberlain L, Elgersma M (2008) Flying fast and low among obstacles: methodology and experiments. Int J Robot Res 27(5):549–574
20. Schouwenaars T, Valenti M, Feron E, How J (2005) Implementation and flight test results of milp-based UAV guidance. In: IEEE aerospace conference, pp 1–13
21. Srinivasan MV, Zhang S, Lehrer M, Collett T (1996) Honeybee navigation en route to the gaoal: visual flight control and odometry. J Exp Biol 199(1):237–244
22. Veksler O (2003) Fast variable window for stereo correspondence using integral image. In: IEEE conf on computer vision and pattern recognition, pp 556–661
23. Viola P, Jones M (2004) Robust real-time face detection. Int J Comput Vis 57(2):137–154
24. Zufferey JC, Floreano D (2006) Fly-inspired visual steering of an ultralight indoor aircraft. IEEE Trans Robot 22(1):137–146

Chapter 11
Design and Implementation of Low-Cost Attitude Quaternion Sensor

Abstract In this chapter, the development of a low-cost attitude sensor is introduced. In the previous chapters, the control system design method of several UAVs/MAVs with rotary wings was shown, and several kinds of controllers were designed. The most important controller is the attitude controller because if attitude control is not achieved, any other control such as velocity and position controls cannot be achieved. For achieving attitude control, it is necessary to measure the attitude of UAV/MAV. Hence, we require an attitude sensor. However, conventional attitude sensors are somewhat expensive and heavy, and they cannot be used for the attitude control of small UAVs and MAVs. Therefore, the design of an attitude estimation algorithm by using low-cost sensors, accelerometers, gyro sensors, and magnetic sensor is introduced. Finally, a low-cost attitude sensor has been developed and evaluated by comparing it with a conventional high-accuracy sensor.

Video Links:

IMU sensor test
http://mec2.tm.chiba-u.jp/monograph/Videos/Chapter11/1.wmv
IMU sensor test by Eagle helicopter
http://mec2.tm.chiba-u.jp/monograph/Videos/Chapter11/2.avi
Fully autonomous hovering1 of Electric power helicopter, Lepton
http://mec2.tm.chiba-u.jp/monograph/Videos/Chapter11/3.avi
Fully autonomous hovering2 of Electric power helicopter, Lepton
http://mec2.tm.chiba-u.jp/monograph/Videos/Chapter11/4.avi

11.1 Introduction

In previous chapters, the control system design of a helicopter-type UAV/MAV and quad-rotor-type MAV was introduced. We designed various types of control systems such as the attitude control system, the horizontal position and velocity control system, and the height control system. Among these control systems, the attitude control system is the most important. In the case of a UAV/MAV with rotary wings,

K. Nonami et al., *Autonomous Flying Robots: Unmanned Aerial Vehicles and Micro Aerial Vehicles*, DOI 10.1007/978-4-431-53856-1_11, © Springer 2010

almost all the control systems can make the driving force to act in all directions by changing the attitude. Hence, if attitude control is not achieved, it is difficult to realize any other control system. To realize attitude control, it is necessary to measure the attitude of the body. Various methods for the measurement of attitude have been suggested. The most popular method uses inertia sensors such as accelerometers and gyro sensors. In this method, static attitude information (the amount a body tilts from the horizontal plane) is obtained from the direction of the gravitational acceleration measured by using an accelerometer. Further, the dynamic attitude movement is obtained by calculating the integral of the angular rate measured by a gyro sensor. Finally, the attitude of the body can be obtained by combining the static attitude information and dynamic attitude movement. Moreover, in the case of heading angles, a method to obtain the heading angle by measuring the geomagnetic force using a magnetic sensor has been derived. In practice, various attitude sensors using the three types of abovementioned sensors are commercially available. However, commercial sensors are relatively expensive and have the characteristic that their weight increases with precision. Hence, they cannot be easily used for small UAVs and MAVs due to the payload limitation. Further, Euler's angle, which has been adopted as the attitude expression by most commercial attitude sensors, has a singular point. Around this singular point, commercial sensors cannot measure the attitude since the singular point is not sufficient for measuring the attitude in three-dimensional space uniquely. Moreover, the accelerometer included in the attitude sensor measures the dynamic acceleration due to movement, although only the gravitational acceleration needs to be measured. Subsequently, they have the disadvantage that a large attitude estimation error is produced when accelerating or decelerating. Such a problem makes it fatal to place a sensor on moving bodies such as UAVs/MAVs, and it is necessary to develop a method to reduce the occurrence of attitude errors by such dynamic accelerations.

In this chapter, we introduce an attitude estimation algorithm by using an extended Kalman filter for attitude sensors that can be developed by using cheap and lightweight sensors; we then reduce the attitude error due to dynamic acceleration.

11.2 Coordinate System and Quaternion

In this section, the coordinate system is described, and the quaternion is introduced.

11.2.1 Definition of Coordinate System

The coordinate systems used in this chapter will be defined now. The first coordinate system is the *r-frame*; it is called the reference frame, and the origin of this frame is placed at an arbitrary point on the ground. The X_r axis indicates magnetic north; Z_r axis, the gravitational direction; and Y_r axis, the vertical direction of the $X_r Z_r$ plane.

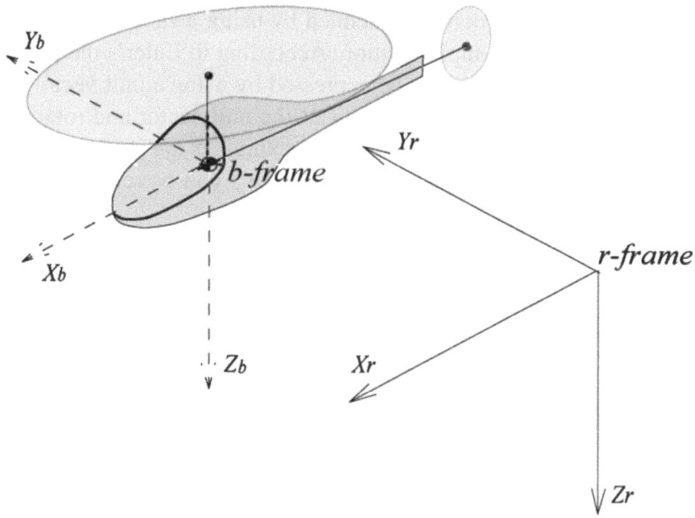

Fig. 11.1 Coordinate system

The second coordinate system is the *b-frame*; it is called the body frame, and the origin is located at the center of gravity of the body. The X_b axis is in the forward direction of the body; Y_b, rightward direction; and Z_b axis, downward direction (Fig. 11.1). In addition, we define the attitude of the *b-frame* for the *r-frame* as the attitude of the helicopter.

11.2.2 Quaternion

The quaternion is introduced now. The quaternion is a number that is expanded from a complex number, and it is redefined for three dimensions. It was devised in 1843 by a British physicist and mathematician William Rowan Hamilton. The quaternion **q** is defined as follows:

$$\mathbf{q} = q_0 + q_1\mathbf{i} + q_2\mathbf{j} + q_3\mathbf{k} \tag{11.1}$$

The quaternion consists of three imaginary numbers and one real number.

i, **j**, and **k** are the imaginary units that satisfy (11.2).

$$\mathbf{i}^2 = \mathbf{j}^2 = \mathbf{k}^2 = -1,\ \mathbf{ij} = \mathbf{k}, \mathbf{jk} = \mathbf{i}, \mathbf{ki} = \mathbf{j} \tag{11.2}$$

In engineering, the quaternion is used for expressing the rotation of a rigid body. Moreover, it has been used in the fields of three-dimensional programming and aerospace because it can express the attitude of the body with a few parameters; it does not have a singular point as in the case of the Euler angle. The quaternion that is used for expressing rotation is defined by using a simple rotation. First, we introduce the definition of simple rotation. According to Euler's theorem, any attitude of the *b-frame* for *r-frame* can be expressed by using a unit vector expressing the rotation axis and rotation angle. A pair of the unit vector and rotation angle is called a simple rotation. A simple rotation that expresses the attitude of the *b-frame* for the *r-frame* is represented by $(\boldsymbol{\lambda}_r^b, \varphi_r^b)$. Here $\boldsymbol{\lambda}_r^b$ can be represented as (11.3), and the norm of $\boldsymbol{\lambda}_r^b$ is equal to 1.

$$\begin{aligned} \boldsymbol{\lambda}_r^b &= [l \;\; m \;\; n]^T \\ \sqrt{l^2 + m^2 + n^2} &= 1 \end{aligned} \tag{11.3}$$

The quaternion $\mathbf{q}_r^b$, which expresses the same attitude as the simple rotation, can be defined by using $\boldsymbol{\lambda}_r^b$ and φ_r^b.

$$\mathbf{q}_r^b = \left[\cos\left(\frac{\phi_r^b}{2}\right), \;\; l \sin\left(\frac{\phi_r^b}{2}\right), \;\; m \sin\left(\frac{\phi_r^b}{2}\right), \;\; n \sin\left(\frac{\phi_r^b}{2}\right)\right]^T \tag{11.4}$$

It is clear that the norm of $\mathbf{q}_r^b$ is equal to 1. Such a quaternion is called the *unit quaternion*.

Next, the sum, difference, and product of the quaternion are defined. In general, the quaternion is represented by a vector representation without an imaginary unit; for example, $\mathbf{q} = [q_0 \;\; q_1 \;\; q_2 \;\; q_3]^T$. Moreover, by using this representation, the sum, difference, and product of a quaternion can be defined by (11.5)–(11.7).

$$\mathbf{q} + \mathbf{p} = [q_0 + p_0 \;\; q_1 + p_1 \;\; q_2 + p_2 \;\; q_3 + p_3]^T \tag{11.5}$$

$$\mathbf{q} - \mathbf{p} = [q_0 - p_0 \;\; q_1 - p_1 \;\; q_2 - p_2 \;\; q_3 - p_3]^T \tag{11.6}$$

$$\mathbf{q} \otimes \mathbf{p} = \begin{bmatrix} q_0 & -q_1 & -q_2 & -q_3 \\ q_1 & q_0 & -q_3 & q_2 \\ q_2 & q_3 & q_0 & -q_1 \\ q_3 & -q_2 & q_1 & q_0 \end{bmatrix} \begin{bmatrix} p_0 \\ p_1 \\ p_2 \\ p_3 \end{bmatrix} \tag{11.7}$$

From (11.7), the commutative law for the quaternion product is not generally formed.

Next, the relation between the quaternion and a coordinate transformation of the vector will be described. A quaternion $\mathbf{q}^*$, which is a conjugate of $\mathbf{q}$, can be defined as follows:

$$\mathbf{q}^* = [q_0 \;\; -q_1 \;\; -q_2 \;\; -q_3]^T \tag{11.8}$$

By using the conjugate quaternion, the coordinate transformation from $\mathbf{r}_r$, which is the vector on the *r-frame* to $\mathbf{r}_b$ (the vector on *b-frame*), is represented as follows:

$$(\mathbf{r}_b)_q = \mathbf{q}_r^{b^*} \otimes (\mathbf{r}_r)_q \otimes \mathbf{q}_r^b \tag{11.9}$$

Here $(\mathbf{r}_b)_q$ and $(\mathbf{r}_r)_q$ imply the quaternion representation of a three-dimensional vector, and each quaternion is defined as (11.10).

$$(\mathbf{r}_b)_q = \begin{bmatrix} 0 & \mathbf{r}_b^T \end{bmatrix}^T, \quad (\mathbf{r}_r)_q = \begin{bmatrix} 0 & \mathbf{r}_r^T \end{bmatrix}^T \tag{11.10}$$

The inverse of the coordinate transformation from $\mathbf{r}_b$ to $\mathbf{r}_r$ is represented as (11.11).

$$(\mathbf{r}_r)_q = \mathbf{q}_r^b \otimes (\mathbf{r}_b)_q \otimes \mathbf{q}_r^{b^*} \tag{11.11}$$

Finally, the time derivatives of the quaternion are introduced. Now, the *b-frame* changes with a very small Δt, and the frame after the change is defined as the *b′-frame*. On the other hand, if the quaternion $\mathbf{q}_r^b$ changes into $\mathbf{q}_r^b(t+\Delta t)$ with the changes in the *b-frame*, in accordance with (11.11), the relation between $\mathbf{r}_{b'}$, which is the vector on *b′-frame*, and $\mathbf{r}_r$ is represented as follows:

$$(\mathbf{r}_r)_q = \mathbf{q}_r^b(t+\Delta t) \otimes (\mathbf{r}_{b'})_q \otimes \mathbf{q}_r^{b^*}(t+\Delta t) \tag{11.12}$$

Next, consider that a simple rotation that expresses the rotational attitude of *the b′-frame* for *b-frame* is $\left(\boldsymbol{\lambda}_b^{b'}, \Delta\varphi\right)$. The quaternion $\Delta\mathbf{q}$ that expresses the same attitude of this simple rotation is defined as (11.13).

$$\begin{aligned} \Delta\mathbf{q} &= \begin{bmatrix} \cos\left(\frac{\Delta\phi}{2}\right) & 0 & 0 & 0 \end{bmatrix}^T + \begin{bmatrix} 0 & \boldsymbol{\lambda}_b^{b'T} \sin\left(\frac{\Delta\phi}{2}\right) \end{bmatrix}^T \\ &\cong [1 \quad 0 \quad 0 \quad 0]^T + \begin{bmatrix} 0 & \boldsymbol{\lambda}_b^{b'T} \left(\frac{\Delta\phi}{2}\right) \end{bmatrix}^T \end{aligned} \tag{11.13}$$

Here $\Delta\varphi$ is very small, and it is assumed that $\sin(\Delta\varphi/2) \cong \Delta\varphi/2$ and $\cos(\Delta\varphi/2) \cong 1$. By using $\Delta\mathbf{q}$, the relation between $\mathbf{r}_b$ and $\mathbf{r}_{b'}$ is represented as follows:

$$(\mathbf{r}_b)_q = \Delta\mathbf{q} \otimes (\mathbf{r}_{b'})_q \otimes \Delta\mathbf{q}^* \tag{11.14}$$

Equation (11.15) is derived by using (11.11), (11.12), and (11.14).

$$\boldsymbol{q}_r^b(t+\Delta t) = \boldsymbol{q}_r^b(t) \otimes \Delta\boldsymbol{q} \tag{11.15}$$

Moreover, by substituting (11.15) into (11.13), we obtain (11.16).

$$\mathbf{q}_r^b(t+\Delta t) - \mathbf{q}_r^b(t) = \mathbf{q}_r^b(t) \otimes \begin{bmatrix} 0 & \boldsymbol{\lambda}_b^{b'T} \left(\frac{\Delta\phi}{2}\right) \end{bmatrix}^T \tag{11.16}$$

By using (11.16), the time derivatives of quaternion $\mathbf{q}_r^b$ can be calculated as follows:

$$\begin{aligned} \dot{\mathbf{q}}_r^b(t) &= \lim_{\Delta t \to 0} \frac{\mathbf{q}_r^b(t+\Delta t) - \mathbf{q}_r^b(t)}{\Delta t} = \lim_{\Delta t \to 0} \frac{\mathbf{q}_r^b(t) \otimes \begin{bmatrix} 0 & \boldsymbol{\lambda}_b^{b'T} \left(\frac{\Delta\phi}{2}\right) \end{bmatrix}^T}{\Delta t} \\ &= \frac{1}{2}\mathbf{q}_r^b(t) \otimes \begin{bmatrix} 0 & \boldsymbol{\lambda}_b^{b'T} \lim_{\Delta t \to 0} \left(\frac{\Delta\phi}{\Delta t}\right) \end{bmatrix}^T \end{aligned} \tag{11.17}$$

Here $\boldsymbol{\lambda}_b^{b'} \lim_{\Delta t \to 0} \left(\frac{\Delta\varphi}{\Delta t}\right)$ implies an angular rate vector of the *b-frame*; we represented this vector as $\boldsymbol{\omega}_b(t)$, and finally, $\dot{\mathbf{q}}_r^b$ can be derived as follows:

$$\dot{\mathbf{q}}_r^b(t) = \frac{1}{2}\mathbf{q}_r^b(t) \otimes (\boldsymbol{\omega}_b)_q(t) \tag{11.18}$$

$(\boldsymbol{\omega}_b)_q$ means the quaternion representation of a three-dimensional vector, similarly to $(\mathbf{r}_b)_q$ and $(\mathbf{r}_r)_q$.

11.3 Attitude and Heading Estimation Algorithms

In this section, we derive the attitude and heading estimation algorithms; this is most important for the development of an attitude sensor. There are many types of algorithms that can estimate the attitude and heading of a rigid body [1–20]; we used one such algorithm called the magnetic force, angular rate, and gravity (MARG) algorithm. In this study, we designed an extended Kalman filter based on the MARG algorithm for the estimation. In this method, attitude and heading are estimated using the magnetic force, angular rate, and gravity. Three-axis magnetic sensors, three-axis gyro sensors, and three-axis accelerometers are required for realizing the MARG algorithm. In the following sections, we introduce the design of an extended Kalman filter based on the MARG algorithm in detail.

11.3.1 Construction of Process Model

For constructing the Kalman filter, a process model of the system is required. In this section, we derive a discrete time process model consisting of the state space equation and observation equation.

First, we derive the state space equation. We apply time derivatives of a quaternion that is represented in (11.18) as the fundamental equation. We can ignore the problems of a singular attitude, which was caused when the Euler angle is used, by using the quaternion. In addition, in the case of the quaternion, the estimated parameters are fewer than in the case of a rotation matrix, and hence, we can reduce the computation time. On the right-hand side of (11.18), $\boldsymbol{\omega}_b$ is output from the gyroscope that is fixed on the body. However, in general, inertia sensors such as gyroscopes and accelerometers have errors caused by various factors. In particular, we have to use a low-cost and low-accuracy gyroscope for realizing the low-cost attitude sensor, and hence, these errors cannot be ignored. For high-accuracy attitude estimation, it is desirable to estimate the error in th gyroscope. Now, let us consider $\boldsymbol{\omega}_{measure} = \begin{bmatrix}\omega_x & \omega_y & \omega_z\end{bmatrix}^T$ as the output of the gyroscope and $\Delta\boldsymbol{\omega} = \begin{bmatrix}\delta\omega_x & \delta\omega_y & \delta\omega_z\end{bmatrix}^T$ as the error that is included in the gyroscope output.

The relationship between these and the true angular rate is represented as follows:

$$\boldsymbol{\omega}_{measure} = \boldsymbol{\omega}_b + \Delta\boldsymbol{\omega} \tag{11.19}$$

Substituting (11.19) into (11.18), we obtained (11.20).

$$\dot{\mathbf{q}}_r^b(t) = \frac{1}{2}\mathbf{q}_r^b(t) \otimes \{(\boldsymbol{\omega}_{measure})_q(t) - (\Delta\boldsymbol{\omega})_q(t)\} \tag{11.20}$$

In general, gyroscope errors consist of the following – bias error, scale factor error, misalignment error, and noise. We can represent these by (11.21).

$$\Delta\boldsymbol{\omega} = \boldsymbol{\omega}_{bias} + \begin{bmatrix} sf_x & 0 & 0 \\ 0 & sf_y & 0 \\ 0 & 0 & sf_z \end{bmatrix} \boldsymbol{\omega}_b + \begin{bmatrix} 0 & ma_{xy} & ma_{xz} \\ ma_{yx} & 0 & ma_{yz} \\ ma_{zx} & ma_{zy} & 0 \end{bmatrix} \boldsymbol{\omega}_b + \boldsymbol{\eta} \tag{11.21}$$

In (11.21), the first term on the right-hand side equation implies bias error; the second term, scale factor error; the third term, misalignment error; and the fourth term, noise. In such errors, the scale factor error and misalignment error are static errors that are caused by individual differences and positioning error of each sensor. Hence, they can be compensated offline, and it is not necessary to estimate them in real time. Hence, only bias error and noise have to be estimated in real time. However, it is impossible to estimate noise, and hence, in this case, we consider $\Delta\boldsymbol{\omega} \cong \boldsymbol{\omega}_{bias}$ and estimate only the bias error. According to [21], it is well known that the bias error of an inertia sensor has the dynamics represented by (11.22).

$$\dot{\boldsymbol{\omega}}_{bias} = \begin{bmatrix} -\beta_x & 0 & 0 \\ 0 & -\beta_y & 0 \\ 0 & 0 & -\beta_z \end{bmatrix} \boldsymbol{\omega}_{bias} + \mathbf{w} \tag{11.22}$$

Here $\mathbf{w} = \begin{bmatrix} w_x & w_y & w_z \end{bmatrix}^T$ is white noise. β_x, β_y, and β_z are constants, and the determination method of these constants is described now.

First, we perform a standstill experiment of approximately 1 h for every sensor and collect data. We then calculate the *Aran variance* from the data, where Aran variance is one of the two sample variances; it is generally used to represent the stability property of the clock. In inertia sensors, the Aran variance is used to quantitatively represent the amount of bias error included in the sensor outputs. We then perform the simulations by using (11.22) after setting the value of β_x, β_y, β_z, and variance of $\mathbf{w}$. Finally, we calculate the Aran variance from the simulation output data. We compare the Aran variances of the experiment and simulation, and we continue to adjust the values of β_x, β_y, and β_z until the two variances match. We consider the state vector as $\mathbf{x} = \left[\mathbf{q}_r^{bT}, \Delta\boldsymbol{\omega}^T\right]^T$; the state space equation is then derived as (11.23).

$$\dot{\mathbf{x}} = \mathbf{f}(\mathbf{x}) + \mathbf{G}\mathbf{u} \tag{11.23}$$

Here, the matrices in the equation are defined as (11.24).

$$\mathbf{f}(\mathbf{x}) = \begin{bmatrix} \frac{1}{2}\mathbf{q}_r^b \otimes (\boldsymbol{\omega}_{measure})_q - \frac{1}{2}\mathbf{q}_r^b \otimes (\boldsymbol{\omega}_{bias})_q \\ \boldsymbol{\beta}\boldsymbol{\omega}_{bias} \end{bmatrix}$$
$$\mathbf{G} = \begin{bmatrix} \mathbf{0}_{4\times 3} \\ \mathbf{I}_{3\times 3} \end{bmatrix}$$
$$\boldsymbol{\beta} = \begin{bmatrix} -\beta_x & 0 & 0 \\ 0 & -\beta_y & 0 \\ 0 & 0 & -\beta_z \end{bmatrix} \tag{11.24}$$

By using the Euler method, we obtain (11.25).

$$\mathbf{x}_{t+1} = \mathrm{x}_t + \mathbf{f}(\mathbf{x}_t)\Delta t + \mathbf{Gu}\Delta t \tag{11.25}$$

Here, $\mathbf{x}_t$ is the state value after t steps and Δt is the sampling time. We consider $\mathbf{f}_t(\mathbf{x}_t) = \mathbf{x}_t + \mathbf{f}(\mathbf{x}_t)\Delta t$ and $\mathbf{G}_t = \mathbf{G}\Delta t$, and the descrete time state space equation is derived as (11.26).

$$\mathbf{x}_{t+1} = \mathbf{f}_t(\mathbf{x}_t) + \mathbf{G}_t\mathbf{u} \tag{11.26}$$

Next, the measurement equation is derived. In this case, the magnetic and gravitational forces can be measured. We consider $\mathbf{g}_r = \begin{bmatrix} 0 & 0 & g \end{bmatrix}^T$ as the gravity force represented as the vector on *r-frame* and $\mathbf{m}_r = \begin{bmatrix} m_n & 0 & m_c \end{bmatrix}$ as the magnetic force represented as the vector on *r-frame*. $\mathbf{g}_b$ and $\mathbf{m}_b$ are then calculated as (11.27).

$$\begin{aligned} (\mathbf{g}_b)_q &= \mathbf{q}_r^{b*}(\mathbf{g}_r)_q\mathbf{q}_r^b \\ (\mathbf{m}_b)_q &= \mathbf{q}_r^{b*}(\mathbf{m}_r)_q\mathbf{q}_r^b \end{aligned} \tag{11.27}$$

$\mathbf{g}_b$ is the gravity force represented on the *b-frame* and $\mathbf{m}_b$ is the magnetic force represented on the *b-frame*. Further, both can be measured by using an accelerometer and magnetic sensor that are fixed on the body. However, the output data of the accelerometer and magnetic sensor have the same error as a gyro sensor. In particular, an accelerometer not only measures gravity but also dynamic linear acceleration that occurs due to the movements of a rigid body, and this dynamic acceleration affects the attitude and heading estimations. Hence, it is necessary to consider these types of errors in the measurement equation. First, we consider the output of accelerometer as $\mathbf{a}_{measure} = \begin{bmatrix} a_x & a_y & a_z \end{bmatrix}^T$, the error in the accelerometer including the dynamic acceleration as $\Delta\mathbf{a} = \begin{bmatrix} \delta a_x & \delta a_y & \delta a_z \end{bmatrix}^T$, the output of the magnetic sensor as $\mathbf{m}_{measure} = \begin{bmatrix} m_x & m_y & m_z \end{bmatrix}^T$, and the error in the magnetic sensor as $\Delta\mathbf{m} = \begin{bmatrix} \delta m_x & \delta m_y & \delta m_z \end{bmatrix}^T$. The relationship among these vectors and $\mathbf{g}_b$ and $\mathbf{m}_b$ can be represented as (11.28).

$$\begin{aligned} \mathbf{a}_{measure} &= \mathbf{g}_b + \Delta\mathbf{a} \\ \mathbf{m}_{measure} &= \mathbf{m}_b + \Delta\mathbf{m} \end{aligned} \tag{11.28}$$

Equation (11.29) is obtained from (11.27) and (11.28).

$$\begin{aligned}(\mathbf{a}_{measure})_q &= \mathbf{q}_r^{b^*}(\mathbf{g}_r)_q\mathbf{q}_r^b + (\Delta\mathbf{a})_q \\ (\mathbf{m}_{measure})_q &= \mathbf{q}_r^{b^*}(\mathbf{m}_r)_q\mathbf{q}_r^b + (\Delta\mathbf{m})_q\end{aligned} \tag{11.29}$$

We define $\mathbf{y}_t = \left[(\mathbf{a}_{measure})_q^T \ (\mathbf{m}_{measure})_q^T\right]^T$ and $\mathbf{v}_t = \left[(\Delta\mathbf{a})_q^T \ (\Delta\mathbf{m})_q^T\right]^T$; the measurement equation that outputs $\mathbf{y}_t$ is represented as (11.30).

$$\mathbf{y}_t = \mathbf{h}_t(\mathbf{x}_t) + \mathbf{v}_t \tag{11.30}$$

Here, vector $\mathbf{h}_t(\mathbf{x}_t)$ is defined as (11.31).

$$\mathbf{h}_t(\mathbf{x}_t) = \begin{bmatrix} \mathbf{q}_r^{b^*}(\mathbf{g}_r)_q\mathbf{q}_r^b \\ \mathbf{q}_r^{b^*}(\mathbf{m}_r)_q\mathbf{q}_r^b \end{bmatrix} \tag{11.31}$$

The discrete time process model of a system that consists of the state space equation represented by (11.26) and the measurement equation represented by (11.30) have been derived.

11.3.2 Extended Kalman Filter

In this section, we show the design of an extended Kalman filter algorithm based on the abovementioned process model. The Kalman filter is mentioned in detail in [21]. Hence, in this book, we will introduce only a simple outline of the Kalman filter and explain our estimation algorithm in detail.

Originally, the Kalman filter was an online estimation algorithm that can estimate the state of a system by using the dynamics of the system, statistical property of noise, the initial value of the system, and the measurement data. The Kalman filter is based on the following four assumptions:

1. Linearity of the system equation
2. Whiteness of the system and measurement noise
3. Gaussian of the noise
4. Quadratic criteria

These assumptions appear to be strictly hard and are not actual. However, a Kalman filter, which is derived by using these assumptions, is available for many real applications. In particular, note the assumption of linearity. The process model represented in (11.26) and (11.30) are both nonlinear equations and does not satisfy the assumption of linearity. However, in such a case, we can also perform the estimation by applying the Kalman filter for an approximate linear model that is obtained by linearizing nonlinear equations. This approximate method is called the Extended Kalman Filter. In the following discussion, the extended Kalman filter algorithm is introduced.

Now, let us consider $\widehat{\mathbf{x}}_{t/t}$ to be the filtered estimate of $\mathbf{x}_t$, which is a state of systems (11.26) and (11.30); $\widehat{\mathbf{x}}_{t/t-1}$, the predicted estimate of $\mathbf{x}_t$; and $\mathbf{F}_t$, the state transition matrix; the measurement matrix $\mathbf{H}_t$ can then be calculated as the partial differential of the vector functions $\mathbf{f}_t(\mathbf{x}_t)$ and $\mathbf{h}_t(\mathbf{x}_t)$. $\mathbf{F}_t$ and $\mathbf{H}_t$ are represented as follows:

$$\mathbf{F}_t = \left(\frac{\partial \mathbf{f}_t(\mathbf{x}_t)}{\partial \mathbf{x}_t}\right)_{\mathbf{x}_t=\widehat{\mathbf{x}}_{t/t}} \qquad \mathbf{H}_t = \left(\frac{\partial \mathbf{h}_t(\mathbf{x}_t)}{\partial \mathbf{x}_t}\right)_{\mathbf{x}_t=\widehat{\mathbf{x}}_{t/t-1}} \tag{11.32}$$

The extended Kalman filter algorithm based on $\mathbf{F}_t$ and $\mathbf{H}_t$ is obtained as (11.33)–(11.37).

$$\widehat{\mathbf{x}}_{t/t-1} = \mathbf{f}_{t-1}(\widehat{\mathbf{x}}_{t-1/t-1}) \tag{11.33}$$

$$\mathbf{P}_{t/t-1} = \mathbf{F}_{t-1}\mathbf{P}_{t/t-1}\mathbf{F}_{t-1}^T + \mathbf{G}_{t-1}\mathbf{Q}_{t-1}\mathbf{G}_{t-1}^T \tag{11.34}$$

$$\mathbf{K}_t = \mathbf{P}_{t/t-1}\mathbf{H}_t^T\left[\mathbf{H}_t\mathbf{P}_{t/t-1}\mathbf{H}_t^T + \mathbf{R}_t\right]^{-1} \tag{11.35}$$

$$\widehat{\mathbf{x}}_{t/t} = \widehat{\mathbf{x}}_{t/t-1} + \mathbf{K}_t\left[\mathbf{y}_t - \mathbf{h}_t(\widehat{\mathbf{x}}_{t/t-1})\right] \tag{11.36}$$

$$\mathbf{P}_{t/t} = \mathbf{P}_{t/t-1} - \mathbf{K}_t\mathbf{H}_t\mathbf{P}_{t/t-1} \tag{11.37}$$

$\mathbf{K}_t$ is the Kalman gain; $\mathbf{P}_{t/t}$ and $\mathbf{P}_{t/t-1}$, the covariance matrix about the estimation error; $\mathbf{Q}_t$, the covariance matrix for the system noise; and $\mathbf{R}_t$, the covariance matrix for the measurement noise. The derivation of each equation is mentioned in detail in [21]. It is guaranteed that we can obtain most probable filtered estimate $\widehat{\mathbf{x}}_{t/t}$ by successive calculations of (11.33)–(11.37). Hence, we can obtain the estimate of the quaternion $\mathbf{q}_r^b$ that expresses the attitude of the body.

11.3.3 Practical Application

In this section, practical applications of the attitude estimation algorithm on a computer are described. Of course, there is a method to calculate (11.33)–(11.37) directly; however, it is not desirable to use an algorithm whose computational complexity increase exponentially with increasing dimensions of the matrices. Moreover, since (11.35) includes calculations of an inverse matrix, it is difficult to calculate by decomposing the matrix, even though $\mathbf{F}_t$ and $\mathbf{H}_t$ have many zero elements. Hence, we use the following algorithm that assumes that a covariance matrix $\mathbf{R}_t$ of the measurement noise, which is a diagonal matrix, is introduced.

11.3.3.1 Assumption

The covariance matrix $\mathbf{R}_t$ depends on the measurement noise p × p symmetric matrix. Moreover, the other matrices shall be divided as follows. For simplicity, the suffix t for the matrix $\mathbf{F}_t$, $\mathbf{G}_t$, $\mathbf{H}_t$, $\mathbf{Q}_t$, and $\mathbf{R}_t$ is omitted.

$$\mathbf{H} = \begin{bmatrix} \mathbf{H}(1) \\ \vdots \\ \mathbf{H}(p) \end{bmatrix}, \quad \mathbf{y} = \begin{bmatrix} y(1) \\ \vdots \\ y(p) \end{bmatrix}, \quad \mathbf{R} = \begin{bmatrix} R(1) & & 0 \\ & \ddots & \\ 0 & & R(p) \end{bmatrix} \tag{11.38}$$

11.3.3.2 Step1: (Initial Value)

Let us consider $\bar{\mathbf{x}} := \hat{\mathbf{x}}_{t/t-1}$ and $\bar{\mathbf{P}} := \mathbf{P}_{t/t-1}$.

11.3.3.3 Step2: (Measurement Update Algorithm)

Considering $\hat{\mathbf{x}} := \bar{\mathbf{x}}$, $\hat{\mathbf{P}} := \bar{\mathbf{P}}$, the following equations are calculated for $i = 1, \ldots, p$ sequentially.

$$\mathbf{f} := \hat{\mathbf{P}}\mathbf{H}(i)^T \tag{11.39}$$

$$\alpha := \mathbf{H}(i)\mathbf{f} + R(i) \tag{11.40}$$

$$\mathbf{K} := \mathbf{f}/\alpha \tag{11.41}$$

$$\widehat{\mathbf{x}} := \widehat{\mathbf{x}} + \mathbf{K}\left[y(i) - \mathbf{h}(i)(\widehat{\mathbf{x}})\right] \tag{11.42}$$

$$\hat{\mathbf{P}} := \hat{\mathbf{P}} - \mathbf{K}\mathbf{f}^T \tag{11.43}$$

11.3.3.4 Step3: (Time Update Algorithm)

Calculate (11.33) and (11.34). This algorithm, which assumes R to be symmetric, is devised such that the inverse matrix calculation does not appear in the calculations of the Kalman gain. Therefore, we can consider that numerous zeros are included in the elements of $\mathbf{F}_t$ and $\mathbf{H}_t$, and we can simplify the algorithm.

11.4 Application and Evaluation

In this section, a small attitude sensor is introduced to demonstrate the application of the estimation algorithm mentioned in the previous section. It comprises low-cost sensors and lightweight micro-computers, and an attitude/heading estimation algorithm is implemented on this sensor. Moreover, we evaluate the sensor by comparing our sensor with a precise conventional attitude sensor.

First, Fig. 11.2 shows an overview of the developed small attitude sensor. This sensor consists of three-axis accelerometers, a three-axis gyroscope, a three-axis magnetic sensor, and a micro computer. The main specifications of this sensor are shown in Table 11.1. According to Table 11.1, the weight of this sensor is very light, and hence, it can be used for various robots and vehicles and not merely UAV/MAV.

Next, the performance of our sensor is compared with that of a conventional sensor. The specifications of a commercial sensor, which is used for the comparison, are shown in Table 11.2. This sensor is often used for movable objects such as aircraft. Hence, the accuracy of this sensor is very high and this sensor is robust for dynamic linear acceleration. However, the weight of this sensor is nearly 50 times that of our attitude sensor, and it is extremely expensive. Hence, this sensor is not sufficient for placing on small UAVs/MAVs. In the evaluation examination, both sensors were

Fig. 11.2 Overview of small attitude sensor

Table 11.1 Specification of small attitude sensor

Weight	15 g	Accelerometer	±3 g
Size	L 60 mm W 40 mm H 14 mm	Magnet sensor	±2 gauss
CPU	SH2–7147	Gyro sensor	±300°/s

Table 11.2 Specification of conventional sensor

Weight	700 g	Accelerometer	±2 g
Size	L 762 mm W 953 mm H 1042 mm	Magnet sensor	±1.5 gauss
CPU		Gyro sensor	±200°/s

placed on the same mount; the experiments were conducted by moving the mount. In addition, the output of the conventional sensor is Euler's angle. Therefore, we plotted it after having converted the estimated quaternion into Euler's angle in order to make it easy to perform comparisons.

Figures 11.3–11.5 show the data of the roll angle, pitch angle, and yaw angle, respectively. On the other hand, Fig. 11.6 shows the data of the roll angle when a dynamic acceleration is expressed in the direction of the Y_b axis when the mount

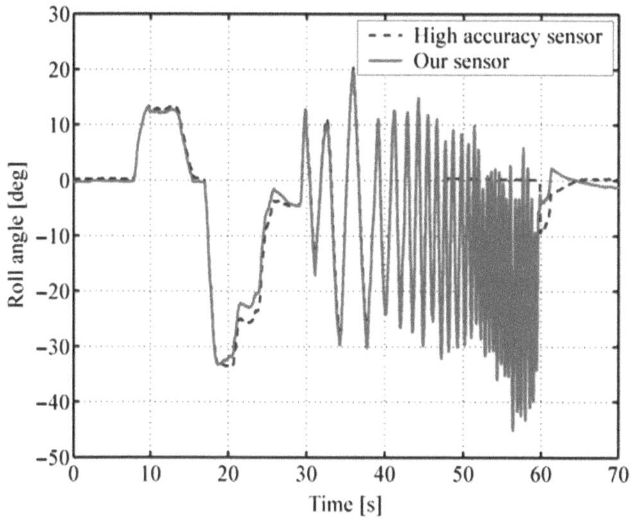

Fig. 11.3 Comparison between conventional sensor and our sensor (roll angle)

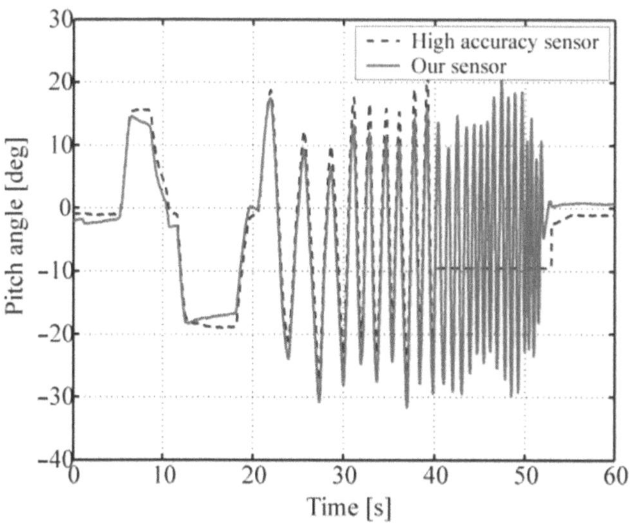

Fig. 11.4 Comparison between conventional sensor and our sensor (pitch angle)

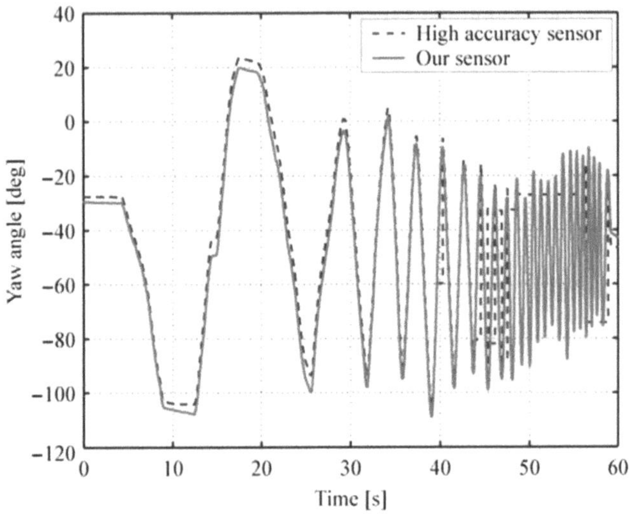

Fig. 11.5 Comparison between conventional sensor and our sensor (yaw angle)

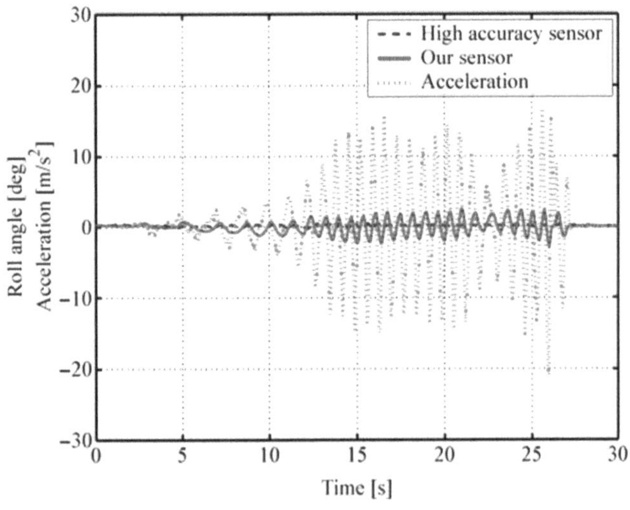

Fig. 11.6 Attitude estimation in dynamic acceleration environment

is horizontal. From Figs. 11.3–11.5, the data of our sensor is in almost complete accordance with that of a high-accuracy attitude sensor, and it is clear that that our sensor can perform accurate estimations. However, there are errors in the data of our sensor in the neighborhood of 5–20 s (in the figure for the pitch angle). This error is due to magnetic agitation, and it is not a particularly major disadvantage in the outdoor environments wherein UAV/MAV flies; however, such magnetism agitations must still be decreased. At several points, the data of the high accuracy sensor has a constant value. The reason for this is mentioned now. High-accuracy attitude sensors cannot estimate the attitude when the angular velocity is beyond 100°/s, and hence, the output data of the sensor becomes constant at high angular velocities. In contrast, our sensor can estimate the attitude when the angular velocity is more than 200°/s. In addition, in the case of the dynamic acceleration, our results are slightly inferior to those of high-accuracy attitude sensors; but our sensor can reduce the attitude error sufficiently when the acceleration is expressed. From the above results, it can be said that our sensor sufficiently meets the specifications to apply it in practice.

11.5 Summary

In this chapter, the algorithm of the attitude sensor based on the extended Kalman filter was introduced as an attitude estimation method that can be realized by using the low cost and lightweight sensors. We used the quaternion in the attitude expression method, and the process model that includes dynamical model of the bias error of gyro sensor was derived. Subsequently, the extended Kalman filter was constructed by using the process model, and the attitude estimation algorithm that reduces the error caused by dynamic acceleration was also realized. Moreover, we inspected the effectiveness of the attitude estimation algorithm that was introduced in this chapter by comparing our sensor with a commercial high-accuracy sensor. From the evaluation examinations, it has been shown that the proposed algorithm can estimate the attitude of a body with an accuracy equal to that of a commercial attitude sensor.

References

1. Lefferts EJ, Markley FL, Shuster MD (1982) Kalman filtering for spacecraft attitude estimation. In: 20th AIAA-1982–70 American Institute of aeronautics and astronautics, aerospace sciences meeting, Orlando, FL, 11–14 Jan 1982, p 17
2. John C, Landis M (2003) Unscented filtering for spacecraft attitude estimation. In: AIAA-2003–5484 AIAA guidance, navigation, and control conference and exhibit, Austin, TX, 11–14 Aug 2003
3. Vaganay J, Aldon MJ, Fournier A (1993) Mobile robot attitude estimation by fusion of inertial data. In: Proceedings of the IEEE international conference on robotics and automation, pp 277–282
4. John LC, Landis M (1997) Predictive filtering for attitude estimation without rate sensors. J Guid Contr Dynam 20(3):522–527
5. Baerveldt AJ, Klang R (1997) Low-cost and low-weight attitude estimation system for an autonomous helicopter. In: Proceedings of the IEEE international conference on intelligent engineering systems (INES) '97, pp 391–395
6. Johnson EN, Calise AJ, Sttigeri R, Watanabe Y, Madyastha V (2004) Approaches to vision-based formation control. In: Proceedings of the IEEE conference on decision and control, pp 1643–1648

7. Psiaki ML, Martel F, Parimal K (1990) Three-axis attitude determination via Kalman filtering of magnetometer data. J Guid Contr Dynam 13(3):506–514
8. Hamel T, Mahony R (2006) Attitude estimation on SO[3] based on direct inertial measurements. In: Proceedings of the IEEE international conference on robotics and automation (ICRA) 2006, pp 2170–2175
9. Mortari D, Pollock TC, Junkins JL (1998) Toward the most accurate attitude determination system using star trackers spaceflight mechanics. In: Proceedings of the AAS/AIAA space flight mechanics meeting, Monterey, CA, USA, 9–11 Feb 1998, pp 839–850
10. Soken HE and Hajiyev C (2009) Adaptive unscented Kalman filter with multiple fading factors for pico satellite attitude estimation. In: Proceedings of 4th international conference on recent advances in space technologies, 2009, RAST '09, pp 541–546
11. Liang X, Weizheng Y, Honglong C, Chengyu J (2009) MEMS-based multi-sensor integrated attitude estimation technology for MAV applications. In: 4th IEEE international conference on nano/micro engineered and molecular systems, 2009, NEMS 2009, pp 1031–1035
12. Young SS (2006) Attitude estimation by multiple-mode Kalman filters. IEEE Trans Ind Electron 53(4):1386–1389
13. Barbour N, Schmidt G (2001) Inertial sensor technology trends. IEEE Sensor J 1(4):332–339
14. Bachmann ER (2000) Inertial and magnetic tracking of limb segment orientation for inserting humans in synthetic environments. Ph.D. thesis, Naval Postgraduate School, Monterrey, CA, USA
15. Sabatini AM, Martelloni C, Scapellato S, Cavallo F (2004) An ambulatory monitoring system for assessing walking features from inertial sensing. In: IMEKO, IEEE, SICE Proc. 2nd international symposium measurement, analysis and modeling of human functions, Genova, Italy, June 2004 (in press)
16. Hall JK, Knoebel NB, McLain TW (2008) Quaternion attitude estimation for miniature air vehicles using a multiplicative extended Kalman filter. In: Position, location and navigation symposium, 2008 IEEE/ION, 5–8 May 2008, pp 1230–1237
17. Yee-Jin C, Jong-Hwan K (2007) Unscented filtering in a unit quaternion space for spacecraft attitude estimation. In: Proceedings of the IEEE international symposium on industrial electronics, 2007, ISIE 2007, pp 66–71
18. Markley FL (2003) Attitude error representation for Kalman filtering. J Guid Contr Dynam 26(2):311–317
19. LaViola JJ (2003) A comparison of unscented and extended Kalman filtering for estimating quaternion motion. In: The proceedings of the 2003 American control conference, IEEE Press, June 2003, pp 2435–2440
20. Wang J, Wilson WJ (1992) 3D relative position and orientation estimation using Kalman filter for robot control. In: Proceedings of the IEEE international conference on robotics and automation, pp 2638–2645
21. Rogers RM (2003) Applied Mathematics in Integrated Navigation System, 2nd edn. AIAA education series

Chapter 12
Vision-Based Navigation and Visual Servoing of Mini Flying Machines

Abstract The design of reliable navigation and control systems for Unmanned Aerial Vehicles (UAVs) based only on visual cues and inertial data has many unsolved challenging problems, ranging from hardware and software development to pure control-theoretical issues. This chapter addresses these issues by developing and implementing an adaptive vision-based autopilot for navigation and control of small and mini rotorcraft UAVs. The proposed autopilot includes a Visual Odometer (VO) for navigation in GPS-denied environments and a nonlinear control system for flight control and target tracking. The VO estimates the rotorcraft ego-motion by identifying and tracking visual features in the environment, using a single camera mounted on-board the vehicle. The VO has been augmented by an adaptive mechanism that fuses optic flow and inertial measurements to determine the range and to recover the 3D position and velocity of the vehicle. The adaptive VO pose estimates are then exploited by a nonlinear hierarchical controller for achieving various navigational tasks including take-off, landing, hovering, trajectory tracking, target tracking, etc. Furthermore, the asymptotic stability of the entire closed-loop system has been established using systems in cascade and adaptive control theories. Experimental flight test data over various ranges of the flight envelope illustrate that the proposed vision-based autopilot performs well and allows a mini rotorcraft UAV to achieve autonomously advanced flight behaviours by using vision.

Video Links:

Vision-based hovering
http://mec2.tm.chiba-u.jp/monograph/Videos/Chapter12/1.wmv
Vision-based precise auto-landing
http://mec2.tm.chiba-u.jp/monograph/Videos/Chapter12/2.wmv
Vision-based moving target tracking
http://mec2.tm.chiba-u.jp/monograph/Videos/Chapter12/3.wmv
Presentation-Vision based hovering
http://mec2.tm.chiba-u.jp/monograph/Videos/Chapter12/4.wmv
Velocity trajectory tracking using optic flow
http://mec2.tm.chiba-u.jp/monograph/Videos/Chapter12/5.wmv

K. Nonami et al., *Autonomous Flying Robots: Unmanned Aerial Vehicles and Micro Aerial Vehicles*, DOI 10.1007/978-4-431-53856-1_12, © Springer 2010

Optic flow based autonomous flight
http://mec2.tm.chiba-u.jp/monograph/Videos/Chapter12/6.wmv
Optic flow based autonomous indoor flight
http://mec2.tm.chiba-u.jp/monograph/Videos/Chapter12/7.wmv

12.1 Introduction

Recently, there is a growing interest in developing fully autonomous UAVs and Micro Air Vehicles (MAVs) for military and civil applications. The design of sensing, navigation and control systems is a crucial step in the development of such autonomous flying machines. Position, altitude and orientation measurements are usually sufficient for the control of UAVs operating at high altitudes. Therefore, conventional avionics that include GPS and IMU provide the required information for flight control and waypoint navigation. On the other hand, mini and micro UAVs are designed to operate at low altitudes in cluttered environments. To achieve realistic missions in such complex environments, the flight controller requires precise estimation of both the vehicle pose and its surrounding environment. Among many sensors for environment mapping and obstacles detection, ultrasonic sensors, Laser Range Finder (LRF), radar and vision have distinct advantages when applied to UAVs. For medium-size UAVs such as the RMAX helicopter, LRF is widely used for obstacles avoidance [33]. We are however, interested in enabling mini UAVs and MAVs to achieve advanced flight behaviors in cluttered environments. The limited payload of these small platforms precludes the use of conventional and standard sensors. Therefore, small cameras, which are available today at low-cost, are attractive sensors for mini UAVs flying near the ground in hazardous environments, Fig. 12.1. Indeed, vision provides a viable and useful solution to sensing on UAVs because cameras are too light to fit the limited payload capabilities of MAVs. Furthermore, visual sensors are passive and contain rich information about the vehicle motion and the environment structure. Unlike GPS, which does not work in the shadow of satellite visibility, vision works in a cluttered urban environment or even in indoor spaces.

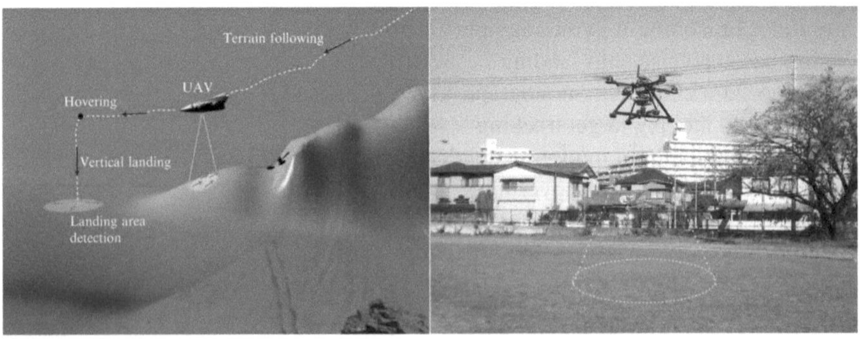

Fig. 12.1 Visual navigation and control of small rotorcraft UAVs using vision

12.1.1 Related Work on Visual Aerial Navigation

Although computer vision has many benefits for UAVs control and guidance, it presents many challenges too. Indeed, the design of reliable and real-time vision algorithms is a complex problem. Moreover, synthesizing robust flight controllers that are based on visual cues is another challenge. Despite these challenging issues, promising results have been obtained using computer vision for aerial vehicles control and navigation. Potential applications of computer vision for UAVs include pose estimation, landing, objects detection and tracking, mapping and localization, obstacles avoidance, etc.

In several UAVs projects, vision is used as a method to sense the aircraft ego-motion. Stereo visual odometers for relative position estimation were implemented in the CMU helicopter [3] and the USC's robotic helicopter. In [23] and [20], structure from motion approach has been applied to recover the ego-motion of autonomous micro air vehicles. Vision has also been applied to recover UAVs pose with respect to some artificial marks like the work presented in [2]. Some researchers have also developed vision systems that can estimate the UAV attitude by detecting the horizon line [9].

The use of vision for autonomous landing has been actively researched. In the BEAR project at the university of California Berkeley, a vision system that uses multiple view geometry has been developed to land an autonomous helicopter on a moving deck [36]. In [32], the authors proposed a vision-based strategy that allows the AVATAR helicopter to land on a slowly moving helipad with known shape. Stereo vision is also used to detect safe landing area and to achieve soft landing [18,42].

There are also some applications of vision for UAVs simultaneous localization and map building (visual SLAM) [6,25,26].

Computer vision is also used as the primary sensor for objects detection and tracking like for windows tracking [29, 30], moving target tracking [19], road following [8], etc.

Most of previous techniques rely on the reconstruction of the vehicle's state vector which is then used in the control loop. It is also important to note that most reported experimental results have been obtained using helicopters that weigh several kilograms.

Recently, many researchers have been interested in using image motion or optic flow for UAVs control and navigation without recovering the explicit motion of the vehicle. These techniques are inspired from insects like honeybees and flies which rely heavily on optic flow for navigation and flight control [39,41].

Barrows [5] designed a VLSI optic flow sensor for use in a MAV, which was tested on a small glider, and later applied for altitude control of a small aircraft [13]. Ruffier and Franceschini [31] developed an optic flow-based autopilot to control a tethered 100-g helicopter. In later work [35], another vision-based system was used for speed control and lateral obstacle avoidance. Another interesting work has been done by Zuffery and Floreano [43] who have developed 30- and 10-g microflyers that can avoid walls using optic flow. A combined optic flow and stereo-based navigation system has been implemented on the AVATAR helicopter by Hrabar et al. [16] to demonstrate obstacles avoidance in an urban environment. In [7], Chahl et al. designed and implemented an optic flow-based system for UAVs landing. In recent work, Garratt and Chahl [11] conducted experiments on a Eagle helicopter and used optic flow and GPS velocity for height estimation, terrain following and lateral motion control.

While most of reviewed works have been done using fixed-wing UAVs or small-scale helicopters, little experimental results have been reported on vision-based control of mini rotorcraft UAVs that weigh less than 1 kg like the quadrotor helicopter. Most existing works on visual servoing of mini quadrotor UAVs have focused on basic stabilization or hovering [2, 10, 14] and indoor flights in structured environments [1, 15, 43].

12.1.2 Description of the Proposed Vision-Based Autopilot

In this chapter, we present an embedded autopilot which uses information extracted from real-time images for navigation and flight control. Our objective in this research is to extend previous findings on vision-based navigation in order to allow mini rotorcraft UAVs to navigate within unknown indoor and outdoor environments and to achieve more advanced flight behaviors like trajectory tracking, precise hovering, vertical landing, target tracking, etc. Furthermore, the system is needed to meet the payload requirements of miniature rotorcraft that weigh less than 1 kg. In order to reach this goal, we have dealt with both theoretical and practical aspects of vision-based control of UAVs. Indeed, we have designed and implemented a vision-based autopilot that includes a real-time vision algorithm for features tracking and image motion integration, an adaptive visual odometer for range and motion estimation, and a hierarchical nonlinear controller for guidance and 3D flight control.

Functionally, the vision algorithm computes in real-time optic flow, tracks visual features and estimates the travelled flight distance in terms of total image displacement (in pixels). These visual measurements are then fused with IMU data in order to overcome the non-desired rotation effects. Indeed, it is difficult to sense UAV translation with vision alone since image displacements also occur with aircraft rotation (translation–rotation ambiguity). Another ambiguity is the scale factor or the unknown range that can not be estimated from visual measurements alone. This ambiguity is resolved by exploiting the adaptive control tools to identify the scale factor or range from optic flow and accelerometers data. The real aircraft position, velocity and height are then recovered. These estimates are then, used by a multipurpose nonlinear controller for autonomous 3D flight control.

Although previous works have shown that visual cues can lead to an autonomous flight, our work has extended this finding in a number of areas. The proposed autopilot is based on only two lightweight sensors, a small single camera and a low-cost IMU without using GPS. A major novelty in our work is that we have demonstrated the possibility and feasibility of recovering the range and aircraft motion from optic flow and IMU data, thereby leading to advanced flight behaviors. Unlike other works, our system has been implemented and demonstrated outdoors and indoors using a miniature unstable rotorcraft that weighs less than 700 g. A final new contribution of this work is that we have developed a 3D flight nonlinear controller and proved its asymptotic stability and robustness even in the presence of estimation errors in the visual odometer.

12.2 Aerial Visual Odometer for Flight Path Integration

The proposed vision system determines the vehicle's relative velocity and 3D position with respect to the initial rotorcraft location or some target which may be stationary or moving. It relies on tracking features appearing in a "target template" initially chosen at the image center. With the computed template location (x_i, y_i) in the image frame, the relative distance (X_i, Y_i) between the rotorcraft and the ground objects appearing in the image template "i" is estimated after compensating the rotation effects using IMU data and recovering the range Z (or height) using an adaptive algorithm. The rotorcraft horizontal motion (X, Y) in the inertial frame is then estimated by accumulating or summing the relative distances (X_i, Y_i) for $i = 1 \ldots n$ as shown in Fig. 12.2.

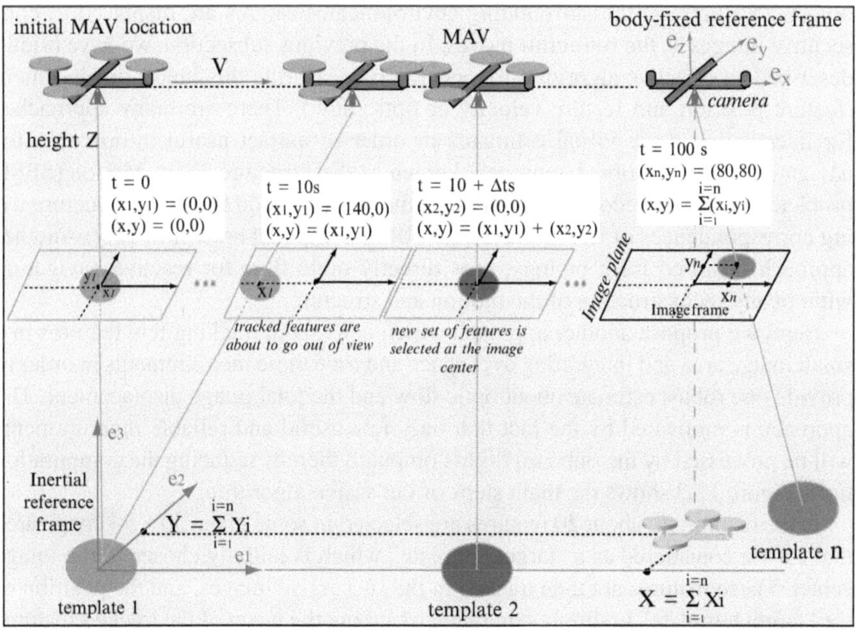

Fig. 12.2 Rotorcraft position estimation by visually tracking ground objects that appear in the camera field of view

The real-time vision algorithm consists of three main stages: features selection and tracking, pseudo velocity/position estimation in terms of pixels, and rotation effects compensation.

12.2.1 Features Selection and Tracking

The tracking algorithm consists of a Lucas–Kanade (LK) tracker which is based on gradient-based optic flow computation. In our implementation, we have used the OpenCV library that contains an efficient implementation of the Shi–Tomasi [37] algorithm for good features selection and the pyramidal LK [28] algorithm for features tracking.

The outputs of this algorithm are the positions $(x_i(t), y_i(t))$ of the tracked features in the image frame. We have slightly modified that algorithm in order to provide also estimates about the optic flow at each feature location.

12.2.2 Estimation of the Rotorcraft Pseudo-motion in the Image Frame

A vision system can estimate rotorcraft pseudo-motion[1] by tracking stationary objects (features) in the surrounding environment. Features are displaced in consecutive images as the rotorcraft moves. In the previous subsection, we have briefly described an effective algorithm for accurately measuring this image displacement (feature position and feature velocity or optic flow). There are many approaches for interpreting these visual estimates in order to extract useful information for navigation. The traditional approach, known as the Structure-From-Motion (SFM) problem consists in recovering both the camera ego-motion and scene structure using correspondences of tracked features [20] or computed optic flow [23]. Another approach, inspired from biology, uses directly optic flow for reactive navigation without any reconstruction of the motion and structure.

Here, we propose another approach which consists in tracking few features in a small image area and integrating over space and time these measurements in order to provide one robust estimate about optic flow and the total image displacement. The approach is motivated by the fact that only few useful and reliable measurements will be processed by the onboard flight computer, thereby reducing the computation time. Figure 12.3 shows the main steps of our vision algorithm.

In the beginning, about 20 features are selected in some given 50×50 image area that can be considered as a "target template" which is initially chosen at the image center. These features are then tracked in the successive images, and the position of the "target template" is simply computed by taking the mean of the tracked features

[1] Pseudo-motion means here, motion in the image frame which is expressed in pixels.

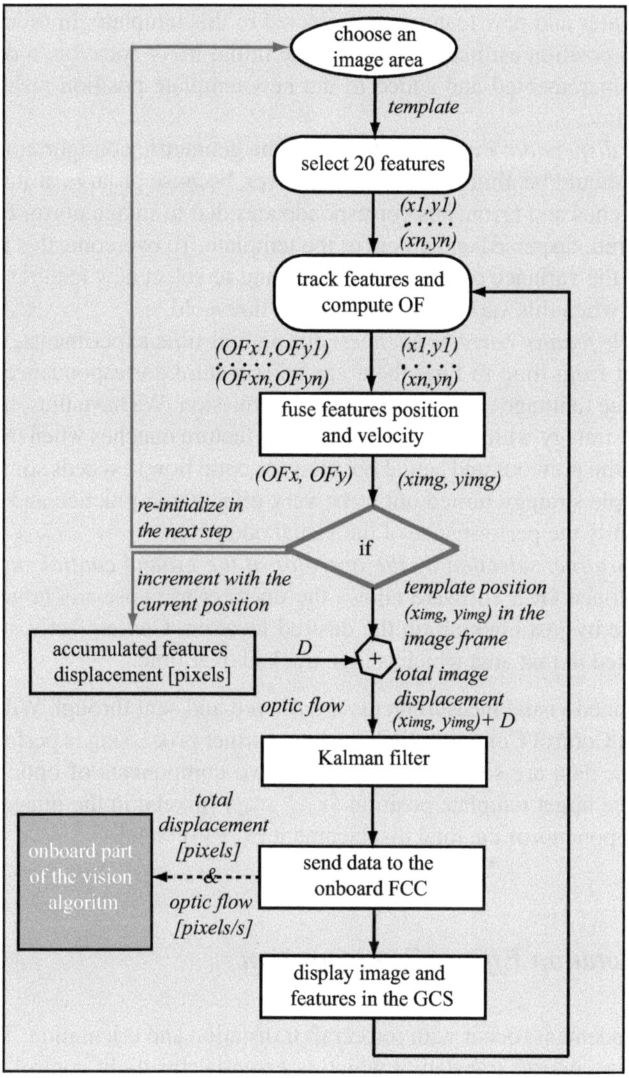

Fig. 12.3 Block diagram of the real-time vision algorithm

position. The "target template" velocity or optic flow is also calculated by taking the mean of the tracked features velocity. As the rotorcraft moves, older features leave the Field Of View (FOV) and new features enter the FOV. Therefore, an effective mechanism is required to deal with features disappearance and appearance and to continue motion estimation. The proposed approach consists in selecting a new set of features under the following conditions:

- *Target template is at the image border*: when the target template is about to go out of view causing features disappearance, a new target template is chosen at the image center and new features are selected in this template. In order to provide a pseudo position estimate relative to the initial MAV location, a displacement offset is incremented and added to the new template position measured in the image frame.

- *Features dispersion*: in an ideal tracking, the geometric configuration of tracked features should be almost constant. However, because of large attitude changes, poor matches and erroneous correspondences due to image noise, features may be scattered, dispersed and go out of the template. To overcome this problem, we compute the variance of features position and re-select new features in the same template when this variance exceeds some threshold.
- *Unreliable feature correspondences*: during real-time experiments, we have noticed that from time to time there are false feature correspondences which are mainly due to image quality and video transmission. We have thus, implemented a simple strategy which consists in rejecting feature matches when the difference between the previous and actual position (or optic flow) exceeds some threshold. This simple strategy turned out to be very effective in practice and it improved significantly the performance of the visual odometer.
- *Arbitrary target selection by the operator at the ground control station (GCS)*: the developed GCS software allows the operator to chose any target or area in the image by just clicking on the desired location. Consequently, new features are selected in that area which is then tracked over time.

The obtained visual measurements are filtered and sent through WiFi to the onboard Flight Control Computer (FCC) where further processing is performed. More precisely, six data are sent to the FCC: the two components of optic flow $(\dot{x}, \dot{y})$ [pixels/s]; the target template position (x_{img}, y_{img}) [pixels] in the image frame; and the two components of the total displacement (x, y) [pixels].

12.2.3 Rotation Effects Compensation

Image displacements occur with rotorcraft translation and orientation. Therefore, in order to sense aircraft translation which is essential for flight control, rotation effects must be eliminated from the measured image displacement and optic flow. Furthermore, this translation–rotation ambiguity is more significant in rotorcraft UAVs since the vehicle translation is a direct result of its attitude change. To overcome this problem and compensate the rotational components of optic flow and image displacement, onboard IMU data (Euler angles (θ, ϕ) and angular rate data $(\omega_x, \omega_y, \omega_z)$) are used.

$$\begin{cases} x_t = x - (-f \ \tan\theta) \\ y_t = y - (f \ \tan\phi) \end{cases} \tag{12.1}$$

and

$$\begin{cases} \dot{x}_t = \dot{x} - \left(\dfrac{x_{img} y_{img}}{f} \omega_x - \dfrac{f^2 + x_{img}^2}{f} \omega_y + y_{img} \omega_z \right) \\ \dot{y}_t = \dot{y} - \left(\dfrac{f^2 + y_{img}^2}{f} \omega_x - \dfrac{x_{img} y_{img}}{f} \omega_y - x_{img} \omega_z \right) \end{cases} \tag{12.2}$$

where f is the camera focal length, (x_t, y_t) and $(\dot{x}_t, \dot{y}_t)$ are the translational components of image displacement and optic flow, respectively.

For effective compensation of the image displacement caused by rotation, the latency problem between the IMU measurements and visual estimates should be addressed. From experimental tests, we have identified a time-delay of about 0.25 s between the IMU and visual data. This delay is mainly due to the fact that images are processed at the ground station and the visual data are sent back to the embedded microprocessor for further processing. To overcome this problem, we applied a low-pass filter to the IMU data to filter the measurements from noise but also to introduce a 0.25 s delay in the IMU data. By using this simple strategy, the filtered (or delayed) IMU measurements were almost identical to vision estimates during pure rotation movements, thereby resulting in effective compensation of rotation effects (see Fig. 12.7).

At this stage, we have visual information $(x_t, y_t, \dot{x}_t, \dot{y}_t)$ about the rotorcraft position and velocity which are expressed in terms of incremented image displacement in pixels and optic flow in pixels per second. The true UAV position and velocity can not be directly deduced because of the range ambiguity. Indeed, the translational image displacement depends on both aircraft translation and relative distance (range) to the perceived objects.

12.3 Adaptive Observer for Range Estimation and UAV Motion Recovery

This section presents an adaptive observer that can estimate the true MAV position in meters and velocity in meters per second using the visual estimates $(x_t, y_t, \dot{x}_t, \dot{y}_t)$ and accelerometers measurements. Furthermore, the proposed estimator allows to recover the range, which is used for height control when the camera is looking downwards or obstacles avoidance for any other camera orientation.

12.3.1 Mathematical Formulation of the Adaptive Visual Observer

Here, we show theoretically that it is possible to recover the range and the real UAV motion using only visual information from a single camera and accelerometers data.

Let us write the relation between the available visual measurements (x_t, y_t [pixels], $\dot{x}_t, \dot{y}_t$ [pixels/s]) and the state variables (X, Y, Z [m], V_x, V_y, V_z [m/s]) expressed in the inertial frame. If the camera is looking downwards, then we can write

$$\begin{cases} x_t = f\dfrac{X}{Z} \\ y_t = f\dfrac{Y}{Z} \end{cases} \tag{12.3}$$

and

$$\begin{cases} \dot{x}_t = f\dfrac{V_x}{Z} + x_{img}\dfrac{V_z}{Z} \\ \dot{y}_t = f\dfrac{V_y}{Z} + y_{img}\dfrac{V_z}{Z} \end{cases} \tag{12.4}$$

It is clear from the above equations that if the range Z (here equivalent to height) can be estimated then the true MAV position and velocity can be recovered using (12.3) and (12.4). As a first step, we propose a real-time identification algorithm that will estimate the range Z under the following assumptions:

1. The height changes are small ($Z \approx cst$ and $V_z \approx 0$) when applying the identifier.
2. The terrain or relief is smooth such that it can be decomposed into flat segments.

In order to satisfy the first assumption, a static pressure sensor was added to the platform for enhancing the vertical motion estimation and control. However, this sensor does not provide height above the ground objects or relative distances to the perceived objects. Like GPS, it indicates the height with respect to the sea level, and when calibrated it can estimate the height with respect to the initial point. We would like to highlight that pressure sensor is very light (few grams), cheep (few dollars) and works in indoor and outdoor environments. Thus, it can be easily integrated into MAV and combined with vision to improve the control of vertical motion.

By considering the previous assumptions, (12.4) becomes

$$\begin{cases} \dot{x}_t \backsimeq f\,\dfrac{V_x}{Z} \\ \dot{y}_t \backsimeq f\,\dfrac{V_y}{Z} \end{cases} \Longrightarrow \begin{cases} \ddot{x}_t \backsimeq f\,\dfrac{a_x}{Z} \\ \ddot{y}_t \backsimeq f\,\dfrac{a_y}{Z} \end{cases} \tag{12.5}$$

where (a_x, a_y) are the MAV linear accelerations expressed in the inertial frame.

Many types of on-line parameter estimation techniques can be applied to estimate Z using the derivatives of optic flow and linear accelerations (see (12.5)). Our odometer is based on the Recursive-Least-Squares (RLS) algorithm which presents many advantages for our application.

12.3.2 Generalities on the Recursive-Least-Squares Algorithm

The choice of this method is essentially motivated by the fact that it is easy to implement and appropriate for real-time applications. In addition, it is robust with respect to noise and it has been widely used in parameter estimation in a recursive and nonrecursive form mainly for discrete-time systems [4, 12, 17, 27].

The RLS method is simple to apply in the case where the unknown parameters appear in a linear form, such as in the linear parametric model:

$$y = b^T \gamma$$

where b^T is a constant vector which contains the unknown parameters and (y, γ) are the available output and input signals.

The RLS algorithm formula are obtained by minimizing some cost function and its expression is as follows [17]:

$$\begin{aligned} \dot{\hat{b}} &= P\ \varepsilon\ \gamma \\ \dot{P} &= \beta P - P\frac{\gamma\gamma^T}{m^2}P, \quad P(0) = P_0 \\ m^2 &= 1 + \gamma^T\gamma \\ \varepsilon &= \frac{y(t) - \hat{b}^T\gamma(t)}{m^2} \end{aligned} \tag{12.6}$$

The stability properties of the RLS algorithm depend on the value of the forgetting factor β.

In our implementation, we have used a robust version of the standard RLS algorithm that includes:

- *RLS with forgetting factor*: (12.6) has been modified to avoid an unbounded growth of he gain matrix $P(t)$ when the forgetting factor $\beta > 0$.
- *RLS with projection*: in many practical problems where b represents the physical parameters of a plant, we may have some a priori knowledge like the upper and/or lower bounds $(b_{\max}, b_{\min})$. Thus, we have modified the RLS algorithm to constrain the estimates search, thereby resulting in the following advantages: (1) speed-up the convergence, (2) reduce large transients, and (3) avoid some singular cases like $\hat{b} = 0$ which is very important for control design.
- *RLS with dead-zone*: the principal idea behind the dead-zone is to monitor the size of the estimation error and adapt only when the estimation error is large relative to the modelling error or disturbance. This means that adaptation or identification process is switched off when the normalized estimation error εm is small.

More details about robust RLS algorithms including their stability analysis can be found in [17] and the references therein.

12.3.3 Application of RLS Algorithm to Range (Height) Estimation

Now, let us apply a robust RLS algorithm to estimate the height Z in (12.5). In a deterministic case, Z can be estimated using one of the following SISO subsystems "$\ddot{x}_t = f\,\frac{a_x}{Z}$" or "$\ddot{y}_t = f\,\frac{a_y}{Z}$". An interesting solution could consist in estimating Z for each SISO subsystem and then, fuse the two obtained estimates $(\hat{Z}_1, \hat{Z}_2)$ in a favorable manner. For more robustness against noise and external disturbances, we have applied the RLS algorithm with projection and dead-zone in order to estimate $\hat{Z}_1, \hat{Z}_2$ from available measurements $(\dot{x}_t, a_x)$, $(\dot{y}_t, a_y)$, respectively. The two estimates $(\hat{Z}_1, \hat{Z}_2)$ have then been fused using a weighted averaging method where the weights depend on the *Persistency Excitation* (PE) property of each signal a_i, $i = x, y$. This idea can be mathematically described as follows:

$$\hat{Z} = \begin{cases} \dfrac{1}{w_1 + w_2}(w_1\hat{Z}_1 + w_2\hat{Z}_2) & \text{if } w_1 + w_2 \neq 0 \\ \hat{Z}_{t-1} & \text{otherwise} \end{cases} \tag{12.7}$$

where the weights (w_1, w_2) are functions of the *PE* condition.

In order to apply the RLS method to estimate Z in (12.5), we need to write the preceding system in the following linear parametric form: $y = b^T\gamma$, and then the application of the modified robust RLS algorithm becomes straightforward. This can be done using the following formula:

$$\begin{cases} y_1 = \dfrac{s\,\dot{x}_t}{(s+\lambda_1)}, & \gamma_1 = \dfrac{a_x}{(s+\lambda_1)}, \quad b_1 = \dfrac{f}{Z} \\ y_2 = \dfrac{s\,\dot{y}_t}{(s+\lambda_2)}, & \gamma_2 = \dfrac{a_y}{(s+\lambda_2)}, \quad b_2 = \dfrac{f}{Z} \end{cases} \tag{12.8}$$

where $\frac{1}{(s+\lambda_1)}$ and $\frac{1}{(s+\lambda_2)}$ are fist-order low-pass filters, which are used to avoid direct signals differentiation.

Theorem 12.1. *For* $i = 1, 2$, *the modified RLS algorithm with projection and dead-zone, applied to system (12.5) and (12.8), guarantees the following properties [17]:*

(a) $\hat{b}_i, \dot{\hat{b}}_i \in \mathcal{L}_\infty$ *(bounded).*
(b) $b_{\min} \leq \hat{b}_i(t) \leq b_{\max}$, $\forall t \geq 0$, *where* $b_{\min} > 0$ *and* $b_{\max} > 0$ *are a priori known lower and upper bounds of the parameter b (projection principle).*
(c) $\lim_{t\to\infty} \hat{b}_i(t) = \bar{b}_i$ *where* $\bar{b}_i$ *is a positive constant.*
(d) If $\gamma_i \in \mathcal{L}_\infty$ *and* γ_i *is* PE, *then the estimate* $\hat{b}$ *converges exponentially to its true value*[2] b.

[2] In the presence of external disturbances, $\tilde{b}_i = b - \hat{b}_i$ converges exponentially to the residual set: $\{\tilde{b}_i / |\tilde{b}_i| \leq c(\rho_0 + \bar{d})\}$, where $\bar{d}$ is the disturbance upper bound, c is a positive constant and ρ_0 characterizes the dead-zone.

Rigorous and detailed proof of this theorem can be found in any book on identification techniques such as [17].

Since the two estimates $(\hat{b}_1, \hat{b}_2)$ satisfy the properties in Theorem 12.1, it is then trivial to show that $\hat{Z}_1 = \frac{f}{\hat{b}_1}$, $\hat{Z}_2 = \frac{f}{\hat{b}_2}$ and $\hat{Z}$ in (12.7) are positive and satisfy also the properties (a)–(d) listed in Theorem 12.1. Once the height Z is identified, the aircraft horizontal position vector (X, Y) and velocity vector (V_x, V_y) can be recovered using (12.3) and (12.5):

$$\begin{cases} \hat{X} = \dfrac{x_t}{\hat{b}} = \hat{Z}\,\dfrac{x_t}{f} \\ \hat{Y} = \dfrac{y_t}{\hat{b}} = \hat{Z}\,\dfrac{y_t}{f} \\ \hat{Z} = \dfrac{f}{\hat{b}} \end{cases} \quad \text{and} \quad \begin{cases} \hat{V}_x = \dfrac{\dot{x}_t}{\hat{b}} = \hat{Z}\,\dfrac{\dot{x}_t}{f} \\ \hat{V}_y = \dfrac{\dot{y}_t}{\hat{b}} = \hat{Z}\,\dfrac{\dot{y}_t}{f} \end{cases} \tag{12.9}$$

Remark 12.1. It is important to note that (x^t, y^t) in (12.9) is the accumulated translational image displacement as shown in Figs. 12.2 and 12.3. Hence, the estimates $(\hat{X}, \hat{Y})$ in (12.9) correspond to the rotorcraft position in the inertial frame provided that the UAV flies at a constant height during path integration. For accurate position estimation even at varying height, it is better to estimate first the relative distances between the MAV and the tracked objects and then to accumulate these relative distances in order to provide an estimate of the MAV position in the inertial frame which is associated to the initial location, Fig. 12.3. This is mathematically equivalent to

$$\begin{cases} \hat{X} = \displaystyle\sum_{i=1}^{i=n} X_i = \sum_{i=1}^{i=n} \hat{Z}_i\,\frac{x_i^t}{f} \\ \hat{Y} = \displaystyle\sum_{i=1}^{i=n} Y_i = \sum_{i=1}^{i=n} \hat{Z}_i\,\frac{y_i^t}{f} \end{cases} \tag{12.10}$$

Proposition 12.1. *The adaptive observer (12.9) is stable and the convergence of the state variables $(\hat{X}, \hat{Y}, \hat{Z}, \hat{V}_x, \hat{V}_y)$ is guaranteed. Furthermore, if the* PE *property is satisfied, then the identification errors $(\tilde{X} = X - \hat{X}, \ \dots \ \tilde{V}_y = V_y - \hat{V}_y)$ converge to zero.*

In fact, the observer stability is a direct consequence of the parameter identifier (RLS algorithm) stability. Indeed, the obtained position and velocity estimates satisfy the following properties:

- Since $\hat{b}(t) \in [b_{\min}, b_{\max}] \rightarrow \bar{b}$, then, from (12.9) we deduce that

$$(\hat{X}, \hat{Y}, \hat{Z}, \hat{V}_x, \hat{V}_y) \longrightarrow (X, Y, Z, V_x, V_y)\, b/\bar{b} \tag{12.11}$$

- Furthermore, if (a_x, a_y) satisfy the *PE* property, then,

$$\begin{cases} \hat{b} \longrightarrow b \Longrightarrow \tilde{b} \longrightarrow 0 \\ (\tilde{X}, \tilde{Y}, \tilde{Z}, \tilde{V}_x, \tilde{V}_y) \longrightarrow (0,0,0,0,0) \end{cases} \tag{12.12}$$

12.3.4 Fusion of Visual Estimates, Inertial and Pressure Sensor Data

This last step of the visual odometer consists in fusing the visual estimates $(\hat{X}, \hat{Y}, \hat{Z}, \hat{V}_x, \hat{V}_y)$, inertial data (a_x, a_y, a_z) and pressure sensor measurements Z_{ps} in order to improve the odometer accuracy and robustness, reduce the noise and estimate the vertical velocity V_z. The data fusion is performed using a linear Kalman filter with choosing (X, Y, Z, V_x, V_y, V_z) as a state vector, $(\hat{X}, \hat{Y}, \hat{Z}, Z_{ps}, \hat{V}_x, \hat{V}_y)$ as a measurement vector and (a_x, a_y, a_z) as an input vector. The implementation of such Kalman filter is straightforward and thus further details are omitted here.

12.4 Nonlinear 3D Flight Controller: Design and Stability

This section presents the design of a nonlinear flight controller and the analysis of the closed-loop system stability. The objective is to design an effective and practical controller that considers plant system nonlinearities and coupling, but also the characteristics and specificities of the adaptive visual odometer estimates. Furthermore, the controller is required to guarantee good flight performance and robustness even in the presence of identification errors.

12.4.1 Rotorcraft Dynamics Modelling

Let us denote by $\xi = (X, Y, Z)$, $\upsilon = (V_x, V_y, V_z)$, $\eta = (\phi, \theta, \psi)$, $\dot{\eta} = (\dot{\phi}, \dot{\theta}, \dot{\psi})$ the position, translational velocity, orientation and angular rate vectors, respectively. Therefore, the dynamics of a rotorcraft UAV such as the quadrotor helicopter, used in this research, can be represented by the following mathematical model (see Chap. 8):

$$\begin{cases} \ddot{\xi} = \frac{1}{m} u R e_3 - g e_3 \\ M(\eta)\ddot{\eta} + C(\eta, \dot{\eta})\dot{\eta} = \Psi(\eta)^T \tau \end{cases} \tag{12.13}$$

where u is the total thrust, τ is the torque vector, R and Ψ are the rotation and Euler matrices, respectively. The pseudo inertial matrix M is defined as $M(\eta) = \Psi(\eta)^T J \Psi(\eta)$, and C is given by $C(\eta, \dot{\eta}) = \Psi(\eta)^T J \dot{\Psi}(\eta) - \Psi(\eta)^T sk(\Psi(\eta)\dot{\eta}) J \Psi(\eta)$.

12.4.2 Flight Controller Design

Control design for rotorcraft UAVs is already a challenging task, especially when it comes to deal with unknown parameters and state variables estimation errors. To cope with these issues, we have proposed a hierarchical flight controller that exploits the structural properties of rotorcraft UAVs model given by (12.13). Our objective is to develop a 3D flight controller that performs well in practice[3] as well as in theory.[4]

The flight controller that we have presented in Chap. 8 proved its effectiveness for GPS-based autonomous flight including waypoint navigation and trajectory tracking. Here, we use the same controller for visual servoing of the rotorcraft using the visual odometer estimates. In Chap. 8, we have described the controller design as well as its stability analysis. The basic idea of that controller is briefly recalled below and its stability and robustness with respect to identification/obsetver errors is given in the next subsection.

To design a practical and hierarchical controller, we have separated the aircraft model into two connected subsystems by decoupling the translational dynamics (outer-loop) and attitude dynamics (inner-loop). The time-scale separation between the two subsystems is made by considering the following transformations (or change of variables):

$$\begin{aligned} \tau &= J\Psi(\eta)\tilde{\tau} + \Psi^{-1}C(\eta,\dot{\eta})\dot{\eta} \\ \mu &= \frac{1}{m}uR(\phi_d,\theta_d,\psi_d)e_3 - ge_3 \end{aligned} \tag{12.14}$$

where $\tilde{\tau}$ is a new torque vector, μ is an intermediary force vector and (ϕ_d,θ_d,ψ_d) are desired roll, pitch and yaw angles.

By defining the tracking errors $\chi = (\xi-\xi_d, \upsilon-\upsilon_d)^T \in \mathbb{R}^6$ and $e = (\eta-\eta_d, \dot{\eta}-\dot{\eta}_d)^T \in \mathbb{R}^6$, replacing η by $\eta_d + e$ in (12.13) and considering (12.14), the system (12.13) can be written in the following form [22, 24]:

$$\begin{cases} \dot{\chi} = \underbrace{A_1\chi + B_1(\mu - \ddot{\xi}_d)}_{f(\chi,\mu,\ddot{\xi}_d)} + \underbrace{\frac{1}{m}u\,H(\eta_d,e_\eta)}_{\Delta(u,\eta_d,e_\eta)} \\ \dot{e} = A_2e + B_2(\tilde{\tau} - \ddot{\eta}_d) \end{cases} \tag{12.15}$$

where $H(\eta_d,e_\eta) = (0,0,0,h_x,h_y,h_z)^T$ is a nonlinear interconnection term. The matrices $A_1 \in \mathbb{R}^{6\times 6}$, $B_1 \in \mathbb{R}^{6\times 3}$, $A_2 \in \mathbb{R}^{6\times 6}$ and $B_2 \in \mathbb{R}^{6\times 3}$ are defined in [22].

The rotorcraft control problem is thus, formulated as the control of two cascaded subsystems which are coupled by a nonlinear term $\Delta(u,\eta_d,e_\eta)$. Some techniques have been proposed in the literature to control systems in cascade [34, 38]. Here,

---

[3] It is easy to implement and guarantees good flight performance.

[4] It considers system nonlinearities/coupling and guarantees the stability of the closed-loop system.

we use partially passivation design to synthesize two stabilizing feedbacks $\mu = \alpha(\hat{\chi}, \ddot{\xi}_d)$, $\tilde{\tau} = \beta(e, \ddot{\eta}_d)$ such that the tracking errors (χ, e) are asymptotically stable for all initial conditions. The idea is to consider $\Delta(u, \eta_d, e_\eta)$ as a disturbance on the χ-subsystem which must be driven to zero, and stabilize independently the χ- and e-subsystems.

Since the χ-subsystem without the coupling $\Delta(.)$ and the e-subsystem in (12.15) are linear, we can use simple linear controllers such as PD or PID. Therefore, we synthesize two control laws

$$\begin{cases} \mu = -K_\chi \hat{\chi} + \ddot{\xi}_d, & K_\chi \in \mathbb{R}^{3\times 6} \\ \tilde{\tau} = -K_e e + \ddot{\eta}_d, & K_e \in \mathbb{R}^{3\times 6} \end{cases} \tag{12.16}$$

such that the matrices $A_\chi = A_1 - B_1 K_\chi$ and $A_e = A_2 - B_2 K_e$ are Hurwitz.

It is important to note that the control $\mu(\hat{\chi})$ is computed using the estimates of the adaptive visual odometer. Indeed, $\hat{\chi} = (\hat{\xi} - \xi_d, \hat{\upsilon} - \upsilon_d)^T$ where $\hat{\xi} = (\hat{X}, \hat{Y}, \hat{Z})^T$ and $\hat{\upsilon} = (\hat{V}_x, \hat{V}_y, \hat{V}_z)^T$.

12.4.3 *Closed-Loop System Stability and Robustness*

In Chap. 8, we have provided a detailed analysis of the closed-loop system in the case of GPS-based flight. Here, we extend previous results to prove the closed-loop stability even in the presence of identification and observation errors in the visual estimates.

By substituting (12.16) into (12.15), the closed-loop system dynamics become

$$\begin{cases} \dot{\chi} = A_1 \chi - B_1 K_\chi \hat{\chi} + \Delta(\hat{\chi}, e) \\ \dot{e} = A_e e \end{cases} \tag{12.17}$$

In this section, the stability of the closed-loop system (12.17) is established, considering the interconnection term $\Delta(\hat{\chi}, e)$ and errors in the state vector estimation (identification and observation errors).

If the visual odometer estimates $(\hat{\xi}, \hat{\upsilon})$ converge to their true values (*no estimation errors*), then the tracking error $\hat{\chi}$ will become χ and system (12.17) can be written as follows:

$$\begin{cases} \dot{\chi} = (A_1 - B_1 K_\chi)\chi + \Delta(\chi, e) = A_\chi \chi + \Delta(\chi, e) \\ \dot{e} = A_e e \end{cases} \tag{12.18}$$

Even for this simplified system it is difficult to prove the stability property because of the complex interconnection term $\Delta(\chi, e)$. However, a detailed demonstration of the stability property for system (12.18) can be found in Chap. 8.

This paper extends the previous results by proving the stability property of system (12.17) even in the presence of identification and estimation errors (i.e., $\hat{b} \longrightarrow \bar{b} \neq b$ and $(\hat{\xi}, \hat{\upsilon}) = (\xi, \upsilon)b/\hat{b}$ which means that $\hat{\chi} \neq \chi$).

First, let us express the tracking error $\hat{\chi} = (\hat{\xi} - \xi_d, \hat{\upsilon} - \upsilon_d)^T$ as a function of the original tracking error $\chi = (\xi - \xi_d, \upsilon - \upsilon_d)^T$:

$$\hat{\chi} = \left(\frac{b}{\hat{b}}\xi - \xi_d, \frac{b}{\hat{b}}\upsilon - \upsilon_d\right)^T = \frac{b}{\hat{b}}\chi + \frac{b - \hat{b}}{\hat{b}}(\xi_d, \upsilon_d)^T \tag{12.19}$$

So, the outer-loop dynamics in (12.17) become:

$$\begin{aligned}\dot{\chi} &= A_1\chi - B_1 K_\chi[\frac{b}{\hat{b}}\chi + \frac{b - \hat{b}}{\hat{b}}(\xi_d, \upsilon_d)^T] + \Delta(\hat{b}, \chi, e) \\ &= \underbrace{(A_1 - \frac{b}{\hat{b}}B_1 K_\chi)}_{\hat{A}_\chi}\chi \underbrace{- \frac{b - \hat{b}}{\hat{b}}B_1 K_\chi(\xi_d, \upsilon_d)^T}_{\varepsilon(\hat{b}, \xi_d, \upsilon_d)} + \Delta(\hat{b}, \chi, e)\end{aligned} \tag{12.20}$$

Now, the closed-loop system (12.17) can be expressed in the following form:

$$\begin{cases} \dot{\chi} = \hat{A}_\chi \chi + \varepsilon(\hat{b}, \xi_d, \upsilon_d) + \Delta(\hat{b}, \chi, e) \\ \dot{e} = A_e e \end{cases} \tag{12.21}$$

The term $\varepsilon(\hat{b}, \xi_d, \upsilon_d)$ is mainly due to the parameter estimate error $b - \hat{b}$ (height estimation error). Indeed, if $\hat{b}$ converges to its true value b, then ε will converge to zero.

According to Sontag's theorem [38], the closed-loop system (12.21) is Globally Asymptotically Stable (GAS) provided that the outer-loop "$\dot{\chi} = \hat{A}_\chi \chi + \varepsilon(\hat{b}, z_d, \upsilon_d)$" and the inner-loop "$\dot{e} = A_e e$" are GAS and the trajectories $(\chi(t), e(t))$ are bounded. We recall below a theorem that we have proposed in Chap. 8 and we will use the prove the stability of system (12.21):

Theorem 12.2. *If the following three conditions hold, then all the solutions $\chi(t)$ and $e(t)$ of (12.21) are bounded.*

A1. The equilibrium point $e = 0$ is GAS and Locally Exponentially Stable (LES).
A2. There exist a positive semi-definite radially unbounded function $V(\chi)$ and positive constants c_1 and c_2 such that for $\|\chi\| \geq c_1$

$$\begin{cases} \frac{\partial V}{\partial \chi}[\hat{A}_\chi \chi + \varepsilon(\hat{b}, \xi_d, \upsilon_d)] \leq 0 \\ \left\| \frac{\partial V}{\partial \chi} \right\| \|\chi\| \leq c_2 V(\chi) \end{cases} \tag{12.22}$$

A3. *There exist a positive constant* c_3 *and one class-*$\mathcal{K}$ *function* $\gamma(.)$, *differentiable at* $e = 0$, *such that*

$$\|\chi\| \geq c_3 \Rightarrow \left\| \Delta(\hat{b}, \chi, e) \right\| \leq \gamma(\|e\|)\ \|\chi\| \tag{12.23}$$

If in addition, $\dot{\chi} = \hat{A}_\chi \chi + \varepsilon(\hat{b}, \xi_d, \upsilon_d)$ *is GAS, then the equilibrium point* $(\chi, e) = (0,0)$ *is GAS.*

The proof of Theorem 12.2 is done in [21].

Proposition 12.2. *The closed-loop system (12.21) is stable and the tracking error* χ *is bounded. Furthermore, if* $\hat{b}(t)$ *converges to b then, the equilibrium point* $(\chi, e) = (0,0)$ *is GAS.*

Proof. In order to prove the stability of the closed-loop system (12.17) or (12.21), we need to prove that the e-subsystem, the χ-subsystem and the coupling term $\Delta(.)$ satisfy the conditions A1, A2 and A3 respectively:

- *A1. e-subsystem stability:* Since the matrix A_e is Hurwitz, then, the e-subsystem (inner-loop) is GES which is stronger than the GAS property.
- *A2.* χ*-subsystem stability:* We are interested here, in the stability of the subsystem $\dot{\chi} = \hat{A}_\chi \chi + \varepsilon(\hat{b}, \xi_d, \upsilon_d)$ without the interconnection term. It is clear from (12.20) that if $\hat{b} \to b$, the system $\dot{\chi} = \hat{A}_\chi \chi + \varepsilon(\hat{b}, \xi_d, \upsilon_d)$ will be GAS. Indeed, for $\hat{b} \to b$, we have $\varepsilon(.) \to 0$ and $\hat{A}_\chi \to A_\chi$, which is Hurwitz. Consequently, the inequalities (12.22) hold.

Now, let us analyze the stability of the χ-subsystem in the presence of parameter estimate errors, i.e., $\hat{b} \to \bar{b} \neq b$. So, in the following, we will show that the term $\varepsilon(\hat{b}, \xi_d, \upsilon_d)$ is bounded, the matrix $\hat{A}_\chi = A_1 - \frac{b}{\hat{b}} B_1 K_\chi$ is Hurwitz and the tracking error $\chi(t)$ is bounded. Furthermore, we will prove that the assumption A2. of Theorem 12.2 is satisfied.

Let us recall the expression of the identification error-related term $\varepsilon(\hat{b}, \xi_d, \upsilon_d) = -\frac{b-\hat{b}}{\hat{b}} B_1 K_\chi (\xi_d, \upsilon_d)^T$. Since $\hat{b}(t) \in [b_{\min}, b_{\max}]$ (see Theorem 12.1) and the desired trajectories $(\xi_d(t), \upsilon_d(t))$ are bounded, then it is trivial to deduce that $\varepsilon(.)$ is also bounded. Then, there exists a positive constant d such that $d = \|\varepsilon(.)\|_\infty$.

We recall that the matrix $A_1 - B_1 K_\chi$ is Hurwitz and the term $\frac{b}{\hat{b}}$ is positive and bounded. Due to the structure of the matrices A_1 and B_1, then it is easy to show that the matrix $\hat{A}_\chi = A_1 - \frac{b}{\hat{b}} B_1 K_\chi$ is also Hurwitz. Therefore, the subsystem $\dot{\chi} = \hat{A}_\chi \chi + \varepsilon(\hat{b}, \xi_d, \upsilon_d)$ is equivalent to the following differential equation: $\dot{\chi} - \hat{A}_\chi \chi = \varepsilon(t)$. The solution of the preceding differential equation is

$$\chi(t) = \chi(0) e^{\hat{A}_\chi t} + \int_0^t e^{\hat{A}_\chi (t-\tau)} \varepsilon(\tau) d\tau \tag{12.24}$$

From (12.24), we can show that after a finite time T, we obtain

$$\|\chi(t)\| \leq \|\varepsilon(t)\| \leq d, \forall t \geq T \tag{12.25}$$

This means that given any initial condition $\chi(0)$, the trajectories $\chi(t)$ of the subsystem "$\dot{\chi} = \hat{A}_{\chi}\chi + \varepsilon(\hat{b}, \xi_d, \upsilon_d)$" converge exponentially to a bounded ball with a radius d. Hence, inequalities (12.22) hold for $\|\chi\| \geq c_2 \triangleq d$.

• *A3. $\Delta(.)$ growth restriction:* Now, let us analyze if the interconnection term $\Delta(.)$ satisfies the growth condition (12.23). By recalling (12.15), we have

$$\|\Delta(\hat{\chi}, e)\| = \frac{1}{m}|u(\hat{\chi})|\|H(\hat{\chi}, e)\| \tag{12.26}$$

where

$$\begin{cases} \|H(\hat{\chi}, e)\| = \sqrt{h_x^2 + h_y^2 + h_z^2} \\ |u(\hat{\chi})| = m\|\mu(\hat{\chi}) + g e_3\| = m\sqrt{\mu_x^2 + \mu_y^2 + (\mu_z + g)^2} \end{cases} \tag{12.27}$$

Lemma 12.1. *Assume that the desired trajectories $(\xi_d(t), \upsilon_d(t))$ and their time-derivatives are bounded. Then, there exist positive constants r, k_1 and k_2 such that the collective thrust $u(\hat{\chi})$ and the coupling term $H(\hat{\chi}, e)$ satisfy the following properties:*

$$\begin{cases} |u(\chi)| \leq \begin{cases} k_1\|\hat{\chi}\|, & \text{for } \|\hat{\chi}\| \geq r \\ k_1 r, & \text{for } \|\hat{\chi}\| < r \end{cases} \\ \|H(\hat{\chi}, e)\| \leq k_2\|e\| \end{cases} \tag{12.28}$$

The proof of Lemma 12.1 is given in [21].

Recalling (12.26) and (12.28), the interconnection term $\Delta(.)$ verifies

$$\|\Delta(\hat{\chi}, e)\| \leq \frac{k_1 k_2}{m}\|e\|\|\hat{\chi}\|, \quad \text{for } \|\hat{\chi}\| \geq r \tag{12.29}$$

Substituting (12.19) into (12.29) and recalling that $\hat{b}$ and (ξ_d, υ_d) are bounded, then one can write

$$\begin{aligned} \|\Delta(\hat{b}, \chi, e)\| = \|\Delta(\hat{\chi}, e)\| &\leq \frac{k_1 k_2}{m}\|e\| \left\| \frac{b}{\hat{b}}\chi + \frac{b - \hat{b}}{\hat{b}}(\xi_d, \upsilon_d)^T \right\| \\ &\leq \frac{k_1 k_2}{m}\|e\|(k_3\|\chi\| + k_4\|\chi\|) \leq \frac{(k_1 k_2)(k_3 + k_4)}{m}\|e\|\|\chi\| \end{aligned} \tag{12.30}$$

for $\|\chi\| \geq 1$, $k_3 = \frac{b}{b_{\min}}$ and k_4 is the upper bound of $\|\frac{b-\hat{b}}{\hat{b}}(\xi_d, \upsilon_d)^T\|$. By defining $\gamma(\|e\|) = \frac{(k_1 k_2)(k_3 + k_4)}{m}\|e\|$, which is a class-$\mathcal{K}$ function, the interconnection term $\Delta(\hat{b}, \chi, e)$ satisfies the growth restriction (12.23), that is, $\Delta(\hat{b}, \chi, e) \leq \gamma(\|e\|)\|\chi\|$ for $\|\chi\| \geq c_3 \triangleq \max(\frac{r - k_4}{k_3}, 1)$.

Finally, the stability of the closed-loop system (12.21) and the boundedness of trajectories $(\chi(t), e(t))$ are direct consequences of Theorem 12.2. Furthermore, if the *PE* property is satisfied, then the estimate $\hat{b}(t)$ converges exponentially to the true value b. Therefore, the χ-subsystem is GAS, thereby ensuring the GAS of the equilibrium point $(\chi, e) = (0, 0)$.

These theoretical results are very interesting because they state that the aircraft position and velocity can be accurately recovered and controlled using only visual cues and IMU data. Furthermore, in the presence of significant external disturbances that may probably induce bounded errors in the parameter estimate $\hat{b}$ (height since $\hat{b} = f/\hat{Z}$), the stability of the connected closed-loop system holds and the position tracking errors remain within a small bounded region (see (12.25)). The tracking errors can be significantly reduced by exciting the system (*PE* property) in order to improve the estimation of the unknown parameter b or height.

12.5 Aerial Robotic Platform and Software Implementation

The proposed adaptive vision-based autopilot was implemented on a quadrotor MAV platform that we have developed at *Chiba University*. This section provides information about the air vehicle, its avionics and real-time architecture, and the implementation of vision and control algorithms.

12.5.1 Description of the Aerial Robotic Platform

Our testbed is based on a quadrotor vehicle which is manufactured by the *Ascending Technologies GmbH* company (Germany). The original kit includes the airframe, four brushless motors, four propellers and some electronics for manual flight. The vehicle is 53 cm rotor-tip to rotor-tip and weighs 400 g including battery. It has a 300-g payload and an endurance of approximately 22 min without payload and 12 min with full payload. The original electronics of the AscTech quadrotor were kept but its attitude controller has been disabled except internal gyros feedback control which is running at 1 kHz.

In order to demonstrate vision-based autonomous flight, we have fitted the quadrotor helicopter with an embedded autopilot, which is designed and built by the authors. As shown in Fig. 12.4, the hardware components that make up the basic flight avionics of our platform include a small micro-controller from *Gumstix Inc.*, the MNAV100CA sensor from *Crossbow Inc.* and a vision system from *Range Video Inc.*

– *Gumstix micro-controller:* the Gumstix micro-controller is a light-weight and powerful computing unit that is very appropriate for our needs and micro UAVs applications in general. Indeed, it weighs only 8 g and has a 400-MHz CPU with 16-MB flash memory and 64-MB SDRAM memory. Two expansion boards (*console-st* and *wifistix*) have been mounted on the Gumstix motherboard, thereby providing

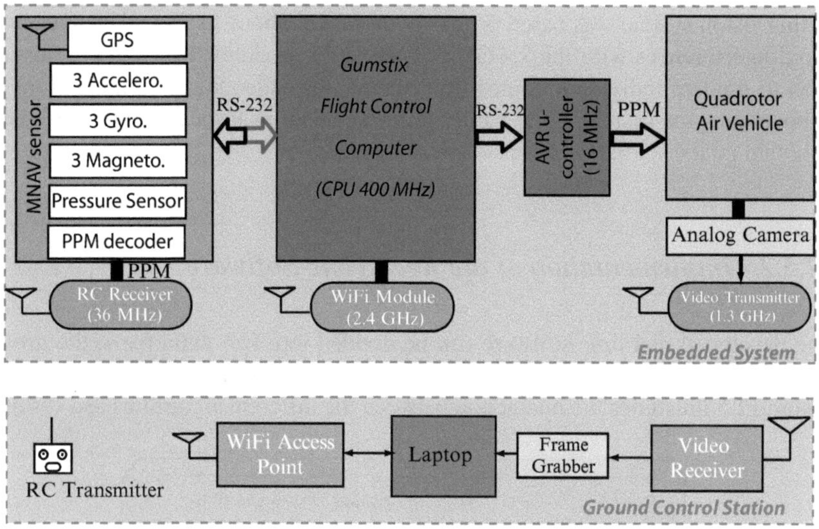

Fig. 12.4 Avionics architecture of our aerial robotic system (GPS is used here for comparison only)

two RS-232 ports and WiFi communication with the GCS. The total weight of the obtained Linux-running computer is about 30 g including interface card, WiFi module and antenna.

The Gumstix-based Flight Control Computer (FCC) performs sensor data acquisition and fusion, implements vision and control algorithms and generates the required control inputs (thrust, pitching torque, rolling torque and yawing torque).

– *AVR micro-controller:* this auxiliary micro-controller has two main tasks:

1. It reads the control inputs from the main FCC and encodes them into a PPM signal which is then used to drive the vehicle motors.
2. It implements the flight termination system which consists in achieving an emergency landing when some hardware or software problem occurs (see [24] for more details).

– *MNAV sensor:* The MNAV100CA sensor is manufactured by *Crossbow* and designed for miniature ground and air vehicles. It includes a digital IMU, a ublox GPS receiver and a pressure sensor in one compact sensor. It is a low-cost (1,500 USD) and light-weight (35 g without GPS antenna) sensor, with a low power consumption, making it ideal for mini and micro UAVs applications. The IMU outputs raw data from three accelerometers, three gyrometers and three magnetometers at the rate of 50 Hz to the FCC. The GPS data are updated at 4 Hz and the static pressure sensor measurements are provided at a rate of 50 Hz. All these sensor data are sent to the FCC through an RS-232 serial link.

– *Imaging system:* for this project, we have used an imaging system from *RangeVideo* that includes an analog camera (KX-171), a 1.3-GHz wireless video transmitter, a hight gain antenna, a video receiver and a grabber card. The choice of this vision system was based primarily on range (about 1 km) and frequency (to avoid interferences with the 2.4 GHz of the WiFi module). The frame grabber is used to digitize individual frames from the analog video signal coming from the onboard imaging system. This enables frame acquisition at speeds up to 25 frames/s to obtain images of 320 × 240 pixels resolution.

12.5.2 Implementation of the Real-Time Software

The developed real-time software can be divided into two main parts, the ground control station software and the embedded navigation and control algorithms. Figure 12.5 illustrates the interaction between the different algorithms and systems.

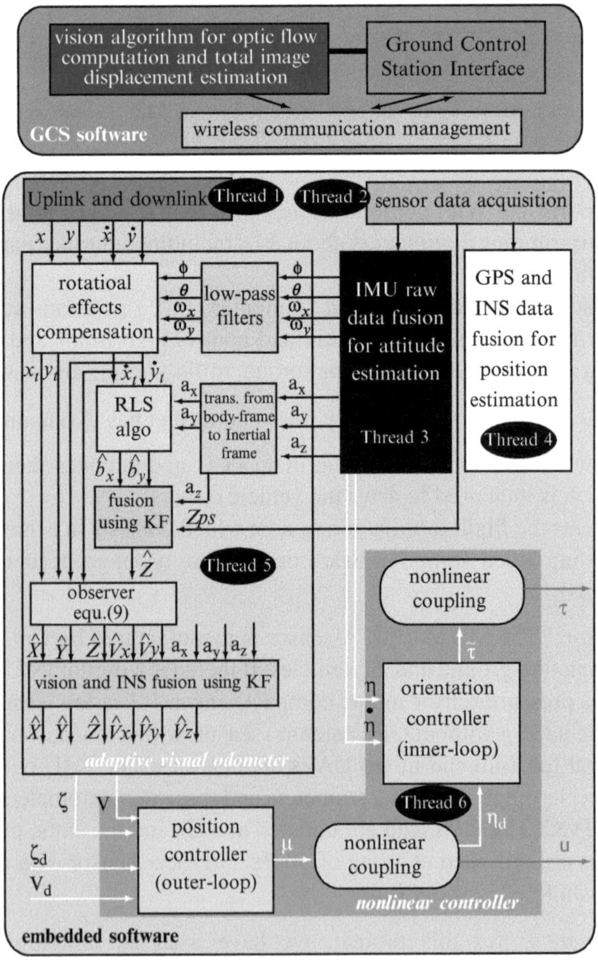

Fig. 12.5 Block diagram of the real-time software

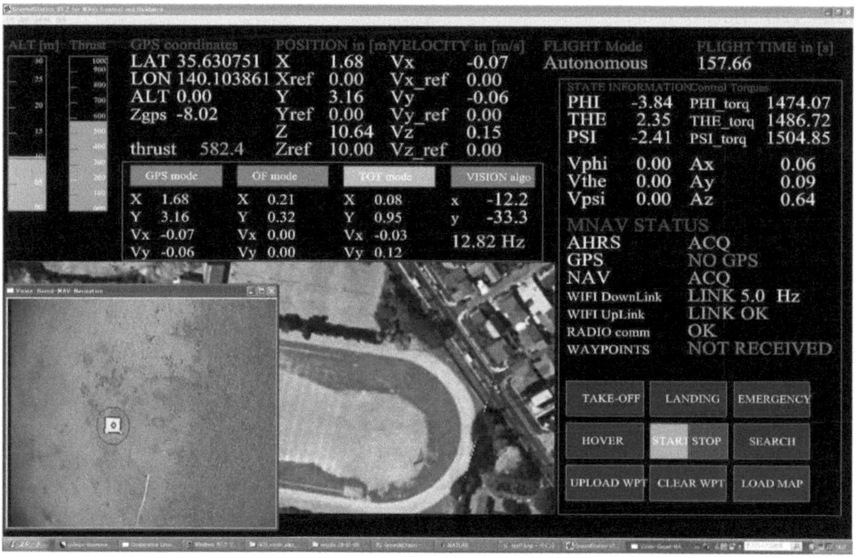

Fig. 12.6 The interactive interface of the ground control station during vision-based autonomous hovering

– *The Ground Control Station (GCS) software:* the GCS software is implemented on a standard laptop using C++, MFC, OpenGL and OpenCV libraries. It includes three different threads that are running at different frequencies. The first thread runs at 5 Hz and handles wireless communication with the onboard FCC for receiving and decoding flight data (telemetry), and sending visual estimates and other navigation commands to the onboard FCC. The communication program uses *socket* and *UDP* protocol. The second thread is also running at 5 Hz and implements the GCS interface that allows to display flight data in real-time and to send flight commands by clicking on the appropriate buttons as shown in Fig. 12.6. The third and last thread implements the first part of the vision algorithm which is described in Sect. 12.2. This vision algorithm runs at 12 Hz and exploits some functions of the OpenCV library.

– *The embedded software:* the adaptive visual odometer described in Sect. 12.3 and the flight controller presented in Sect. 12.4 are implemented on the *Gumstix* FCC using multi-thread programming. Other navigation algorithms for sensor data acquisition and fusion are implemented in the onboard FCC. In total, the embedded software is composed of six different threads that are running at different frequencies: (1) thread 1 for communication with the GCS (10 Hz); (2) thread 2 for sensor data acquisition (50 Hz); (3) thread 3 for attitude estimation (50 Hz); (4) thread 4 for GPS-INS fusion and position estimation (10 Hz); (5) thread 5 that implements the adaptive visual odometer (10 Hz); (6) guidance and control algorithms (50 Hz). Figure 12.5 shows the different programs that are running on the GCS laptop and the onboard micro-controller.

Table 12.1 Control system gains

Parameter	Value	Parameter	Value
kp_x, kp_y	0.6	kp_ϕ, kp_θ	28
ki_x, ki_y	0.02	ki_ϕ, ki_θ	0.5
kd_x, kd_y	0.8	kd_ϕ, kd_θ	1
kp_z	0.8	kp_ψ	3
ki_z	0.03	ki_ψ	0.05
kd_z	1	kd_ψ	0.2

The values of the nonlinear controller gains are shown in Table 12.1. The parameters of the RLS algorithm used for height estimation are chosen as follow: $\beta = 0.4$, $P(0) = 5$, $P_{\max} = 10{,}000$, $Z_{\min} = 0.5\,\text{m}$, $Z_{\max} = 20\,\text{m}$, $dead - zone = 0.1$, $\lambda_1 = \lambda_2 = 0.1$, $Z(0) = 0.5\,\text{m}$.

12.6 Experimental Results of Vision-Based Flights

The performance of the adaptive vision-based autopilot was demonstrated in real flights using the quadrotor MAV described in Sect. 12.5. We have performed various outdoor flight tests under autonomous control for take-off, landing, hovering, trajectory tracking, stationary and moving target tracking. All the outdoor flight tests, described in this section, have been conducted at the play-ground of the Chiba University Campus which was not prepared in any way for flight experiments. As it can be seen in Fig. 12.6 and flight videos, the field is almost homogenous without rich texture.

Remark 12.2. For the experimental flight tests, described in this section, the flight controller used the adaptive visual odometer estimates for position, height and velocity feedback. GPS measurements are recorded during flights and plotted here for comparison only. It is important to note that GPS data can not be considered as the *ground-truth* since we are using a low-cost GPS with $\pm 2\,\text{m}$ accuracy for horizontal position and $\pm 5\,\text{m}$ error for height.

12.6.1 Static Tests for Rotation Effects Compensation and Height Estimation

These static tests aim at demonstrating the effectiveness of the approach (12.1)–(12.2) for compensating the rotational component of optic flow and image displacement using IMU data. We also analyze the performance of the adaptive visual odometer for height estimation using optic flow and accelerations.

We have conducted two static tests under different conditions. In *test A*, we have moved the MAV by hand in the following sequence: (1) vertical movement from 0

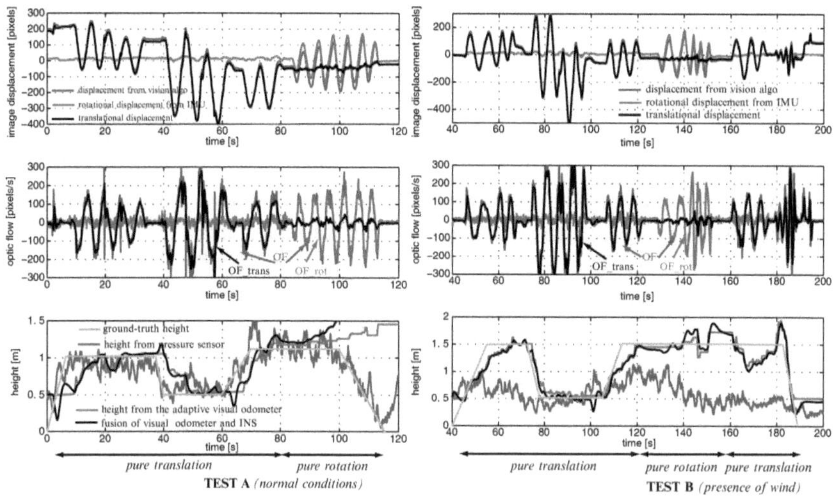

Fig. 12.7 Static tests showing rotation effects compensation and comparison of pressure sensor and visual odometer for height estimation

to 1 m height; (2) horizontal movement along the x-axis at different heights (1, 0.5, 1 m); (3) pure rotational movement (pitch movement); and (4) vertical landing. In *test B*, we have repeated almost the same maneuvers with two differences: (1) the MAV is moved along the x-axis (zigzag) when performing vertical landing; (2) a small fan or ventilator is used to simulate wind.

For both tests, we have plotted in Fig. 12.7, the total image displacement and OF computed by the vision algorithm (blue line), the rotational image motion obtained from IMU data (red line), and the resulted translational image motion after compensation (black line). We can clearly see in Fig. 12.7 that rotation effects are effectively cancelled.

Concerning height estimation by the pressure sensor and the adaptive visual odometer, several remarks can be made:

- The pressure sensor works well in good weather conditions and it is sensitive to wind and temperature.
- The adaptive visual odometer is able to estimate the height when the translational OF is not very small.
- Compared to the pressure sensor, the visual odometer takes some time to converge to the true value when the MAV experiences vertical motion (height changes).

These tests prove the feasibility and possibility of estimating the height using optic flow and IMU data. For more robustness and accuracy, the height estimate used in flight tests is obtained by fusing the odometer measurement, the pressure sensor data and INS using a linear Kalman filter.

12.6.2 *Outdoor Autonomous Hovering with Automatic Take-off and Landing*

The objective of this experiment is to check the robustness and accuracy of the developed vision-based autopilot for achieving stationary flights in natural environments with poor texture. The MAV is tasked to take-off autonomously, to hover at a desired altitude of 5 m and then to achieve a vertical auto-landing.

As shown in Fig. 12.8, this task was achieved successfully with good performance. Indeed, the MAV maintained its 3D position with good accuracy (± 2 m maximum error) using the adaptive visual odometer estimates. The small errors in position control are mainly due to wind which was about 3.5 m/s during the flight test. We can also see on Fig. 12.8 that reference height trajectories are tracked accurately (± 1 m maximum error) during take-off, hovering and landing phases. The inner-loop controller performs also well and tracks the reference angles.

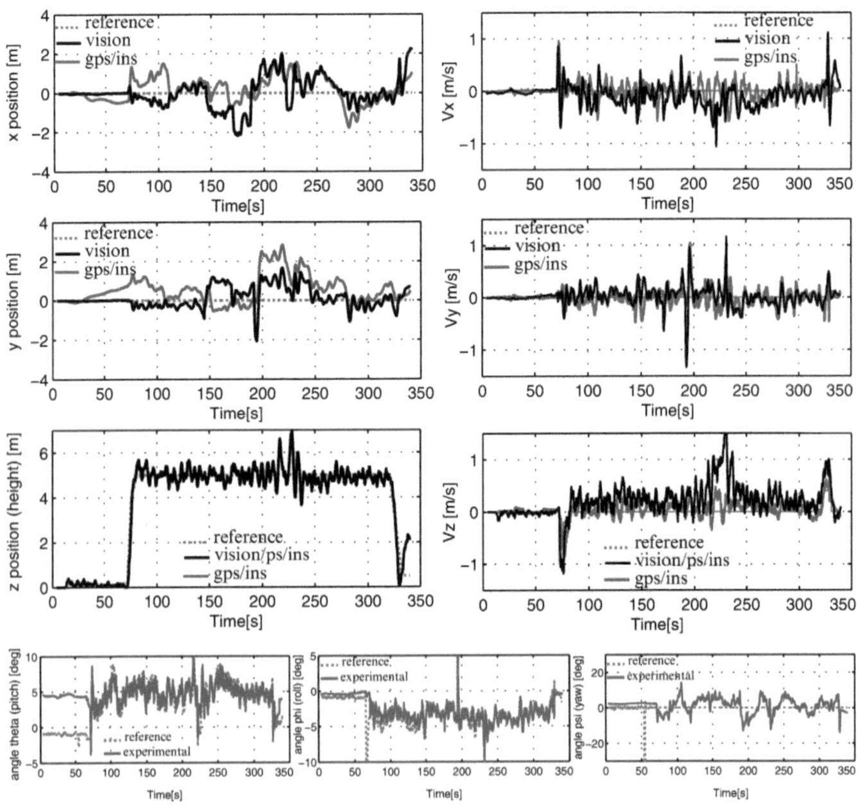

Fig. 12.8 Outdoor autonomous hovering with automatic takeoff and landing using the visual odometer

We conclude from this test that all the components of the proposed vision-based autopilot (vision algorithm, adaptive visual odometer, nonlinear controller) perform stably and robustly despite the textureless environment.

Video clips of autonomous vision-based hovering can be found at http://mec2.tm.chiba-u.jp/monograph/Videos/Chapter12/1.wmv.

12.6.3 Automatic Take-off, Accurate Hovering and Precise Auto-landing on Some Arbitrary Target

As described in Sect. 12.2, the image area where optic flow and image displacement are computed is initially chosen at the image center. However, the developed GCS and embedded softwares allow to choose this target-template at any location of the image by just selecting the desired area/object on the image. This flight test consists in exploiting this useful characteristic to achieve an accurate hovering above some designated ground target and to perform a precise auto-landing on it.

The rotorcraft was put on a small box of about 50 cm × 70 cm which is used as a target. The take-off procedure is launched from the GCS and the target is selected when it appeared in the camera FOV (about 1 m height during take-off). When the MAV reached the desired height of 10 m, it performed an accurate hovering by detecting the target and keeping it at the image center (see Fig. 12.9). Finally, the auto-landing procedure is activated and the MAV executed descent flight while controlling its horizontal position to keep the target at the image center. The MAV landed at 25 cm from the target, but it can be seen from the video that the MAV was exactly on the target at 30 cm height and then landed just near the target. This is due

Fig. 12.9 Rotorcraft during vision-based autonomous hovering above a stationary target

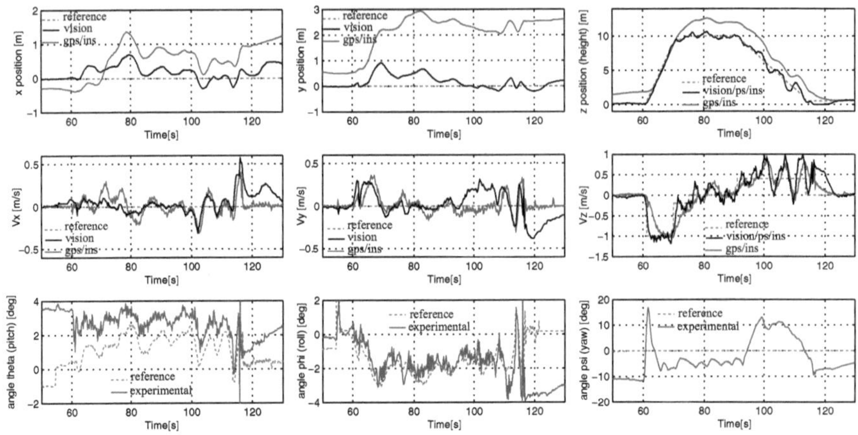

Fig. 12.10 Accurate hovering and precise auto-landing on some designated ground target

to very large image displacements when the MAV is at few centimeters from the target or ground. One approach to solve this problem could consist in deactivating the visual odometer and decrementing the thrust when the aircraft is under some height (50 cm for example).

Figure 12.10 shows the obtained MAV trajectories (position, height, velocity, orientation). The relative horizontal position between the MAV and the target was regulated to zero with about ±0.5 m maximum error. The height is also estimated and controlled accurately. The MAV was very stable even at 10 m height. Indeed, as shown in Fig. 12.10, the horizontal velocities, pitch and roll angles are stabilized and kept very small.

The good performance of this flight test can be checked by seeing the associated video clip at http://mec2.tm.chiba-u.jp/monograph/Videos/Chapter12/2.wmv.

12.6.4 Tracking a Moving Ground Target with Automatic Take-off and Auto-landing

Here, we explore the possibility of our vision-control system to track a ground moving target. For this experiment, we have used a small cart (see Fig. 12.12) as a target and placed it at about 20 m from the GCS.

First, the MAV performed automatic take-off from the target and hovered above the target (6 m height) for nearly 100 s. Then, the target is continuously moved towards the GCS by pulling some wire attached to the target. The control objective is thus, to keep the moving target at the image center by controlling the relative position between the MAV and the target to zero.

Figure 12.11 shows that the target is accurately tracked even when it is moving. The GPS ground-track on the first graph shows that the MAV flied about 20 m

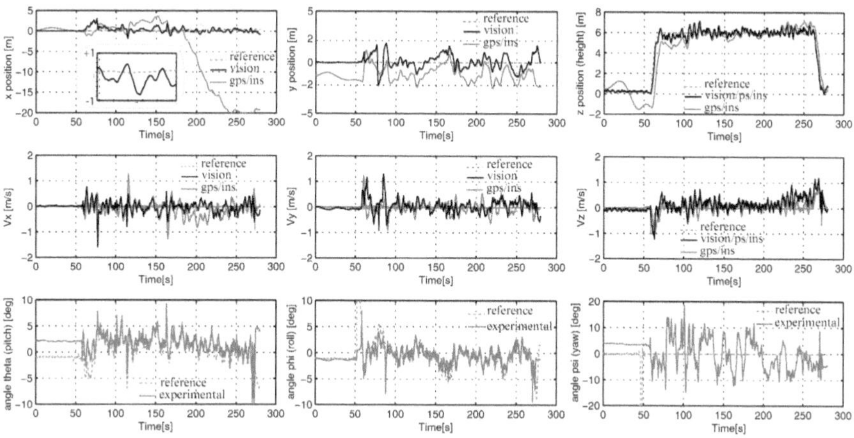

Fig. 12.11 Application of the vision-based autopilot for ground moving target tracking

Fig. 12.12 (a) Onboard image showing the detection and tracking of a moving target; (b) MAV tracking the moving target

(which corresponds also to target movement) while controlling the relative position between the MAV and the target to zero with ± 1 m maximum error during tracking.

Figure 12.12a shows the tracked target on an image captured by the onboard camera and displayed at the GCS. We can also see in Fig. 12.12b the rotorcraft tracking the moving target.

Video clip of this flight test is available at http://mec2.tm.chiba-u.jp/monograph/Videos/Chapter12/3.wmv.

12.6.5 *Velocity-Based Control for Trajectory Tracking Using Vision*

This flight test involves a velocity control scheme. It aims at evaluating and demonstrating the ability of the MAV to achieve hovering flight and velocity trajectory tracking by relying on velocities computed from optic flow without position feedback. After automatic take-off, the MAV is tasked to hover and then to achieve autonomous translational flight by tracking some reference trajectories, sent in real-time from the GCS. The commands for this test were: take-off, fly left, stop, fly forward, stop, fly right, stop, fly backward, stop, hover, land.

From Fig. 12.13, the rotorcraft can be seen to clearly respond to commands and to track reference velocity trajectories. Although the closed-loop control of horizontal position is not used in this test, the MAV seems to track also the position reference trajectories with small drifts. In fact, the position reference trajectories are obtained by integrating over time the velocity reference trajectories.

Despite the poor texture (see Fig. 12.14), the vision algorithm and adaptive visual odometer were able to track features, compute optic flow and recover the MAV motion parameters.

When doing these tests, the GPS signal was not available because of some technical problem of the GPS antenna connector.

This flight behavior is very useful and needed for many real-world applications where GPS signal is not available. Indeed, autonomous velocity control is sufficient to achieve many realistic tasks by just sending high-level commands.

A video clip of this flight test and other tests can be found at http://mec2.tm.chiba-u.jp/monograph/Videos/Chapter12/5.wmv.

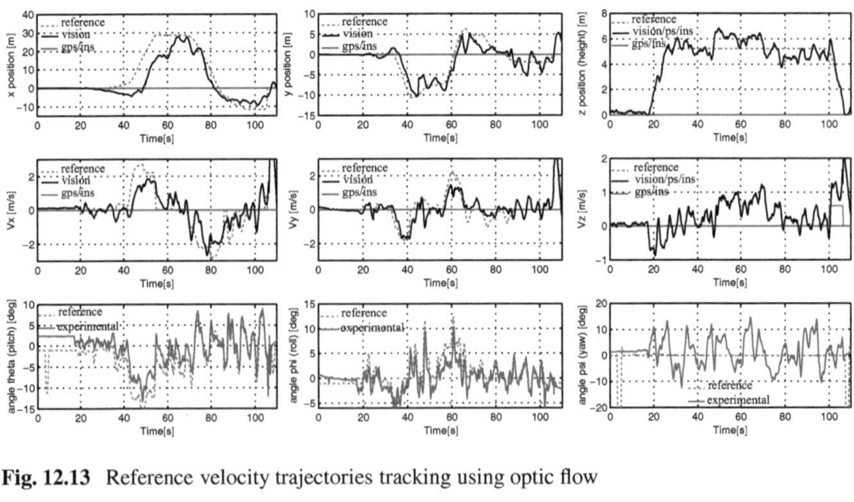

Fig. 12.13 Reference velocity trajectories tracking using optic flow

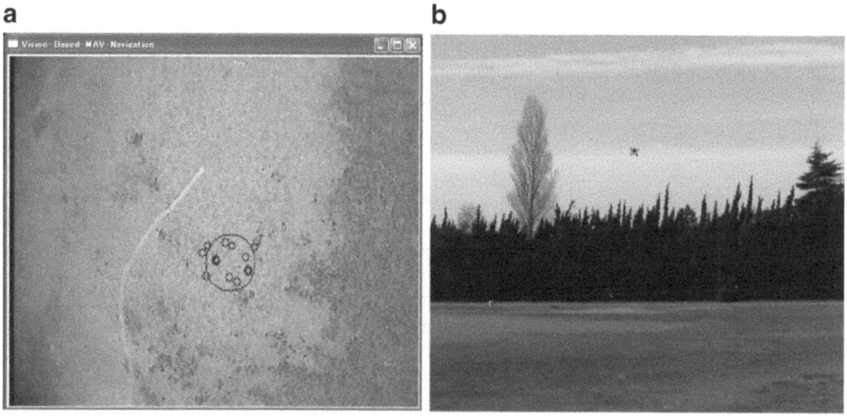

Fig. 12.14 Autonomous vision-based translational flight. (**a**) Onboard image showing features tracking in a textureless environment; (**b**) rotorcraft during vision-based trajectory tracking

12.6.6 Position-Based Control for Trajectory Tracking Using Visual Estimates

In this experiment, both position and velocity estimates are used by the controller to track arbitrary position and velocity trajectories, pre-programmed in the onboard FCC or sent by the operator from the GCS. The reference trajectories, shown in Fig. 12.15, consist in vertical climb (take-off) to an altitude of 6 m, 12-m sideward flight and 16-m forward flight, simultaneous backward and sideward flight to return to the starting point. The height is then, reduced to 3 m and small displacement commands are given.[5]

The obtained results, shown in Fig. 12.15 are very satisfactory. Indeed, the reference position, height, velocity and attitude trajectories are well tracked. The MAV was stable along the flight course.

12.6.7 GPS-Based Waypoint Navigation and Comparison with the Visual Odometer Estimates

In this last test, we have performed waypoint navigation using GPS data for horizontal movement control and adaptive visual odometer estimates for height control. A set of four waypoints were chosen by just clicking the desired locations on the 2D map of the GCS interface (see Fig. 12.6). The MAV should then, pass the assigned

[5] In fact, height is reduced to 3 m to avoid damaging the platform in case where the MAV crashes because of empty battery (there was no charged battery for this test).

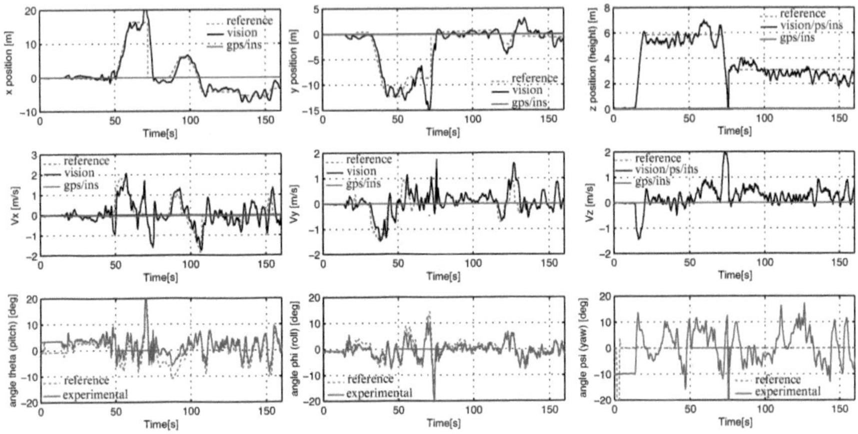

Fig. 12.15 Reference position trajectories tracking using visual odometer estimates

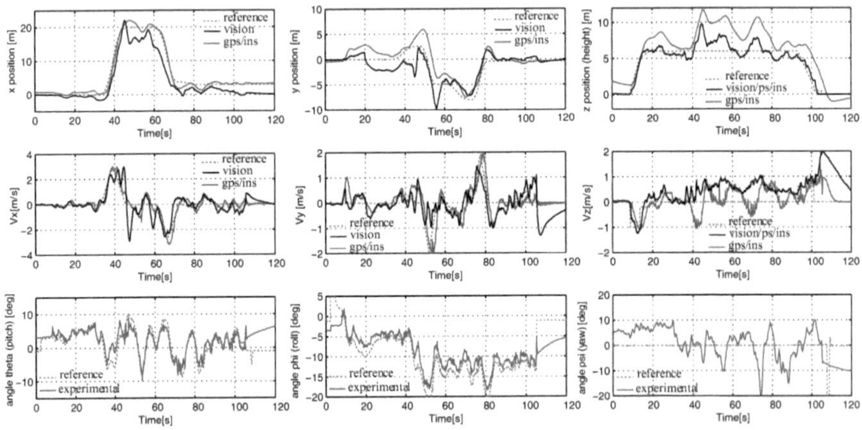

Fig. 12.16 GPS waypoint navigation and comparison with the visual odometer estimates

waypoints in a given sequence. The objective of this test is to compare the GPS and the visual odometer for estimating the horizontal position during waypoint navigation. The obtained results are shown in Fig. 12.16.

One can see that the reference trajectories are tracked and the mission is accomplished. It is also important to note that during this translational flight, the adaptive visual odometer was able to estimate the MAV position or the travelled flight distance despite the textureless environment. Therefore, vision-based waypoint navigation is possible by the current system when combined with some landmarks recognition algorithm.

12.6.8 Discussion

The reported results over different flight scenarios testify on the effectiveness and robustness of the developed vision-based autopilot. However, the proposed system can be further improved especially the adaptive visual odometer. Here, we discuss briefly the main weaknesses of the proposed visual odometer and some ideas to address these issues.

Inaccuracies in position estimation are mainly due to two factors, estimation errors growing in long range navigation and identification errors in the adaptive process.

– *Growing estimation errors in Visual Odometry (VO):* VO pose estimates are produced by tracking visual features in the environment and integrating or accumulating their image displacement. Although VO is an incremental process, our results have shown that positioning accuracy to within a few percent of measured ground truth is possible, over distances of about 100 m. However, the performance of the VO is expected to degrade for longer flight distances due to the accumulation of unbounded errors. Furthermore, the poor texture of some terrain surfaces leads to poorer performance since VO is an exteroceptive sensing method that relies on features tracking.

To enable accurate and robust long range navigation with VO, this latter can be combined with other terrain relative navigation techniques like landmarks recognition. An interesting approach could consist in performing landmarks matching and recognition in a moment-to-moment fashion (or periodically) to estimate the absolute position, and integrating the flight path between landmarks to estimate the relative position. Therefore, computation of distance flown is re-commenced whenever a prominent landmark is encountered. Re-setting the VO at each landmark facilitates accurate long-range navigation by preventing excessive accumulation of odometric errors. Indeed, some animals like bees seem to use landmark-based cues as well as visual path integration to navigate to a goal [40].

In this paper, we have concentrated on the VO for estimating the vehicle's position relative to some known location like the initial location or recognized landmark.

– *Range estimation errors in the adaptive process:* The visual odometer presented here requires an approximate measure of height above the ground. A RLS algorithm has been used to estimate the height by fusing the derivatives of optic flow and IMU accelerations data. This process suffers from the general well-known drawbacks of identification algorithms like noise and poor input–output signals (non-excited system). During real-time experiments, we have noticed performance degradation in height estimation by the RLS algorithm when the rotorcraft is hovering or flying at very low speeds. In this case, input and output signals are small compared to the noise level and do not contain sufficient information to estimate the height. As stated in Sect. 12.3, this is related to the persistent excitation (PE) property which is not satisfied in this case.

This issue has been handled by switching off the adaptation process when the input and output signals are small (stationary or slow flights). In this case, height estimate is primary provided by the pressure sensor as a propagation of previous estimates using a kinematic Kalman filter. Although the obtained results are satisfactory, range estimation process can be further improved by exciting the system (small lateral oscillations or zigzag flight) from time to time and/or using more robust identification algorithms.

12.7 Summary

In this chapter, we have presented a vision-based autopilot which is designed for miniature UAVs and MAVs flying at low altitudes. The developed system is based on a downward-looking camera and relies heavily on visual cues to achieve various navigation tasks in unknown environments. Unlike some existing bio-inspired flight controllers which are usually used for reactive navigation, the proposed vision-based autopilot extends optic flow-based control capabilities to complex navigational tasks such as accurate hovering, arbitrary trajectory tracking, stationary and moving target tracking, etc. These capabilities are mainly due to the incorporated adaptive visual odometer that allows the estimation of the height and the recovery of the rotorcraft position and velocity. A practical nonlinear control system is also designed and used for flight control. The combined visual odometer and flight controller result in an effective autopilot that is easy to implement while guaranteeing the asymptotic stability of the entire closed-loop system.

The experimental flight tests, performed outdoors under realistic conditions, have shown the good performance of the proposed adaptive vision-control system. Indeed, experimental results over various ranges of the flight envelope illustrate that the proposed vision-based autopilot allowed a small quadrotor MAV to achieve automatic take-off, accurate hovering, precise auto-landing, trajectory tracking and moving target detection and tracking. These results demonstrate that visual odometry can enable UAVs and MAVs to navigate in unknown environments where global localization is impossible.

As future work, we plan to combine the visual odometer with other navigation algorithms, including landmarks recognition and vision-based obstacle avoidance. This should enable visual odometry to operate in long range and goal-oriented navigation. We are also implementing a modified version of the system which operates in real-time on-board the MAV.

References

1. Ahrens S, Levine D, Andrews G, How JP (2009) Vision-based guidance and control of a hovering vehicle in unknown, GPS-denied environments. In: Proceedings of the IEEE international conference on robotics and automation, Kobe, Japan, pp 2643–2648
2. Altug E, Ostrowski JP, Taylor CJ (2005) Control of a quadrotor helicopter using dual camera visual feedback. Int J Robot Res 24(5):329–341

3. Amidi O, Kanade T, Fujita K (1999) A visual odometer for autonomous helicopter flight. Rob Auton Syst 28(2–3):185–193
4. Astrom KJ, Wittenmark B (1989) Adaptive control. Addison-Wesley, Reading, MA
5. Barrows GL (1999) Mixed-mode VLSI optic flow sensors for micro air vehicles. PhD Dissertation, Department of Electrical Engineering, University of Maryland
6. Caballero F, Merino L, Ferruz J, Ollero A (2009) Vision-based odometry and slam for medium and high altitude flying UAVs. J Intell Rob Syst 54:137–161
7. Chahl J, Srinivasan M, Zhang S (2004) Landing strategies in honeybees and applications to uninhabited airborne vehicles. Int J Robot Res 23(2):101–110
8. Frew E, McGee T, ZuWhan K, Xiao X, Jackson S, Morimoto M, Rathinam S, Padial J, Sengupta R (2004) Vision-based road-following using a small autonomous aircraft. In: Proceedings of the IEEE aerospace conference, vol 5, pp 3006–3015
9. Ettinger SM, Nechyba MC, Ifju PG, Waszak M (2002) Towards flight autonomy: vision-based horizon detection for micro air vehicles. In: Proceedings of the Florida conference on recent advances in robotics, Miami, FL
10. Fowers S, Lee DJ, Tippetts B, Lillywhite K, Dennis A, Archibald J (2007) Vision aided stabilization and the development of a quad-rotor micro UAV. In: Proceedings of the IEEE international symposium on computational intelligence in robotics and automation, Florida, USA, pp 143–148
11. Garratt MA, Chahl JS (2008) Vision-based terrain following for an unmanned rotorcraft. J Field Robot 25(4–5):284–301
12. Goodwin GC, Sin KC (1984) Adaptive filtering prediction and control. Prentice Hall, Englewood Cliffs, NJ
13. Green WE, Oh PY, Sevcik K, Barrows G (2003) Autonomous landing for indoor flying robots using optic flow. In: Proceedings of the 2003 ASME international mechanical engineering congress, Washington, 15–21 November 2003
14. Guenard N, Hamel T, Mahony R (2008) A practical visual servo control for a unmanned aerial vehicle. IEEE Trans Robot 24(2):331–341
15. Herisse B, Russotto FX, Hamel T, Mahony R (2008) Hovering flight and vertical landing control of a VTOL unmanned aerial vehicle using optical flow. In: Proceedings of the IEEE/RSJ international conference on intelligent robots and systems, Nice, France, pp 801–806
16. Hrabar S, Sukhatme GS, Corke P, Usher K, Roberts J (2005) Combined optic-flow and stereo-based navigation of urban canyons for a UAV. In: Proceedings of the IEEE international conference on intelligent robots and systems, Canada, pp 302–309
17. Ioannou P, Sun J (1996) Robust adaptive control. Prentice Hall, Englewood Cliffs, NJ
18. Johnson A, Montgomery J, Matthies L (2005) Vision guided landing of an autonomous helicopter in hazardous terrain. In: Proceedings of the IEEE international conference on robotics and automation, Barcelona, Spain, pp 4470–4475
19. Johnson EN, Calise AJ, Watanabe Y, Ha J, Neidhoefer JC (2007) Real-time vision-based relative aircraft navigation. AIAA J Aerosp Comput Inform Commun 4:707–738
20. Kanade T, Amidi O, Ke Q (2004) Real-time and 3D vision for autonomous small and micro air vehicles. In: Proceedings of the 43rd IEEE conference on decision and control, Atlantis, Bahamas, pp 1655–1662
21. Kendoul F (2007) Modelling and control of unmanned aerial vehicles, and development of a vision-based autopilot for small rotorcraft navigation. PhD Thesis Report, CNRS Heudiasyc Laboratory, University of Technology of Compiegne, France
22. Kendoul F, Fantoni I, Lozano R (2008) Adaptive vision-based controller for small rotorcraft UAVS control and guidance. In: Proceedings of the 17th IFAC world congress, Seoul, Korea, pp 797–802
23. Kendoul F, Fantoni I, Nonami K (2009) Optic flow-based vision system for autonomous 3D localization and control of small aerial vehicles. Rob Auton Syst 57:591–602
24. Kendoul F, Zhenyu Y, Nonami K (2009) Embedded autopilot for accurate waypoint navigation and trajectory tracking: application to miniature rotorcraft UAVS. In: Proceedings of the IEEE international conference on robotics and automation, Kobe, Japan, pp 2884–2890

25. Kima J, Sukkarieh S (2007) Real-time implementation of airborne inertial-SLAM. Rob Auton Syst 55:62–71
26. Lacroix S, Jung I, Soueres P, Hygounenc E, Berry J (2002) The autonomous blimp project of LAAS/CNRS: current status ànd research challenges. In: Proceedings of the workshop WS6 aerial robotics, IEEE/RSJ international conference on intelligent robots and systems, Lausanne, Switzerland, pp 35–42
27. Landau I, Lozano R, M'Saad M (1998) Adaptive control. Communications and control engineering. Springer, Berlin
28. Lucas B, Kanade T (1981) An iterative image registration technique with an application to stereo vision. In: Proceedings of the DARPA IU workshop, pp 121–130
29. Mejias L, Saripalli S, Campoy P, Sukhatme G (2006) Visual servoing of an autonomous helicopter in urban areas using feature tracking. J Field Robot 23(3/4):185–199
30. Proctor AA, Johnson EN, Apker TB (2006) Vision-only control and guidance for aircraft. J Field Robot 23(10):863–890
31. Ruffier F, Franceschini N (2005) Optic flow regulation: the key to aircraft automatic guidance. Rob Auton Syst 50(4):177–194
32. Saripalli S, Montgomery J, Sukhatme G (2003) Visually-guided landing of an unmanned aerial vehicle. IEEE Trans Robot Autom 19(3):371–381
33. Scherer S, Singh S, Chamberlain L, Elgersma M (2008) Flying fast and low among obstacles: methodology and experiments. Int J Robot Res 27(5):549–574
34. Sepulcre R, Jankovic M, Kokotovic P (1997) Constructive nonlinear control. Communications and control engineering series. Springer, Berlin
35. Serres J, Dray D, Ruffier F, Franceschini N (2008) A vision-based autopilot for a miniature air vehicle: joint speed control and lateral obstacle avoidance. Auton Robots 25:103–122
36. Shakernia O, Sharp C, Vidal R, Shim D, Ma Y, Sastry S (2002) Multiple view motion estimation and control for landing an unmanned aerial vehicle. In: Proceedings of the IEEE conference on robotics and automation, vol 3, pp 2793–2798
37. Shi J, Tomasi C (1994) Good features to track. In: Proceedings of the IEEE conference on computer vision and pattern recognition, Seattle, WA, USA, pp 593–600
38. Sontag E (1988) Smooth stabilization implies coprime factorization. IEEE Trans Autom Contr 34:435–443
39. Srinivasan MV, Zhang S, Lehrer M, Collett T (1996) Honeybee navigation en route to the gaoal: visual flight control and odometry. J Exp Biol 199(1):237–244
40. Srinivasan MV, Zhang S, Bidwell N (1997) Visually mediated odometry in honeybees. J Exp Biol 200:2513–2522
41. Tammero L, Dickinson M (2002) The influence of visual landscape on the free flight behavior of the fruit fly drosophila melanogaster. J Exp Biol 205:327–343
42. Yu Z, Nonami K, Shin J, Celestino D (2007) 3D vision based landing control of a small scale autonomous helicopter. Int J Adv Robot Syst 4(1):51–56
43. Zufferey JC, Floreano D (2006) Fly-inspired visual steering of an ultralight indoor aircraft. IEEE Trans Robot 22(1):137–146

Chapter 13
Autonomous Indoor Flight and Precise Automated-Landing Using Infrared and Ultrasonic Sensors

Abstract The accuracy of small and low cost GPS is insufficient to provide data for precisely landing micro air vehicles (MAVs). This study shows how a MAV can land on a small targeted landing site by using rangefinders rather than imprecise GPS data. This chapter describes a proposed movable range finding sensor system for measuring the environment and an algorithm of position measurement. This range finding system consists of four infrared (IR) rangefinders, four servo motors and one ultrasonic rangefinder. In order to measure the MAV's position, the sensor system vertically swings each IR rangefinder using the servo motors and these IR sensors detect the edge of landing target. And this sensor system calculates the position from the measured edge direction and the ultrasonic altitude. Additionally, experiments of autonomous hovering over a 52 × 52 cm table and autonomous landings were carried out indoors using the proposed sensor system. Our experiments succeeded, and as a result, the MAV kept flying horizontally within a 17.9 cm radius circle, and then it landed on the table from a height of 50 cm. The IR rangefinders were selected because of the payload limitations of small MAVs. If the MAV's payload capacity were higher than a laser rangefinder would be used since lasers can operate better in sunlight than IR sensors and also they tend to have longer scanning ranges and are more accurate.

Video Links:

Pinpoint_Landing_of_Quad-rotor_MAV by IR and Ultrasonic sensors
Landing Target: 60 × 60 cm table
http://mec2.tm.chiba-u.jp/monograph/Videos/Chapter13/1.wmv
Pinpoint_Landing_of_Quad-rotor_MAV by IR and Ultrasonic sensors (1)
Landing Target: 52 × 52 cm table
http://mec2.tm.chiba-u.jp/monograph/Videos/Chapter13/2.wmv

13.1 Introduction

In order to minimize the damages caused by disasters such as earthquakes and typhoons, it's crucial to collect information to warn the public before these catastrophes occur. Advisory warnings and alerts are used to warn people to avoid the loss

K. Nonami et al., *Autonomous Flying Robots: Unmanned Aerial Vehicles and Micro Aerial Vehicles*, DOI 10.1007/978-4-431-53856-1_13, © Springer 2010

of lives and properties. And information gathering and communication are effective in preventing secondary disasters and also for drawing up recovery plans.

Today, manned helicopters are widely used during disasters since they have a wide field of view, high mobility and are ideal for rescuing people. However, a manned helicopter has limitation of flying altitude at low altitudes due to regulation. In addition, since they can cost over a million dollars each they are very expensive to deploy widely. Currently, there are limited means of collecting disaster information at low altitudes. However, the development of MAVs is expected to enable low altitude information gathering as MAVs become widely deployed.

In previous research, our group studied external vision-based autonomous control of 12.3 g coaxial-rotor helicopter μFR by Wang [1] and external stereovision-based autonomous control of coaxial-rotor helicopter X.R.B by Ohira [2]. However, in order to develop a practical MAV, a quad-rotor MAV which can carry much higher payloads was chosen as our research platform.

Recent research on quad-rotor MAVs by Bouabdallah et al. has developed a micro quad-rotor helicopter. They built a model of the quad-rotor's attitude, and succeeded in the control of altitude [3]. Hoffmann et al. has been developing a test bed of multi-vehicle control systems in the STARMAC Project [4]. Altuğ et al. has succeeded in a vision-based control of height and yaw by an external vision system [5]. Additionally, a quad-rotor such as the X-3D BL which was designed by Ascending Technologies GmbH [6] and the Draganflyer which was designed by Draganfly Innovations Inc. [7] have both been chosen for many studies because of their high flight performance.

The key objective is to provide autonomous flight capability for MAVs beyond visual range operation in order to collect information during disasters. Also automated takeoff and landing are needed, as well as obstacle avoidance, obstacle detection and even self-charging of the MAVs may be required depending on the situation. In addition, another current key objective is to develop robots which can fly autonomously in an unknown indoor environment where GPS is unavailable.

This research focused on developing an autonomous flying robot which can help gather information during rescue-efforts after natural disasters including earthquakes and fires. In related work, Wang has succeeded flying a quad-rotor MAV outdoors using GPS. However, its position control includes a few meters of errors [8], since it relies on the limited accuracy of GPS feedback. In addition, GPS is unable to detect obstacles. So if only GPS is used, it's impossible to land on a specified target or, and search around objects including buildings. This study proposed a range finding system which consists of four movable IR rangefinders and one ultrasonic rangefinder. And it's capable of detecting objects in various directions. This chapter describes a measurement algorithm of this proposed sensor system. And it also describes the experimental results of hovering control and the automated landings of a quad-rotor MAV.

13.2 System Configuration

In this section, we describe the experimental platform, a proposed movable range finding system and a block diagram of the system configuration.

13.2.1 Description of the Experimental Platform

Our platform is based on a quad-rotor type helicopter from Ascending Technologies GmBH which is called the X-3D-BL. This MAV, which is shown in Fig. 13.1, is equipped with our autopilot unit (see Chap. 3), a communication unit and the proposed range finding system all of which are mounted on the X-3D-BL. Our MAV's specifications are shown in Table 13.1.

13.2.2 Movable Range Finding System

Figure 13.2 shows the proposed movable range finding unit for auto-landing in this study. This unit consists of one ultrasonic rangefinder and four IR rangefinders and the servo motors. The IR rangefinder is able to move within a 120° range using a servo motor for each rangefinder and it has a measuring range from 1 to 5.5 m.

Fig. 13.1 Our autonomous quad-rotor flying robot for precise landing

Table 13.1 Specifications of our flying robot

Item	Value	Unit
Mass of X-3D BL	275	g
Mass of equipment	316	g
Mass of battery	148	g
Total Mass	739	g
Maximum width (with propellers)	540	mm
Maximum width (without propellers)	380	mm
Height	206	mm
Battery voltage	11.1	V
Battery capacity	2,100	mAh

Fig. 13.2 The IR (*on the left*) and ultrasonic range finding units

These IR sensors are installed on each end of the MAV's legs as shown in Fig. 13.1. The measuring range of the ultrasonic rangefinder is up to 6 m and the width of its beam is 50°. It is attached on the bottom of the MAV and it's pointed downwards. The resulting detection area of this MAV is shown in Fig. 13.3.

13.2.3 MAV Operation System

Finally, Fig. 13.4 shows the current system configuration.

13.3 Principle of Position Measurement

By tilting the IR rangefinder to the horizontal direction, the proposed range finding system can measure the horizontal position by using objects which are in the horizontal plane, such as walls. These objects provide reference points to calculate

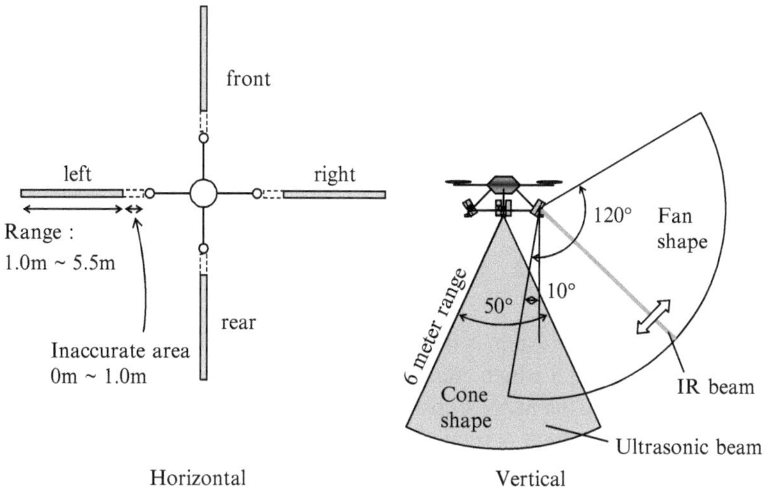

Fig. 13.3 Detection area

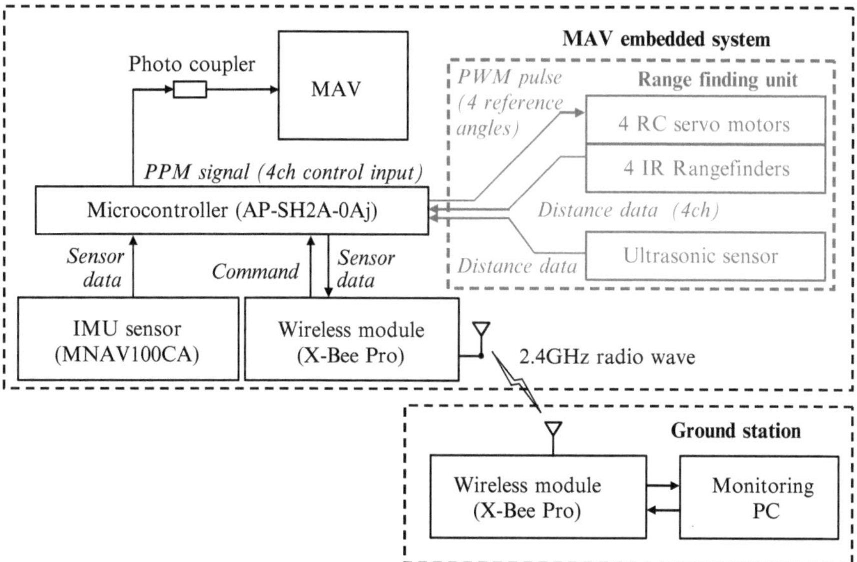

Fig. 13.4 Block diagram of the system configuration

the MAV's position. However, in these experiments, no walls were nearby so we assumed there were no objects in horizontal plane. So we proposed an algorithm which would measure the horizontal position by using the edge detection of an object beneath the MAV such as a car or a table.

Recent studies on helicopter auto-landings have focused on single rotor type helicopters [9–12] and some quad-rotor systems [13, 14]. These studies adopted vision-base approaches for auto-landings which used stereo-vision cameras. However, the CPU capability and payload capacity of MAVs are limited. Therefore this study used lightweight rangefinders and uncomplicated method for automatic-landings of MAVs. In this section, an algorithm of position measurement for automated-landings is described.

13.3.1 Basic Principle

This automated landing system is designed for precision landings without GPS on a rectangular shaped target which could represent a rescue vehicle.

In order to measure the MAV's position, this algorithm detects the edge of object. Each IR rangefinder, which is rotated by a servo motor, is used for edge detection. First, we consider an MAV flying above the object. As the IR-rangefinder is tilted by the servo motor, the measured distance spikes downward when the IR beam strides the edge of object such as in Fig. 13.5. $\tilde{\alpha}$ is defined as an angle between the IR beam line and the Z axis during this moment. Additionally, the ultrasonic rangefinder can measure the altitude h between object and MAV at same time.

The horizontal distance from edge to center of MAV, L, is calculated as follows where the width of MAV is w and also using the measured $\tilde{\alpha}$ and h.

$$L = h \tan \tilde{\alpha} + \frac{1}{2} w \tag{13.1}$$

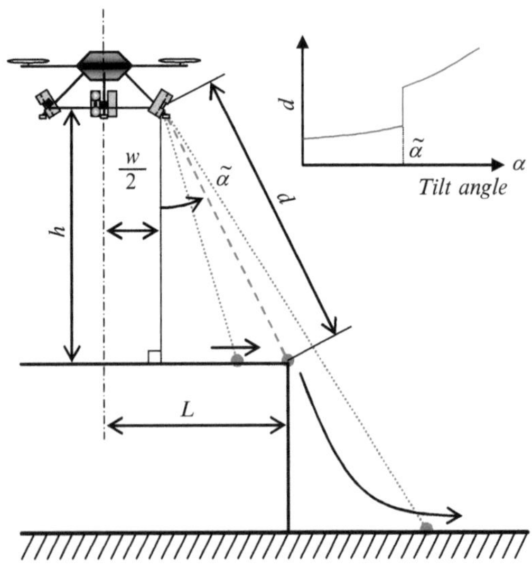

Fig. 13.5 Distance measurement

13.3.2 Definition of Coordinate System

At first, we must define a coordinate system which will also be used for flight control. A local coordinate system MAV-frame is defined for the MAV. The origin for these coordinates is defined as the center of gravity of the MAV. The MAV's-fixed X axis is defined as direction of the front rotor and Y axis is defined as direction of the right rotor. The Z axis is defined as direction which is perpendicular to the X and Y axis, and it points in the direction beneath the MAV. And, the landing target frame is defined as the landing target as shown in Fig. 13.6.

In addition, the rotation about the X axis is defined as the Roll angle ϕ, the rotation about the Y axis is defined as the Pitch angle θ, and the rotation about the Z axis is defined as the Yaw angle ψ. Their angles are defined as the Euler angle between the MAV's frame and the targeted landing frame.

13.3.3 Edge Detection

By tilting each servo motor in order to detect $\tilde{\alpha}$, the angle is made to converge at the edge where the measured distance jumps. For example as shown in Fig. 13.7, the servo motor tilts the IR rangefinder to an upper point when an IR spots the landing point. Otherwise, the servo motor tilts the IR rangefinder to a lower position. If it is able to do this, it is able to hold the motor motion around the edge.

In this study, a reference angle to the servo motor is defined as follows.

$$\alpha_{ref}^{k} = \alpha_{ref}^{k-1} + \frac{V_s T_s \cos^2 \hat{\alpha}^k}{h^k} \mathrm{sgn}\left(d_t^k - d^k\right) \tag{13.2}$$

Where, suffix k is a step number on the time axis. And V_s is a given number about tilting speed, T_s is control cycle (20 ms), $\hat{\alpha}$ is the estimated angle of the servo motor, sgn is a sign function, d_t is a threshold value for deciding a tilting direction and d is the measured distance by IR rangefinder. In the case of a constant swing speed by servo motors, moving velocity of the target spot depends on the altitude h of MAV and the IR beam direction α. Therefore, (13.2) can be compensated by $\cos^2 \hat{\alpha}^k / h^k$.

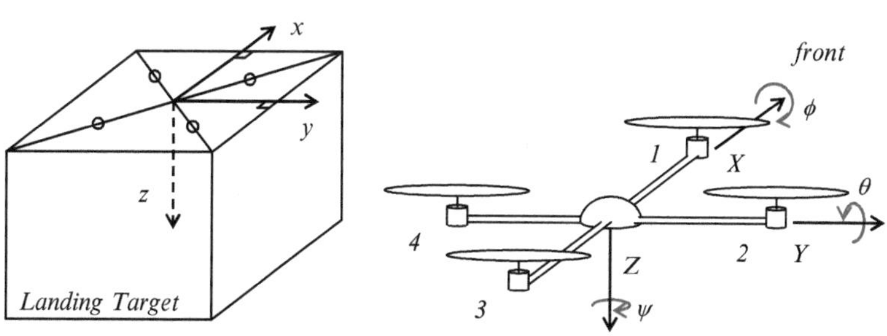

Fig. 13.6 The coordinate system

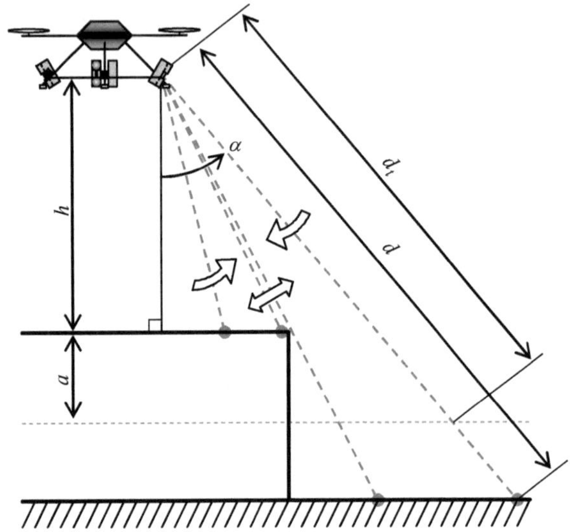

Fig. 13.7 Edge detection

There is no sensor for measuring the servo motor angle α in this system. Because of this, the system calculates an estimated angle $\hat{\alpha}$ and uses it in place of the true value α. The following equation is an updating rule of $\hat{\alpha}$ which is used in this system, and the servo dynamics is approximated to first-order delay.

$$\hat{\alpha}^k = A_s\hat{\alpha}^{k-1} + (1 - A_s)\,\alpha_{ref}^{k-1} \tag{13.3}$$

The difference between d_t as the threshold value and d decides the rotational direction of servo motor. First, a parallel plane to the ground is assumed in the bottom distance $h + a$ of the MAV as shown in Fig. 13.7. Assuming d_t is the distance from the MAV to the cross-point of the IR beam and above-mentioned plane, the value is as follows.

$$d_t = \frac{h + a}{\cos\hat{\alpha}} \tag{13.4}$$

A motor angle when it detected the edge is defined as $\tilde{\alpha}$ and used for calculation of MAV's position. It is updated as follows.

$$\tilde{\alpha}^k = \begin{cases} \hat{\alpha}^k & \operatorname{sgn}\left(d_t^k - d^k\right) \neq \operatorname{sgn}\left(d_t^{k-1} - d^{k-1}\right) \\ \tilde{\alpha}^{k-1} & \textit{otherwise} \end{cases} \tag{13.5}$$

Edge detection and tracking is enabled by the above rules. Additionally, if (13.2) is replaced by

$$\alpha_{ref}^k = \alpha_{ref}^{k-1} + c\frac{V_s T_s \cos^2\hat{\alpha}^k}{h^k}\operatorname{sgn}\left(d_t^k - d^k\right) \tag{13.6}$$

at the time of $\operatorname{sgn}\left(d_t^k - d^k\right) \neq \operatorname{sgn}\left(d_t^{k-1} - d^{k-1}\right)$, the updating speed of $\tilde{\alpha}$ rises, where, c is a given number.

13.3.4 Position Calculation

This system executes the edge detection in four directions: forward, right, backward and left. Therefore, four variables are calculated in each direction. And so, four suffixes from 1 to 4 are marked for each variable clockwise from the front of the MAV as shown in Fig. 13.2.

For example, by using (13.1), the center position of the MAV is calculated as follows.

$$x = \frac{h}{2}\left(\tan\tilde{\alpha}_3 - \tan\tilde{\alpha}_1\right) \tag{13.7}$$

$$y = \frac{h}{2}\left(\tan\tilde{\alpha}_4 - \tan\tilde{\alpha}_2\right) \tag{13.8}$$

Where, the origin is the center of landing point. This equation does not include the width of the landing point. That is, the position calculation doesn't require the width of landing point.

However, (13.7) and (13.8) did not include the attitude of the MAV. This influence is not negligible if h is high. In fact, the position control has a tendency to be unstable by using this equation. And so, actual calculation of the position includes this influence even though, an actual equation is simplified by assuming a low attitude. Because of this, the equation becomes too complicated. In addition, a distance measurement from the ultrasonic rangefinder is almost stationary if the attitude is low. Therefore, h is assumed to be unaffected by attitude. For these reasons, the center position of the MAV is calculated by following equations.

$$x = \frac{1}{2}\left\{\left(h - \frac{w\theta}{2}\right)\tan\left(\tilde{\alpha}_3 - \theta\right) - \left(h + \frac{w\theta}{2}\right)\tan\left(\tilde{\alpha}_1 + \theta\right)\right\} \tag{13.9}$$

$$y = \frac{1}{2}\left\{\left(h + \frac{w\phi}{2}\right)\tan\left(\tilde{\alpha}_4 + \phi\right) - \left(h - \frac{w\phi}{2}\right)\tan\left(\tilde{\alpha}_2 - \phi\right)\right\} \tag{13.10}$$

Parameters for the edge detection algorithm and the position calculation are shown in Table 13.2.

Table 13.2 Values of the edge detection algorithm

Symbol	Unit	Value	
V_s	m/s	0.87	Given value (>0)
T_s	s	0.02	Equal to the control cycle
A_s		0.846	Depending on the servo motor and T_s
a	m	0.6	Given value (>0)
c		2.0	Given value (>1)
w	m	0.26	Depend on MAV size

13.4 Modeling and Controller Design

In this section, we describe the design of a position controller and modeling. Additionally, we describe the altitude and heading control in order to implement fully autonomous control.

13.4.1 Configuration of the Control System

In this study, we separate the MAV's dynamics into four independent systems as in Chap. 3. These four systems are: (1) $\theta - x$, the position control system which includes attitude θ and position x, (2) $\phi - y$, the position control system which includes attitude ϕ and position y, (3) the heading control system which includes the attitude ψ and (4) the altitude control system which includes position z, where $\theta - x$ and $\phi - y$ control system have the same structure due to the quad-rotor's symmetry.

For $\theta - x$ and $\phi - y$ control, first an inner-loop control system with feedback attitude and angular velocity is designed. Next as shown in Fig. 13.8, an outer-loop control system which calculates the position, attitude and angular velocity is designed outside the inner-loop system. Additionally, ψ and z are controlled by a single control loop as shown in Fig. 13.9.

In addition, attitude is calculated by the attitude and heading reference systems (AHRS) [15] shown in Figs. 13.8 and 13.9a at this system.

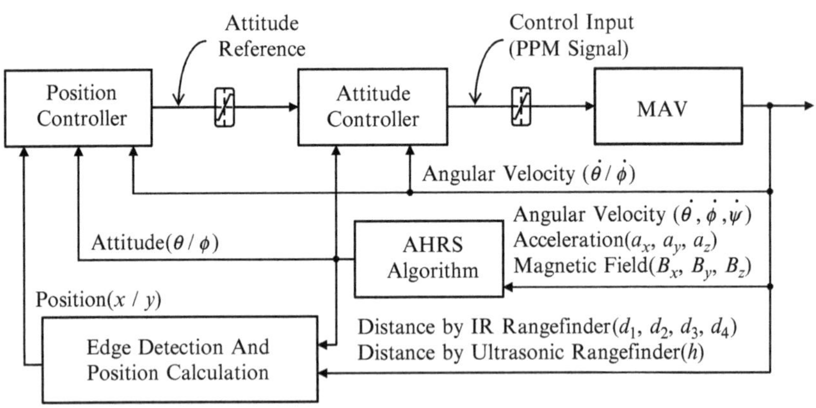

Fig. 13.8 The block diagram of the x and y control loop

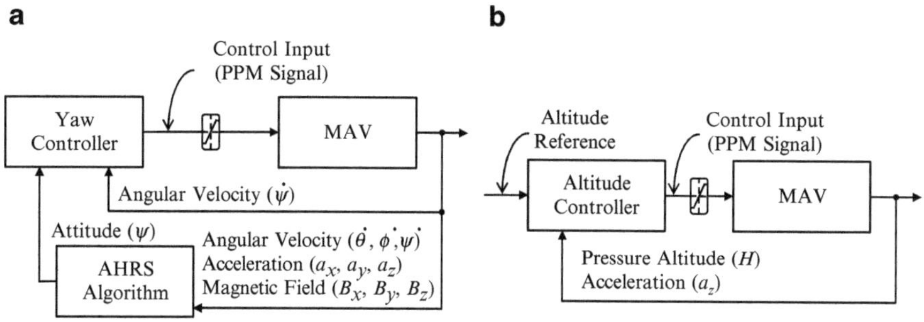

Fig. 13.9 Block diagram: (**a**) yaw (heading) control loop, (**b**) z (altitude) control loop

13.4.2 Modeling

13.4.2.1 Attitude Model (θ, ϕ)

The transfer functions between the control input u_θ, u_ϕ and measured angular velocity $\dot{\theta}$, $\dot{\phi}$, are assumed to be second order delay as Chap. 3. And the transfer function from the angular velocity to attitude is regarded as an integrator. So our attitude model is as follows.

$$\frac{\theta(s)}{u_\theta(s)} = \frac{\phi(s)}{u_\phi(s)} = \frac{K\omega_n^2}{s\left(s^2 + 2\xi\omega_n s + \omega_n^2\right)} \tag{13.11}$$

In this system, θ, ϕ, $\dot{\theta}$ and $\dot{\phi}$ are measurable.

13.4.2.2 Position Model (x, y)

At first, attitude of the MAV is assumed to be a stabilized system by control of Fig. 13.8. In addition, this output is assumed to be able to follow the attitude reference in designing the position model. Since the order of this attitude closed system is high, a transfer function from attitude reference to attitude is designed as the dominant poles. These dominant poles are one pair of conjugate poles. Therefore, a transfer function of attitude response is as follows.

$$\frac{\theta(s)}{\theta_{ref}(s)} = \frac{\phi(s)}{\phi_{ref}(s)} = \frac{Re^2 + Im^2}{s^2 + 2Res + Re^2 + Im^2} \tag{13.12}$$

Next, a transfer function from the attitude to the position is based on the following model where the horizontal thrust is the horizontal component of the lift force mg [3].

$$\begin{cases} \ddot{x} = -g\left(\cos\phi\sin\theta\cos\psi + \sin\phi\sin\psi\right) \\ \ddot{y} = -g\left(\cos\phi\sin\theta\sin\psi - \sin\phi\cos\psi\right) \end{cases} \tag{13.13}$$

By linearization of their equation of motion, the actual model is calculated as follows.

$$\frac{x(s)}{\theta(s)} = -\frac{y(s)}{\phi(s)} = -\frac{g}{s^2} \tag{13.14}$$

Where g is acceleration of gravity. In this system, $x, y, \theta, \phi, \dot{\theta}$ and $\dot{\phi}$ are measurable.

13.4.2.3 Altitude Model (z)

The transfer function between the control input u_z and the measured acceleration is assumed to be first order delay. And the transfer function from the acceleration to the position is regarded as double integrator. So our altitude model is as follows.

$$\frac{z(s)}{u_z(s)} = \frac{K_z}{s^2\left(T_z s + 1\right)} \tag{13.15}$$

In this system, z and $\dot{z}$ are measurable.

13.4.2.4 Heading Model (ψ)

The transfer function between the control input u_ψ and the measured yaw angular velocity $\dot{\psi}$ is assumed to be a second order delay which has two real poles. And the transfer function from the angular velocity to the attitude is regard as integrator. So our heading model is as follows.

$$\frac{\psi(s)}{u_\psi(s)} = \frac{K_\psi}{s\left(T_{\psi 1}s + 1\right)\left(T_{\psi 2}s + 1\right)} \tag{13.16}$$

In this system, ψ and $\dot{\psi}$ are measurable.

13.4.3 Parameter Identification

The procedure of parameter identification is as follows.

1. Fly the MAV by manual operation and acquire the system outputs and manual inputs

2. Assume the parameter values. And simulate the system output by using a system model, with parameters and input series of 1
3. Compare the simulated output and the acquired output. If these do not match, try 2 and 3 again

As a result of this procedure, the following parameters were obtained as shown in Table 13.3. However, *Re* and *Im* are given as the dominant poles of the attitude closed loop system such as Fig. 13.10.

Table 13.3 Values of identified parameters

Symbol	Value
K	3.94×10^{-3}
ζ	1.01
ω_n	15.4
Re	3.94
Im	3.96
K_z	1.90×10^{-2}
T_z	1.15×10^{-1}
K_ψ	9.50×10^{-3}
$T_{\psi 1}$	1.00
$T_{\psi 2}$	0.60

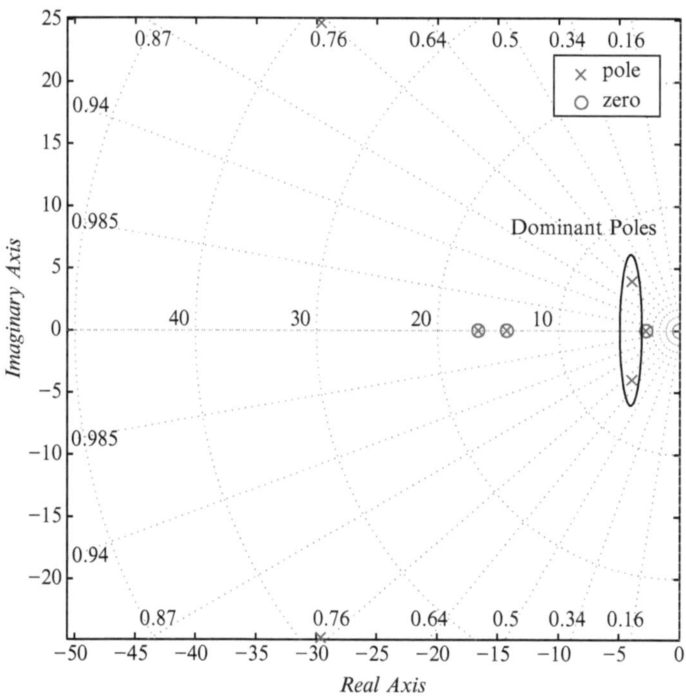

Fig. 13.10 Pole zero placement of the attitude closed-loop system (including the attitude observer)

13.4.4 Controller

Each controller was designed for each model. Our controller is based on optimal state feedback. Additionally, in order to create a servo system, an extended system is designed. In order to design an extended system, we define a state variable x_r as the integral of the error of the difference between n-th output $y_n = C_n x$ and the reference r where the suffix n is an index of the output vector which we want to follow the input, and C_n is n-th row of matrix C. An extended system is calculated as follows.

$$\frac{d}{dt}\begin{bmatrix} x \\ x_r \end{bmatrix} = \begin{bmatrix} A & 0 \\ -C_n & 0 \end{bmatrix}\begin{bmatrix} x \\ x_r \end{bmatrix} + \begin{bmatrix} B \\ 0 \end{bmatrix} u + \begin{bmatrix} 0 \\ I \end{bmatrix} r \qquad (13.17)$$

By applying optimal control theory to this, a feedback gain F, which is able to converge y_n to r, is calculated.

The state feedback controller needs all state variables. However, this system is unable to measure some state variables. Therefore, unmeasurable states are estimated by a Kalman filter in this system.

13.5 Experiments

In order to test the efficacy of the proposed sensor system, a hovering experiment and automated landing experiment were carried out. These experiments used a 52 cm-square table. The relative positions of the table and MAV are shown in Fig. 13.11. The origin of the coordinate system is defined as center of table.

13.5.1 Autonomous Hovering Experiment

In this experiment, first the MAV was flown manually above the center of table. And the heading of the MAV ψ was conformed to the table. For this, operator

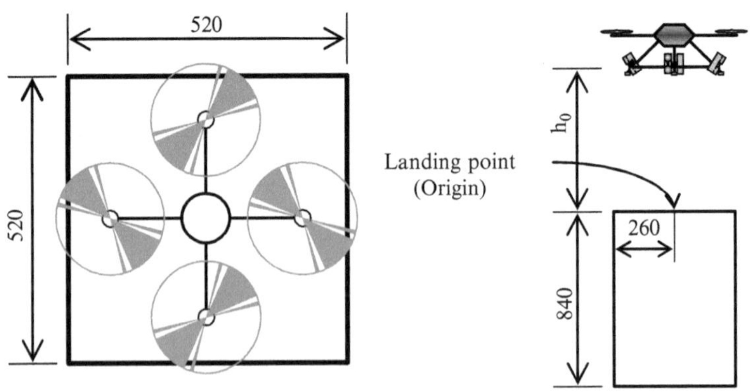

Fig. 13.11 Experimental configuration

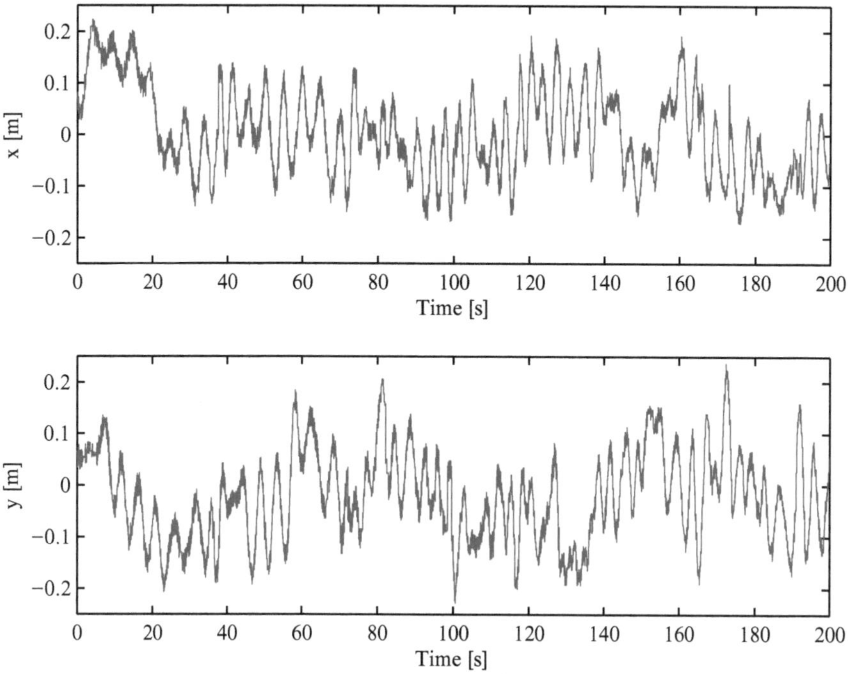

Fig. 13.12 Time variation of the position (raw data)

adjusted the MAV's heading in order that MAV's X and Y axes is perpendicular and/or parallel to each side of table such as in Fig. 13.11. Next, after checking that the movable range finding system was working, control was switched from manual to autonomous. During autonomous control, the horizontal x and y positions were controlled to keep them at zero, and the altitude z and the heading ψ were controlled to keep the initial state at the moment autonomous control began.

The results of these tests are shown in Fig. 13.12 which measured the horizontal position using the movable range finding system. There were from 5 to 6 Hz band noises in this position measurement because the sensor system was based on swing measurements. Since this controller filtered this band's noises using a Kalman filter, the influence on the performance of this control was small. According to this graph, hovering was accurate to ± 20 cm, however the time variation oscillated. Its frequency was about 0.22 Hz, and was difficult to reduce by tuning the horizontal position controller. Unremoved attitude coupling in position measurement and modeling errors caused this oscillation.

The horizontal trajectory of the MAV is shown in Fig. 13.13. This is made from a horizontal position shown in Fig. 13.12 and this data was corrected using a Kalman filter. A circle in this figure is drawn in order to include 95% of the data, and its radius is 17.9 cm. An "*" mark is at the center of the circle and it is the average of all data, where, the position is $x = 5$ mm and $y = 8$ mm. So this shows that autonomous hovering successfully occurred near the center of table.

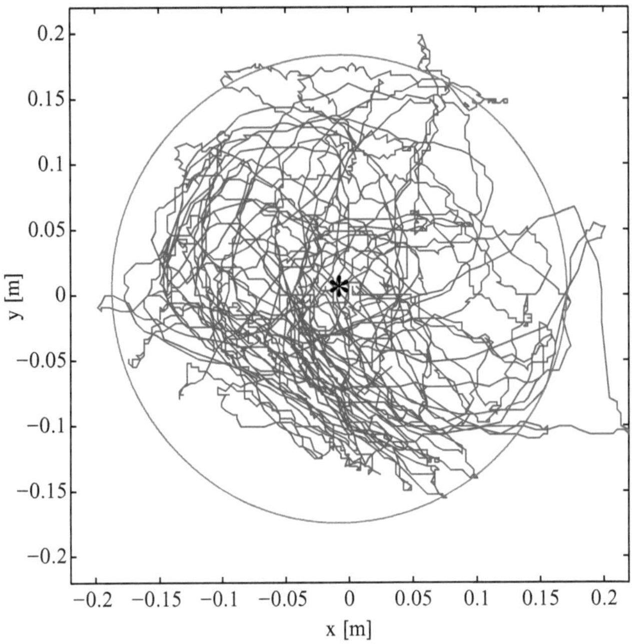

Fig. 13.13 2D plot of the position (with Kalman filter)

In addition, the relative altitude between the MAV and table, which was measured by ultrasonic rangefinder, averaged 85 cm.

13.5.2 Automated Landing Experiment

In this experiment, autonomous control of the horizontal position and the heading were carried out the same as the hovering experiment. On the other hand, the altitude control used the following rules.

- Hold the altitude reference for an initial 20 s
- After the initial 20 s, descend the altitude reference with 0.001 m/step speed when an energy function is less than a constant value ε. (1 step = 0.02 s)
- If the altitude measured by the ultrasonic rangefinder is less than 40 cm, then land the MAV by decreasing the control input without feedback control

The first step should be to wait for convergence of the initial transient response. The second step is to improve the landing relativity by waiting until convergence of the horizontal motion where the horizontal motion was influenced during the descent. Assuming the state variables are $\boldsymbol{x}_x = \{\hat{\dot{\theta}}, \hat{\theta}, \hat{\dot{x}}, \hat{x}, x_r\}$ and $x_y = \{\hat{\dot{\phi}}, \hat{\phi}, \hat{\dot{y}}, \hat{y}, y_r\}$, an energy function is defined as the following positive definite quadratic form.

$$V\left(\boldsymbol{x}_x, \boldsymbol{x}_y\right) = \boldsymbol{x}_x^T L_0 \boldsymbol{x}_x + \boldsymbol{x}_y^T L_0 \boldsymbol{x}_y \tag{13.18}$$

Where x_r and y_r are integrated value of deviation in (13.17), and other state variables are estimated values by a Kalman filter. L_0 is defined as the following positive definite solution which satisfies that the time derivative of V is always negative.

$$(A_i - B_i F) L_0 + L_0 (A_i - B_i F)^T + I = 0 \tag{13.19}$$

Where A_i and B_i are A and B matrix in extended system (13.17) of horizontal position control, and F matrix is a state feedback gain of its extended system. The third step is to prevent a fault measurement during low altitudes where a measurement of IR rangefinder fails.

The procedures for this experiment are identical to the hovering experiment. The condition of the descent ε is chosen as 20 which was based on previous experimental results.

The experimental results are shown in Fig. 13.14 using the measured horizontal distance from the movable range finding system. In the current landing algorithm, landing accuracy depends on the position control accuracy. Therefore, according to the results of this experiment, the landing accuracy was about ±10 cm. Since hovering altitude in these tests were lower than the hovering experiment, the accuracy of horizontal position control is higher than the hovering experiment.

The experimental results are shown in Fig. 13.15 where the MAV's flying altitude was measured by the pressure sensor and the ultrasonic rangefinder. The pressure

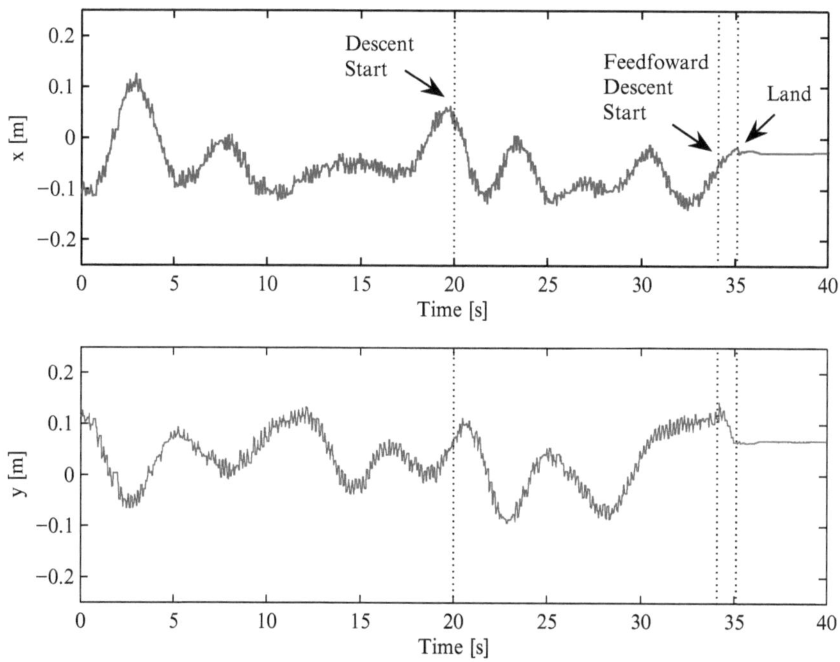

Fig. 13.14 Time variation of the position (raw data)

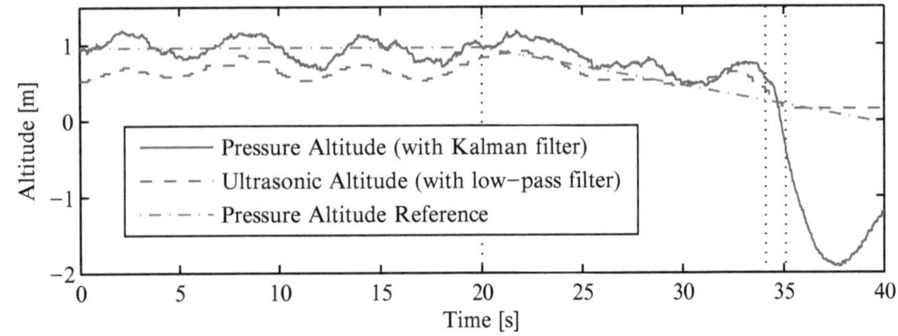

Fig. 13.15 Time variation of the altitude

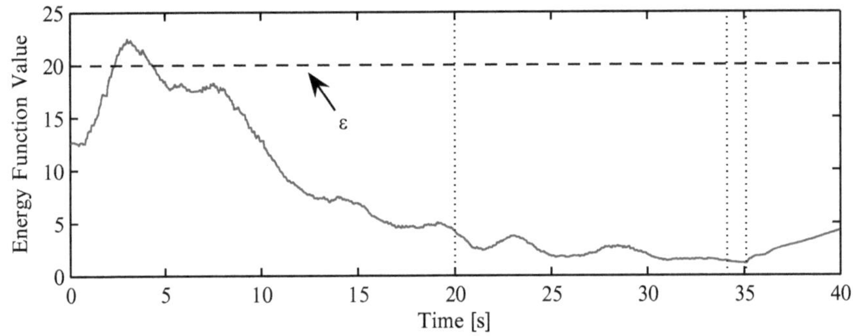

Fig. 13.16 Time variation of the energy function

altitude was corrected using a Kalman filter, and the ultrasonic altitude was adjusted using a low-pass filter. Since these two measurements have different mean-values, an offset in pressure altitude of this figure has been removed. Altitude control is based on pressure altitude, but the accuracy of the ultrasonic altitude is higher than the pressure altitude. Because an altitude based on ultrasonic rangefinder might be sometimes affected by bump between table and floor, we chose the pressure sensor for altitude control. According to the ultrasonic altitude, autonomous altitude control started at above 0.5 m high as the beginning of control. And after ascending to 1 m altitude, the MAV began to descend. Since pressure altitude decreased after 20 s, it indicates that attitude of MAV followed the reference. Pressure altitude was overshooting after 34 s. This was due to the feed forward control input for altitude control which is generated at the final stage of landing. Since landing is completed immediately, this overshooting is not a matter of concern for controlling the landing.

An experimental result shown in Fig. 13.16 is an energy function value of horizontal position control. At 20 s, the value of the energy function was well converged and it remained small until the MAV landed. So the descending speed was constant because of so quite behavior of the MAV.

13.6 Summary

In order to measure a MAV's position without GPS, this chapter presented a movable range finding system, which consists of servo motors, and IR and ultrasonic rangefinders. And this chapter described this sensor system and its algorithm for position measurement. In addition, it discussed the design of a control system, and the hovering and landing experiments which were conducted. Results are summarized as follows.

- In the hovering experiment, our MAV successfully flew over the 52×52 cm table and maintained a horizontal position within a 17.9 cm radius circle. The mean-value of the flying altitude was 85 cm
- In the landing experiment, our MAV could land on a table which was about the same size as the MAV by maintaining ± 10 cm horizontal position accuracy. During these landings, the initial altitude was 50 cm and maximum altitude was 1 m
- The closed-loop system of the position control oscillated at 0.22 Hz. Therefore for future research, we will design a new controller which can more robustly handle both the modeling errors and also improve the measurement algorithms.

After solving this oscillating problem, our goal will be to achieve high-altitude hovering and outdoor precision landings using the combination of GPS and the movable range finding system. Moreover, utilizing the capability of our sensor's versatility could lead to obstacle avoidance, mapping and indoor localization.

References

1. Wang W, Hirata M, Nonami K, Miyazawa O (2007) Modeling and autonomous control of microminiature ultra-small scale/light weight micro-flying robot. J Robot Soc Jpn 25(1):121–129
2. Ohira Y, Wang W, Nonami K (2008) Modeling and autonomous control of small co-axial helicopter using stereo vision camera. J Robot Soc Jpn 26(6):175–182
3. Bouabdallah S, Murrieri P, Siegwart R (2004) Design and control of an indoor micro quadrotor. In: Proceedings of the 2004 IEEE international conference on robotics & automation, New Orleans, LA, April 2004, pp 4393–4398
4. Hoffmann G, Rajnarayan DG, Waslander SL, Dostal D, Jang JS, Tomlin CJ (2004) The Stanford testbed of autonomous rotorcraft for multi agent control (STARMAC). In: Proceedings of the 23rd digital avionics systems conference, Salt Lake City, UT, November 2004
5. Altuğ E, Ostrowski JP, Mahony R 2002 Control of a quadrotor helicopter using visual feedback. In: Proceedings of the IEEE international conference on robotics and automation, Washington, DC, May 2002, pp 72–77
6. Ascending Technologies GmbH, http://www.asctec.de/main/index.php?id=14
7. Draganfly Innovations Inc., http://www.draganfly.com/
8. Wang W, Suzuki S, Nonami K, Iwakura D, Pebrianti D, Zhu L, Guo D, Kendoul F (2008) Fully autonomous quad-rotor MAV and flying performance with complete embedded system. In: Proceedings of the 9th international conference on motion and vibration control, Munich, Germany, September 2008, ID.1301
9. Srikanth S, Montgomery JF, Sukhatme GS (2002) Vision-based autonomous landing of an unmanned aerial vehicle. In: Proceedings of the IEEE international conference on robotics and automation, Washington DC, Virginia, USA, 2002, pp 2799–2804

10. Srikanth S, Montgomery JF, Sukhatme GS (2003) Visually guided landing of an unmanned aerial vehicle. IEEE Trans Robot Autom 19(3):371–381
11. Johnson A, Montgomery J, Matthies L (2005) Vision guided landing of an autonomous helicopter in hazardous terrain. In: Proceedings of the IEEE international conference on robotics and automation, Barcelona, Spain, 2005, pp 3966–3971
12. Yu Z, Nonami K, Shin J, Celestino D (2007) 3D vision based landing control of a small scale autonomous helicopter. Int J Adv Robotic Syst 4(1):51–56
13. Frietsch N, Meister O, Schlaile C, Seibold J, Trommer GF (2008) Vision based hovering and landing system for a VTOL-MAV with geo-localization capabilities. In: AIAA guidance, navigation and control conference and exhibit, Honolulu, Hawaii, USA, August 2008
14. Herisse B, Russotto F-X, Hamel T, Mahony R (2008) Hovering flight and vertical landing control of a VTOL unmanned aerial vehicle using optical flow. In: Proceedings of the IEEE/RSJ international conference on intelligent robots and systems, Nice, France, September 2008, pp 801–806
15. MNAV Autopilot, http://sourceforge.net/project/showfiles.php?group_id=146680

Index

MIX
Papier aus verantwortungsvollen Quellen
Paper from responsible sources
FSC® C105338

If you have any concerns about our products,
you can contact us on
ProductSafety@springernature.com

In case Publisher is established outside the EU,
the EU authorized representative is:
Springer Nature Customer Service Center GmbH
Europaplatz 3, 69115 Heidelberg, Germany

Printed by Libri Plureos GmbH
in Hamburg, Germany